Multimodality Imaging, Volume 2

Heart, lungs and peripheral organs

Online at: https://doi.org/10.1088/978-0-7503-2352-9

Multimodality Imaging, Volume 2

Heart, lungs and peripheral organs

Edited by
Mainak Biswas
Kalinga Institute of Industrial Technology, Bhubaneshwar, India

Jasjit S Suri
AtheroPoint LLC, Sacramento, CA, United States of America

IOP Publishing, Bristol, UK

ISBN 978-0-7503-2352-9 (ebook)
ISBN 978-0-7503-2350-5 (print)
ISBN 978-0-7503-2353-6 (myPrint)
ISBN 978-0-7503-2351-2 (mobi)

DOI 10.1088/978-0-7503-2352-9

Version: 20231201

IOP ebooks

British Library Cataloguing-in-Publication Data: A catalogue record for this book is available from the British Library.

Published by IOP Publishing, wholly owned by The Institute of Physics, London

IOP Publishing, No.2 The Distillery, Glassfields, Avon Street, Bristol, BS2 0GR, UK

US Office: IOP Publishing, Inc., 190 North Independence Mall West, Suite 601, Philadelphia, PA 19106, USA

To my late parents, my wife and daughter.

—Mainak Biswas

To all my collaborators around the world.

—Jasjit S Suri

Contents

Part II An overview of deep learning and its applications in cardiovascular diseases

Preface

Deep learning and artificial intelligence (AI) is now as good for use as a general tool as other medical equipment and software. Having been deployed for some time now, it has been observed that inaccurate diagnosis cases (both false positives and false negatives) are decreasing with the usage of AI/DL aided tools. Medical practitioners are increasingly using diagnosis tools embedded with AI for coming to accurate decisions. With coming-of-age of AI platforms such as Google Health, DeepMind and OpenAI, the diagnosis will become more affordable and accessible than ever before. This book will look into some aspects of usage of deep learning for effective treatment.

Purpose

The book is written in the post-COVID-19 era and therefore a major section of this book is dedicated to deep learning (DL) and artificial intelligence (AI) applications in COVID-19 and respiratory diseases. This book discusses the effect on organs such as the brain and heart and how in the long term, AI will be able to detect the damage caused to vital organs. One single chapter is also dedicated to tuberculosis. AI also has made significant advances in the area of detection of cardiovascular diseases using multiple medical imaging modalities such as MRI, CT, and ultrasound. This book covers multiple areas where DL/AI technologies have been critical in accurate characterization of diseases.

Content and organization

The content of this book is divided into two sections: AI/DL in COVID-19/respiratory diseases and other cardiovascular diseases. In the first part, four chapters are dedicated to COVID-19 and one chapter to tuberculosis. In the second part, three chapters are dedicated to AI/DL in cardiovascular diseases. The description of each chapter in the first part is given as follows: Chapter 1 discusses the four pathways through which COVID-19 affects the heart and the brain, and how AI-assisted medical imaging can detect and diagnose the damage caused. Chapter 2 discusses critical AI technologies that have been applied to detection of COVID-19 induced Acute Respiratory Distress Syndrome (ARDS). In Chapter 3, eight pruned deep-learning models for COVID-19 CT lung segmentation and lesion localization are discussed in detail. An inter-variability study of the results of lung segmentation is done in Chapter 4. In Chapter 5, a segmentation study of tuberculosis-infected lung images using deep learning is discussed.

In the second part, there are three chapters. Chapter 6 presents a study of different applications of AI/DL in cardiovascular ultrasound. Chapter 7 discusses different segmentation/characterization studies related with atherosclerosis. Finally, Chapter 8 talks about different techniques for carotid disease management.

Editor biographies

Mainak Biswas

Mainak Biswas, PhD, is a computer scientist with specialization in the application of machine learning and deep learning in the biomedical domain. His research is inspired by providing an effective solution for computer-aided diagnosis for diverse diseases. His PhD specialization was in application of advanced machine learning and deep learning in complex tissue characterization and segmentation from ultrasound images of liver and carotid arteries. His other interests are development of advanced machine-learning architectures and early warning systems for risk estimation of both symptomatic and asymptomatic patients at high risk of CVDs. He also has keen interest in development of new metrics for machine learning algorithms based on statistical mechanics. He has published and presented more than 30 papers in the area of characterization and segmentation of ultrasound images through machine and deep learning platforms. His H-index is 13, and he has more than 1000 citations on his research. One of his review papers 'The Present and Future of Deep Learning in Radiology' has approximately 200 citations. Dr Mainak Biswas completed his BTech from the Government College of Engineering and Ceramic Technology under West Bengal University of Technology, Kolkata, MTech from Jadavpur University and his PhD from the National Institute of Technology Goa, India. Currently, he is serving as Associate Professor at Vignan's Foundation for Science, Technology and Research.

Jasjit S Suri

Jasjit S Suri (PhD, MBA, FIEEE, FAIMBE, FAIUM, FSVM, FAPVS) is an innovator, a visionary, a scientist, and an internationally known world leader. He has spent over 30 years in the field of biomedical engineering/sciences, software and hardware engineering and its management. During his career in biomedical industry/imaging, he has had an upstream growth and responsibilities from scientific engineer, scientist, manager, Director of R&D, Senior Director, Vice President, Chief Technology Officer (CTO), CEO level positions in industries like Siemens Medical Systems, Philips Medical Systems, Fisher Imaging Corporation and Eigen Inc., Global Biomedical Technologies Inc., AtheroPoint™, respectively, and managed up to a maximum of 100 people.

Dr Suri is a pioneer in the area of artificial intelligence (AI) and has published over 100+ papers in international journals covering several fields such as vascular, coronary, prostate, mammography, diabetes, and COVID-19 CT lung areas. He has developed products and worked extensively in the areas of breast, mammography, orthopedics (spine), neurology (brain), angiography (blood vessels), atherosclerosis (plaque), ophthalmology (eye), urology (prostate and ovarian), image-guided surgery (for neurology and orthopedics) and several kinds of biomedical devices from inception phase to commercialization, including 510(K)/FDA clearances. Under his leadership he has obtained over 5 FDA clearances in urology, angiography and image-guided surgery product lines ranging from 1000 page to 5000 page submissions. He has conducted *in vivo* and *ex vivo* validations on biomedical devices and surgery systems. Dr Suri has developed several collaboration programs between university–industry partnerships. He has managed funds ranging up to $10 million dollars. He has very successfully built IP portfolios during his career bringing attraction for larger OEMs spin-offs. Dr Suri has submitted over 100 US/European inventions, 20 trademarks, 50 books and over 750+ peer-reviewed Google Scholar articles, over 300+ journal articles with the National Library of Medicine (NLM), Washington DC, and currently holds over 38 000+ citations and an H-index of 90. Dr Suri has conducted over 50 national and international seminars around the globe. He received his Masters from the University of Illinois, Chicago, Doctorate from University of Washington, Seattle, and Masters in Business Administration (MBA) from Weatherhead School of Management, Case Western Reserve University (CWRU), Cleveland.

Dr Suri was crowned with Director General's President's gold medal (in 1980); one of the youngest Fellows of the American Institute of Medical and Biological Engineering (AIMBE, 2004) for his outstanding contributions in health imaging, a recipient of Marquis Life Time Achievement Award (2018) for his outstanding contributions in healthcare, Fellow of American Institute of Ultrasound in Medicine, Fellow of Asia Pacific Vascular Society, Fellow of Society of Vascular Medicine, and Fellow of IEEE, all for exceptional contributions. He believes in 'getting a job done' using his strengths of innovation, strategic partnerships and strong team collaborations by bringing cross-functional and multi-disciplinary teams together both in-house and outsourcing relationships. Dr Suri currently lives in California, USA.

List of contributors

Mohit Agrawal
Bennett University, Greater Noida, Uttar Pradesh, India

Sushant Agrawal
Excelra, Hyrerabad, India

Amita Banerjee
Kalinga Institute of Industrial Technology, Bhubaneshwar, India

Radhakrishn Birla
Kalinga Institute of Industrial Technology, Bhubaneshwar, India

Swastika Bishnoi
Kalinga Institute of Industrial Technology, Bhubaneshwar, India

Mainak Biswas
Kalinga Institute of Industrial Technology, Bhubaneshwar, India

Gautam Chugh
Kalinga Institute of Industrial Technology, Bhubaneshwar, India

Sujoy Datta
Kalinga Institute of Industrial Technology, Bhubaneshwar, India

Pankaj Kumar Jain
Washington University School of Medicine, Washington, USA

Ashutosh Jha
Kalinga Institute of Industrial Technology, Bhubaneshwar, India

Piyush Kumar
Kalinga Institute of Industrial Technology, Bhubaneshwar, India

Venkateshh Moningi
Kalinga Institute of Industrial Technology, Bhubaneshwar, India

Harshit Sharma
Kalinga Institute of Industrial Technology, Bhubaneshwar, India

Riddhika Shringi
Kalinga Institute of Industrial Technology, Bhubaneshwar, India

Mayank Singhal
Kalinga Institute of Industrial Technology, Bhubaneshwar, India

Jasjit Suri
AtheroPoint LLC, Sacramento, CA, USA

Part I

An overview of deep learning and its applications in COVID-19 and tuberculosis

IOP Publishing

Multimodality Imaging, Volume 2
Heart, lungs and peripheral organs
Mainak Biswas and Jasjit S Suri

Chapter 1

An overview of AI applications in medical imaging for COVID-19-related brain and heart injuries

Harshit Sharma, Radhakrishn Birla, Mainak Biswas and Jasjit S Suri

Artificial intelligence (AI) has significantly impacted the field of medicine, especially radiology, in recent years. The COVID-19 pandemic has caused a devastating impact, with over 416 million people infected and more than 5.8 million lives lost as of February 23, 2022. Although there have been approximately 228,391 publications on COVID-19, only a few articles have focused on the influence of AI and medical imaging on infected patients with comorbidities.

A comprehensive study has recently been conducted to investigate the various pathways that lead to heart and brain damage in individuals who have contracted COVID-19. This study has provided valuable insights into the importance of medical imaging in the management of patients with comorbid conditions, utilizing statistical data on COVID-19 symptoms. Common symptoms associated with COVID-19 include hypoxia, arrhythmias, plaque rupture, coronary thrombosis, encephalitis, ischemia, inflammation, venous and lung injury, as well as thromboembolism. The research primarily focuses on the application of AI in identifying specific tissues in COVID-19 patients and assessing the severity of their illness through the analysis of medical images. Given the limited medical resources available to governments worldwide in the fight against the pandemic, the use of image-based AI has become increasingly essential for the detection and diagnosis of COVID-19.

The integration of imaging and AI-based tissue classification, along with preliminary test probability and COVID-19 symptoms, has revealed a promising method to evaluate the potential danger posed by patients with comorbidities. Techniques like these can play a crucial role in monitoring and enhancing the healthcare system during and after the epidemic. Keywords such as COVID-19, comorbidity, pathophysiology, heart, brain, lung, imaging, artificial intelligence, and risk assessment have been identified as important factors in this context.

doi:10.1088/978-0-7503-2352-9ch1

1.1 Introduction

In December 2019, a new coronavirus called 'severe acute respiratory distress syndrome coronavirus 2' (SARS-CoV-2) was identified in Wuhan, the capital of Hubei Province in the People's Republic of China [1]. Initially, the Chinese government referred to the illness caused by the viral infection as 'new coronavirus pneumonia' (NCP), while the World Health Organization (WHO) named it 'coronavirus disease 2019' (COVID-19). A global public health emergency was declared on January 30, 2020 [2]. The primary mode of transmission for SARS-CoV-2 is believed to be by means of respiratory droplets or nasal secretions [3]. Interhuman transmission was first observed by Jasper Fuk-Woo Chan *et al* during their investigation at the University of Hong Kong-Shenzhen Hospital [4]. As of July 16, 2022, the pandemic had spread to more than 200 countries, resulting in over 416 million infections and 5.8 million deaths due to its high contagion rate (Ro = 2.7) [5], as shown in figure 1.1.

Recent studies have revealed that individuals with preexisting conditions face a higher risk of severe consequences due to COVID-19 [6–10]. In a specific study focused on COVID-19 patients, diabetic individuals (48, 24.9%) exhibited significantly higher mortality rates (81.3% vs. 47.6%) and ICU hospitalization rates (66.7% vs. 41.4%) compared to non-diabetic individuals (145, 75.1%) [11]. Diabetic individuals also experienced severe inflammatory reactions and coagulopathy in the heart, liver, and kidneys. Infected individuals with chronic disorders such as diabetes, renal disease, dyslipidemia, hypertension, cardiovascular diseases, and chronic obstructive pulmonary disease (COPD) had a higher prevalence of heart and

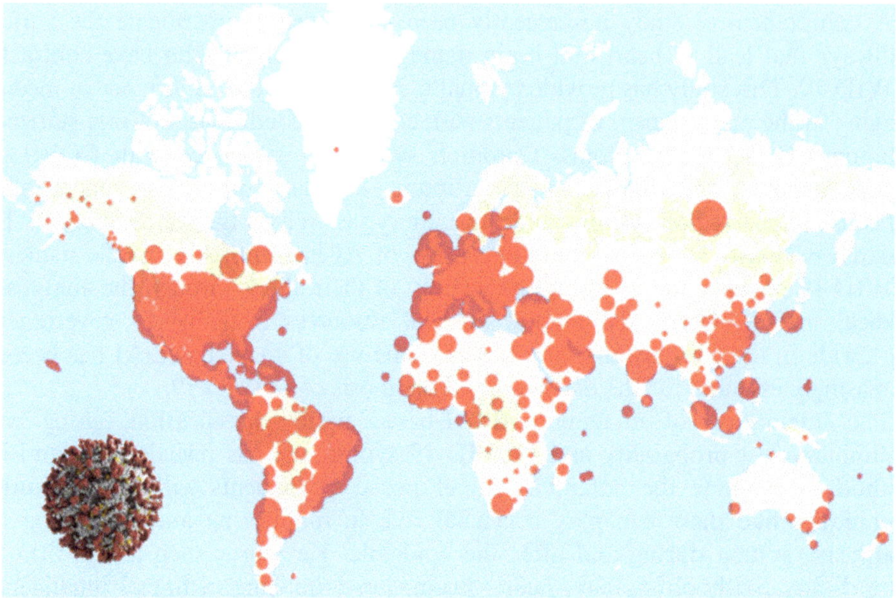

Figure 1.1. COVID-19 is distributed over 213 countries on a world map (courtesy: John Hopkins University).

brain (H&B) damage [12–16]. SARS-CoV-2 has been found to infect the thin lining of the epithelial cells that line the arteries, leading to atherosclerosis and arterial inflammatory disease, which are significant contributors to cardiovascular diseases (CVDs) and H&B damage [12, 17–21]. This may be attributed to a decrease in the production of angiotensin-converting enzyme 2 (ACE2), which results in endothelial dysfunction and exacerbates existing atherosclerosis [22, 23]. When individuals with comorbidities undergo image screening, it has been observed that they exhibit a wide range of preliminary test probabilities (PTPs) for COVID-19, ranging from mild to severe [24]. Conventional cardiovascular risk factors (CCVRFs), such as imaging techniques of the heart or alternative indicators used as substitutes for assessing coronary artery disease (e.g., carotid artery disease), are closely associated with comorbid patients. COVID-19 severity prediction models can benefit from the incorporation of both biomarkers and imaging [25–30]. Figure 1.2 illustrates the connections between SARS-CoV-2 and comorbidities, as well as the survival rates of COVID-19 individuals with and without diabetes.

The expression of ACE2 can lead to scarring and potential artery rupture [31–34]. Therefore, it is essential to evaluate CCVRF alongside imaging in individuals with COVID-19 and other comorbidities [35]. In stage two of the disease, when patients are severely affected by COVID-19, there is a higher risk of heart damage or the release of troponin T (TnT). Imaging has proven to be valuable in keeping track of tissue scarring caused by COVID-19 [35–39].

Different imaging modalities such as magnetic resonance imaging (MRI), computed tomography (CT), and ultrasound can be employed to detect COVID-19 symptoms in patients [40–44]. These imaging techniques offer the advantage of visualizing the scar tissue caused by the disease. However, a drawback is their inability to provide a 'risk assessment' on their own. Artificial intelligence (AI) technologies have the potential to leverage information from imaging modalities and generate more precise predictions, enabling accurate identification of tissues and disease processes [45–51]. The combination of AI and medical imaging (MI) has demonstrated significant advancements in diagnosis, risk stratification, rapid patient

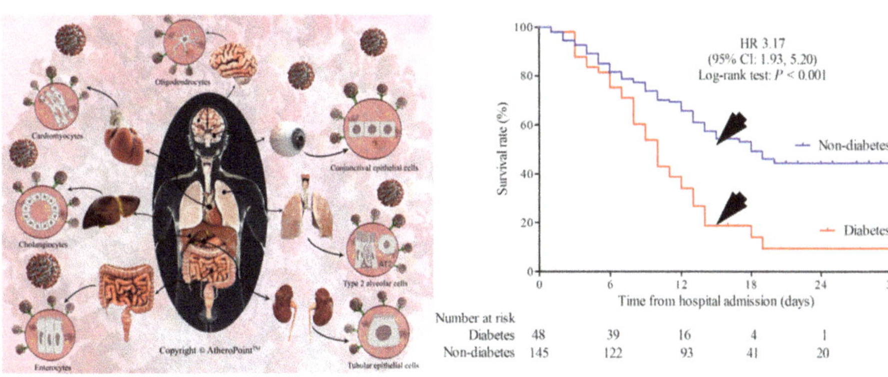

Figure 1.2. (a) SARS-CoV-2 and its link with other comorbidities, and (b) COVID-19 diabetes and non-diabetic patients' mortality rates compared (with permission to reprint [11]).

evaluation, disease monitoring, and early intervention [40, 48, 52–57]. Consequently, this review focuses on the utilization of AI-based tissue characterization through medical imaging in comorbid patients affected by COVID-19.

The chapter is organized as follows: section 1.2 examines the physiological mechanisms underlying the four pathways that result in heart and brain injuries. Section 1.3 presents an overview of the justification for utilizing imaging in the context of the COVID-19 pandemic. Section 1.4 provides an in-depth exploration of utilizing AI-based tissue characterization for risk assessment. Ultimately, the paper concludes with a thorough critical analysis.

1.2 SAR-CoV-2 pathophysiology in the context of heart and brain injury

Numerous studies indicate that SARS-CoV-2 relies on the ACE2 receptor for cell entry, achieved by binding to the spike protein (S protein) on the cell surface [58–60] (see figure 1.2). ACE1 and ACE2 are carboxypeptidase enzymes that are structurally similar but have distinct roles in the renin-angiotensin-aldosterone system (RAAS) [61]. The ACE2 is found in various cardiac cells, including, astrocytes (brain cells), enterocytes and type 2 pneumocytes [15, 61–63], and is recognized as a contributing factor to extrapulmonary complications. Figure 1.3 provides a comprehensive visual representation of how SARS-CoV-2 induces cardiac and brain damage through four

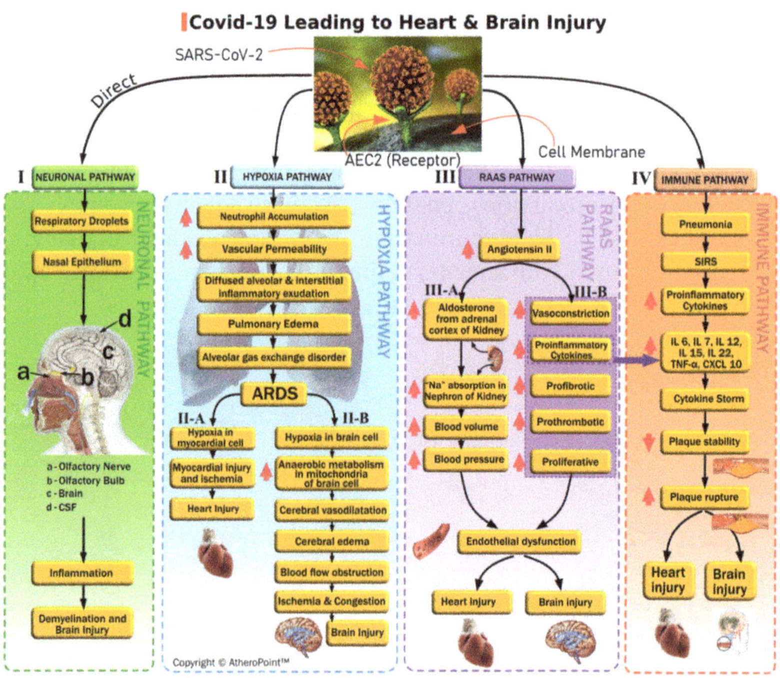

Figure 1.3. We have shown in four pathways how COVID-19 can cause brain and heart injury. Brain image in pathway I: http://debuglies.com/2020/01/23/olfactory-disturbances-have-implications-in-mental-and-emotional-well-being-health/ (courtesy of Debug Lies).

distinct paths: (i) the RAAS pathway, (ii) the immune pathway, (iii) the neural pathway, and (iv) the hypoxia pathway. These pathways will be further discussed, along with the resulting injuries, which may encompass infectious toxic encephalopathy, acute cerebrovascular diseases and viral encephalitis.

(i) The neural pathway (figure 1.3, the first pathway): Recent epidemiological investigations have highlighted genomic similarities between MERS, SARS-CoV-1, and SARS-CoV-2 [6, 64, 65]. Prior research has demonstrated that coronaviruses, including SARS-CoV-1, have the ability to enter the brain and directly infect it [66, 67]. In figure 1.3, the sagittal brain image representing the neural pathway illustrates the olfactory nerve and bulb, labeled as 'a' and 'b,' respectively [68–70]. It has been observed that individuals infected with SARS-CoV-2 may experience symptoms such as dysgeusia (taste loss) and anosmia (loss of smell) [64, 71–73]. Furthermore, a mouse experiment where the olfactory bulb was surgically removed demonstrated a limitation of CoV within the central nervous system (CNS) [74]. These findings suggest that the neural pathway could be one of the potential routes for SARS-CoV-2.

(ii) The hypoxia pathway (figure 1.3, the second pathway): Following the entry of the coronavirus into lung parenchyma cells, there is a reduction in ACE2 levels, leading to the accumulation of neutrophils, increased vascular permeability, and the release of diffuse alveolar and interstitial exudates. This process contributes to the development of acute respiratory distress syndrome (ARDS) and pulmonary edema [75]. ARDS is described by significant irregularities in the composition of blood gases, causing an imbalance of oxygen and carbon dioxide and leading to decreased blood oxygen levels [76, 77]. Prolonged hypoxia can induce myocardial ischemia and cardiac damage [78, 79] (see figure 1.3, pathway II-A). In the brain, hypoxia is the primary cause of cerebral vasodilation, edema, and reduced blood flow due to increased anaerobic metabolism in brain cell mitochondria. This can lead to cerebral ischemia and acute cerebrovascular disorders, such as acute ischemic stroke [71, 80] (see figure 1.3, pathway II-B).

(iii) The RAAS pathway (figure 1.3, the third pathway): The renin-angiotensin-aldosterone system (RAAS) pathway plays a critical role in regulating blood pressure and electrolyte balance. Disruption of this pathway can contribute to the development of cardiovascular disorders [15]. Prior to SARS-CoV-2 invasion, angiotensin I (Ang I) is converted to angiotensin II (Ang II) by ACE1. Ang II causes vasoconstriction and possesses pro-inflammatory, prothrombotic, and proliferative properties that can negatively impact the hemostasis and vascular tone [77, 80]. Conversely, ACE2 counteracts the effects of Ang II by converting it to Ang (1–7), which has mitigating effects [75, 78]. Both ACE2 and Ang (1–7) have protective effects on the cardiovascular and cerebrovascular systems [61]. SARS-CoV-2 infection disrupts the RAAS, leading to injuries in the heart and brain through two distinct pathways. The primary mechanism involves an increase in Ang II levels due to a decrease in ACE2 levels (figure 1.3, pathway III-A). Firstly, elevated Ang II levels stimulate the adrenal cortex in the kidney, resulting in increased aldosterone production. Aldosterone, a steroid hormone, facilitates the reabsorption of sodium and water in the distal tubule and collecting duct of the nephron [81]. This leads to an increase in blood volume and raises blood pressure, causing endothelial dysfunction and subsequent damage to the heart and brain [82]. Secondly, elevated Ang II levels and decreased ACE2 levels contribute to endothelial dysfunction in arterial walls, which can be observed in arterial wall images [21] (see figure 1.3, pathway III-B). High levels of Ang II can also trigger the release of pro-inflammatory cytokines, contributing to a cytokine storm.

(iv) The immune pathway (figure 1.3, the fourth pathway): In recent studies, SARS-CoV-2 viral pneumonia has been linked to an elevated inflammatory response known as a 'cytokine storm' [7, 77, 83, 84]. Advanced stages of severe COVID-19 are characterized by increased levels of inflammatory cytokines, which can contribute to multiple organ failure [85–87]. Inflammatory markers such as IL-6, IL-7, IL-12, IL-15, IL-22, TNF-α, and CXCL-10 have been associated with plaque destabilization. This increased inflammation can potentially lead to plaque rupture and subsequent damage to the heart and brain [37, 68–70, 80, 85–87, 89–91].

1.3 The role of imaging in patients with comorbidities and COVID-19

COVID-19 leads to significant damage to the heart and brain through four pathways (neuronal, hypoxia, RAAS, and immunological), as discussed earlier. This highlights the need for increased utilization of medical imaging (MI) to expedite assessments, differential diagnoses, and patient management [92] with appropriate safety measures. The choice of imaging modality depends on symptom severity, with consideration for portability and invasiveness. Portable and non-invasive ultrasound imaging in B-Mode is suitable for low-risk individuals, while x-rays, magnetic resonance imaging (MRI), and computed tomography (CT) are non-portable and can be used for patients with a medium risk level [40, 41]. Invasive imaging techniques like intravascular ultrasonography (IVUS) and ventriculography are reserved for life-threatening situations [42, 43, 98–100]. Ultrasound is particularly advantageous due to its rapidity, reproducibility, cost-effectiveness, radiation-free nature, and portability. It can be performed in isolation, minimizing the risk of COVID-19 transmission [101, 102].

Throughout the early and later stages of the pandemic, various imaging modalities have proven effective. X-ray imaging of the lungs has revealed different patterns, signaling the advancement of COVID-19 at different stages and aiding in treatment planning [103]. Chest CT scans have shown lung involvement in nearly 86% of COVID-19 patients, affecting at least one lobe [104]. Chest MRI scans have revealed pulmonary tissue consolidation, diffusion-restricted areas, and lung injury in COVID-19 patients [105]. MRI examinations of recovered patients have identified myocardial edema and late gadolinium augmentation, indicating long-term cardiac damage requiring ongoing care even after recovery [106]. MR scans of COVID-19 patients have demonstrated myocardial inflammation, highlighting cardiac damage caused by the cytokine storm associated with the infection (Pathway IV) [107]. Studies have also investigated the impact of COVID-19 on the brain, with MRI images showing hemorrhagic rim enhancing lesions in specific brain regions [108] (figure 1.4). Abnormal findings have been observed in brain MRI scans of COVID-19 patients, and combined CT and ultrasound investigations have revealed liver disease and gallbladder abnormalities [109, 110]. Recent MRI scans of COVID-19 patients' olfactory bulbs have revealed inflammatory occlusion caused by the interaction between SARS-CoV-2 and the ACE2 protein expressed in the olfactory epithelium, resulting in the loss of olfactory function [111].

Invasive imaging techniques are employed to ascertain the diagnosis of individuals with COVID-19 with significant comorbidities. One trial utilized intravascular ultrasonography (IVUS) in combination with stenting for a COVID-19 patient who experienced a myocardial infarction [112] (figure 1.5). Precautions regarding invasive imaging techniques are further explained in section 1.5. Another study employed ventriculography to detect takotsubo syndrome, a type of cardiac damage associated with COVID-19 [113]. In several trials, MI of COVID-19 patients played a critical role in assessing tissue damage and determining the severity of infection, even in the absence of obvious signs [39, 114]. Therefore, MI is recommended for evaluating the degree of damage to cardiac and cerebral tissues in individuals with

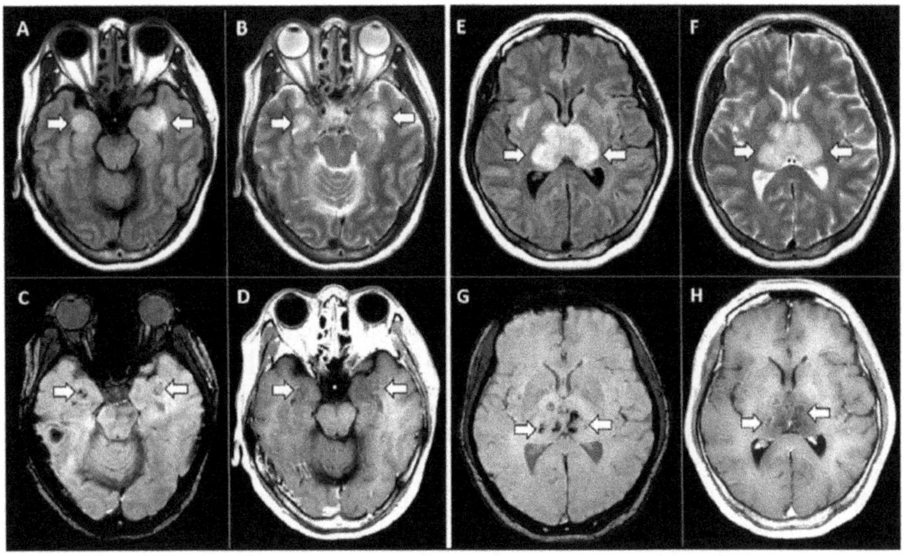

Figure 1.4. The MRI scan of a patient with COVID-19 showed evidence of bleeding. T2 FLAIR hyperintensity was observed in the paired medial temporal lobes and thalami (A, B, E, F), and the hemorrhage was identified by a hypointense signal intensity on susceptibility-weighted images (C, G). Additionally, postcontrast imaging revealed rim enhancement (D, H) (reprinted with permission [108]).

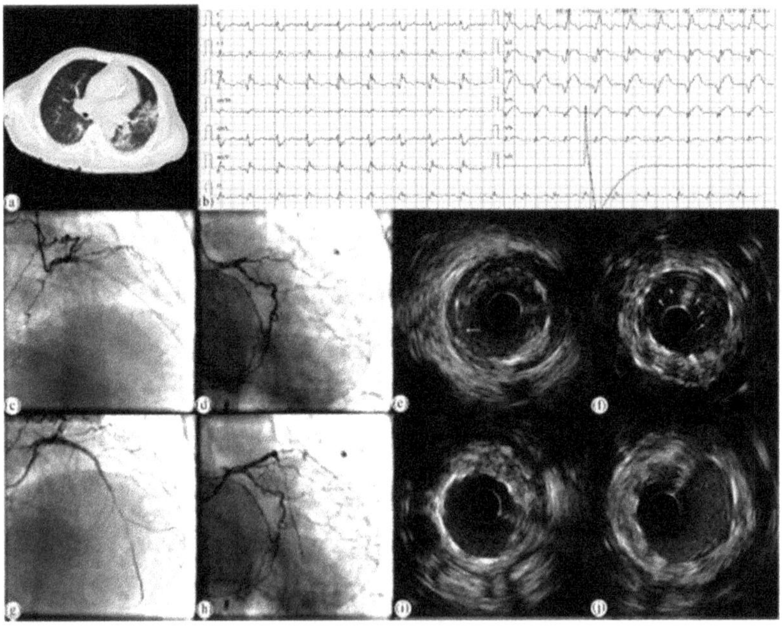

Figure 1.5. In a COVID-19 patient with myocardial infarction, both chest CT and intravascular ultrasound (IVUS) were utilized for diagnostic purposes. The findings from these imaging techniques are as follows: (a) Chest CT scan revealed localized fibrinous exudative alterations, which are associated with viral pneumonia. (b) ECG data showed ST-segment elevations in leads V1–V5 when the patient experienced chest pain. (c, d) Coronary angiography (CAG) indicated occlusion in the proximal segment of the left anterior descending artery (LAD). (e, f) Blood flow in the LAD was restored after the placement of two drug-eluting stents (DESs). (g) IVUS revealed a dissection distal to the stent in the LAD, specifically from the 7–12 o'clock position. (h) A low echogenic shadow with dispersed increased echogenic flicker was observed, indicating the presence of a thrombus. (i) The dissection was no longer visible after the DES was implanted and the stent was adequately inflated. (j) The thrombus disappeared following the intervention. These findings were obtained from a published study and are reprinted with permission [112].

COVID-19 throughout their lifetime. Individuals with preexisting medical conditions who have contracted COVID-19 are particularly vulnerable and should undergo MI examination from the time of diagnosis. MI can also be beneficial for COVID-19 patients with deep vein thrombosis (DVT). An analysis found that patients suffering from COVID-19 and DVT had a worse prognosis compared to those lacking DVT. The DVT group had a higher rate of ICU admission (18.2%), lower rate of discharge (48.5%), and higher mortality rate (38.5%) [115].

However, the evaluation, diagnosis, and monitoring processes for myocardial infarction imaging can be challenging due to the exponential nature of the pandemic, limited medical resources, and a shortage of radiologists. These factors contribute to time-consuming processes and a higher risk of errors [116–118]. To address these challenges, the utilization of artificial intelligence (AI) in medical imaging (MI) for tissue characterization can offer valuable support. AI-based systems have the potential to be scaled up to meet the demands of the pandemic, facilitating rapid MI assessments and diagnoses during the COVID-19 outbreak [119–121].

Based on the severity of symptoms and patient presentation, AI-driven assessments have the capability to classify or categorize the risk level into different categories, including zero-risk, low, low-medium, high-medium, low-high, or high-high risk [120, 122], as illustrated in figure 1.6. The choice of imaging modality depends on the assessed risk level. For zero-risk patients, no imaging is necessary.

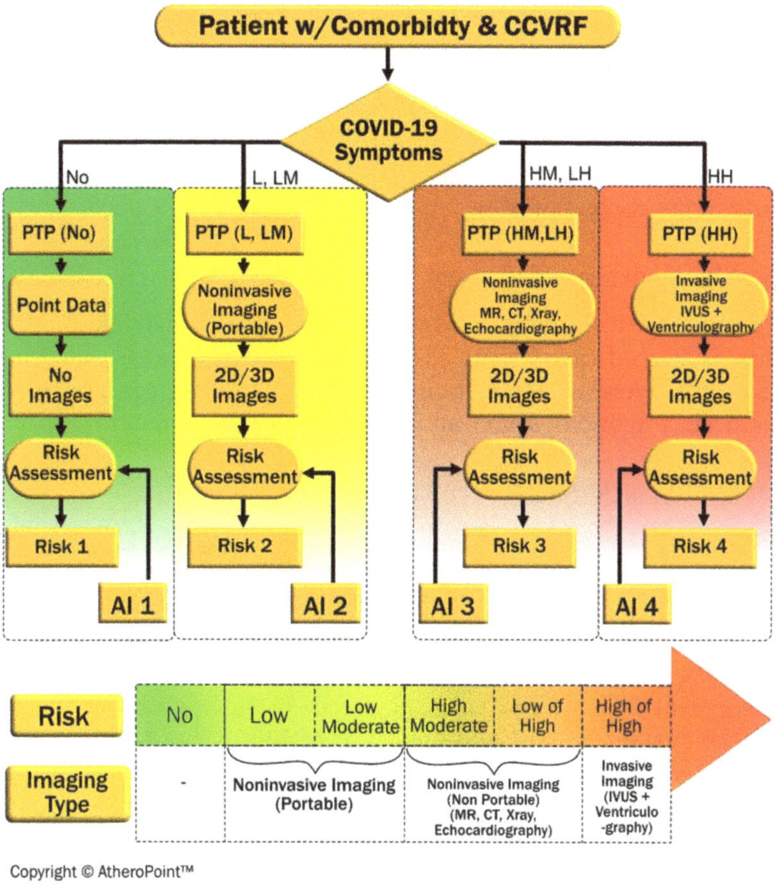

Figure 1.6. AI-based risk assessment plays a crucial role in managing comorbidity patients with COVID-19, offering valuable insights and aiding in healthcare administration [54, 55, 127–129]. The implementation of AI in healthcare encompasses various systems that enable accurate decision-making in patient monitoring, diagnosis, management, and treatment. In the field of medical imaging, artificial intelligence has gained significant importance due to the abundant volume of three-dimensional data accessible and the necessity to characterize and quantify diseases utilizing imaging observations [130–132]. Tissue imaging and classification are particularly vital as they directly impact decisions regarding the severity of COVID-19 in patients [133–135]. The key advantage of AI technologies lies in their ability to be trained to emulate the cognitive actions of physicians, allowing for the prediction of disease severity in asymptomatic patients. Several machine learning (ML)-based approaches have effectively utilized AI to combat COVID-19 within a short timeframe [136, 137].

Portable imaging modalities are suitable for patients at zero-risk and low-medium risk levels. Non-portable imaging techniques such as MRI, CT, x-ray, and echocardiography are appropriate for high-medium and low-high risk patients. Invasive imaging methods like IVUS and ventriculography are reserved for high-high risk patients. Precise evaluation of diagnostic results and patient categorization into specific risk groups can be achieved through pre-test probability (PTP) assessment [123–126]. Non-imaging biomarkers can be utilized by AI-based algorithms for risk assessment in zero-risk patients. Patients with low risk may undergo portable 2D/3D imaging modalities, such as ultrasound, while non-portable and invasive 2D/3D imaging modalities are suitable for low-medium risk patients. High-high risk patients may require invasive imaging techniques like ventriculography and IVUS. AI-driven MI plays a crucial role in assessing the risk level based on data obtained from multiple 2D/3D scans, and treatment decisions can be made accordingly. The subsequent section will focus on deep learning (DL)-based MI, particularly in the context of ultrasound scans for COVID-19 patients.

1.4 Machine learning and deep learning for tissue classification

The exponential rise in the number of patients during the pandemic and the limited availability of trained radiologists have presented challenges in achieving timely diagnoses. Nonetheless, the integration of AI and related technologies in healthcare holds significant promise in significantly reducing diagnosis times [119].

1.4.1 ML and DL architectures

The machine learning process consists of two stages. In stage I, various attributes from the images of COVID lesions are extracted and processed by a machine learning (ML) model to produce offline parameters. These parameters are then modified by test lesion photos, leading to intelligent categorization or inference. Figure 1.7 illustrates a typical machine learning system used for predicting risk class. The development of a CUSIP (image-based phenotype) relies on the event equivalent gold standard (EEGS) model [57, 138, 139]. Deep learning (DL) functions similarly to the visual cortex, employing multiple neural layers directly applied to tissue images for feature extraction and classification purposes [54]. Convolutional neural networks (CNN) [140], as shown in figure 1.8, are a common type of deep learning network used for medical image classification. Convolution and max-pooling operations are employed to extract features and carry out characterizations. Both ML and DL utilize a supervised learning method is employed, in which models are trained using preexisting data.

The previous sections have discussed how COVID-19 spreads through four distinct pathways and can cause damage to the heart and blood vessels (H&B). Myocardial infarction (MI) can be used to assess the level of tissue damage in these pathways, aiding healthcare professionals in developing appropriate treatment strategies for patients. The use of AI models for tissue classification based on medical images has been widely employed, both during the pandemic and in routine healthcare settings. In the subsequent sections, we will present a proposed approach

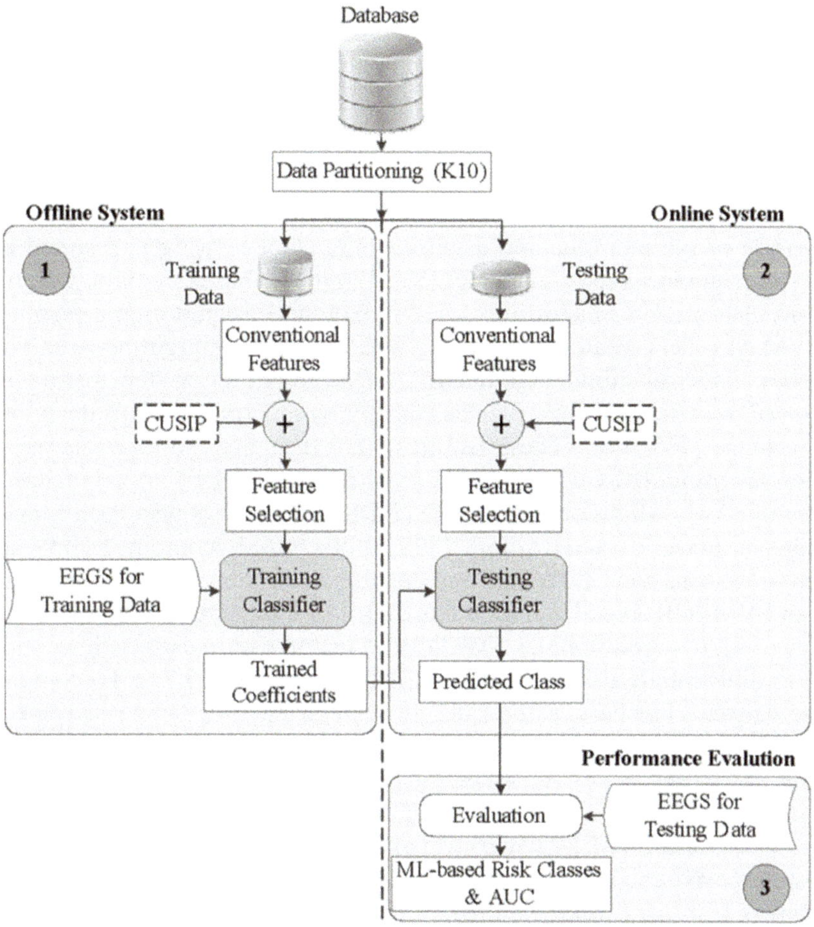

Figure 1.7. Classic ML model utilizing EEGS model.

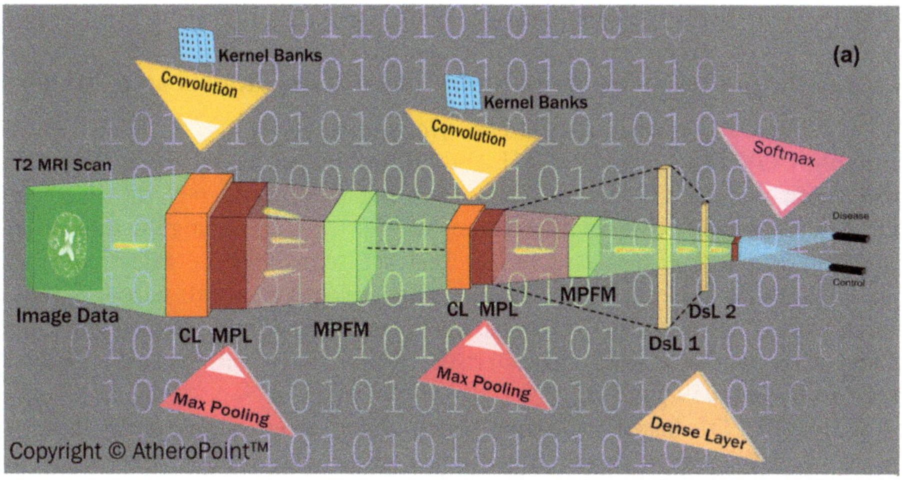

Figure 1.8. A convolution neural network (courtesy of AtheroPoint™, CA, USA).

for describing tissue classification using deep learning (DL) and provide specific examples of AI applications for each organ.

1.4.2 Tissue characterization ML system for stroke risk stratification

There are two primary types of AI-based approaches: (i) ML-based and (ii) DL-based methods [63, 141, 142]. ML-based techniques have been developed for the classification of symptomatic and asymptomatic plaques using ultrasound images. For instance, support vector machines (SVM), an ML-based method, were utilized to classify 346 carotid images into symptomatic and asymptomatic plaques [143, 144]. SVM classifiers create a hyperplane with the largest margin between points of two classes, known as support vectors. In the feature extraction step, texture analysis is employed to extract features such as entropy, symmetry, standard deviation, and run percentage [145, 146]. These features were then used to characterize plaque tissue lesions using SVM with a radial basis function (RBF) kernel, achieving an accuracy of 82.4%. Higher-order spectra (HOS) analysis has also been found to be significant in tissue characterization [130]. Another study combined HOS, discrete wavelet transformations (DWTs), and texture data from 146 patient images to create an SVM-RBF-based classifier [46, 130, 146–148]. This classifier achieved an accuracy of 91.7%. Additionally, DWT-based features were used with second-order kernels to differentiate tissues, resulting in an accuracy of 83.7%. To compare and evaluate various classifiers, a total of 346 scans from two distinct carotid plaque datasets (Portugal and the United Kingdom) were utilized. Various classifiers, including fuzzy classifier [154], k-nearest neighbor [152], radial basis probabilistic neural network [150], decision tree [151], Gaussian mixture model [149], naive Bayes classifier [153], and SVM [45] fuzzy classifier [154], were evaluated. The primary features employed encompassed trace transform [155], fuzzy gray level co-occurrence matrix [156], and fuzzy run-length matrix [157]. In the Portugal cohort, the fuzzy classifier attained the highest accuracy of 93.1%, while both the NBC and SVM-RBF kernels exhibited comparable performance at 85.3%. These AI models for plaque classification have been applied in various approaches for cardiovascular disease (CVD) risk stratification [27, 28, 158].

1.4.3 Vessel characterization, measurement, and risk stratification using ML/DL

1.4.3.1 Chest CT and liver disease classification using AI

During the COVID-19 pandemic, ML and DL techniques have been utilized for the classification of lung CT images, demonstrating varying degrees of effectiveness [159–164]. Kang *et al* achieved an accuracy of 95.5% by employing the utilization of representation learning to characterize chest CT scans without infection of data from COVID-19 patients [168]. Wang *et al* developed a DL-based system to differentiate CT scans of COVID-19 patients from those of non-infected individuals, yielding a receiver operating characteristic (ROC) curve with an area under the curve (AUC) of 0.959 [169].

Additionally, DL-based radiomics using shear wave elastography has been applied to distinguish diseased (fatty liver) ultrasound images and assess liver fibrosis stages with an impressive accuracy of 100% [170–172]. This technique proves valuable for the identification and categorization of COVID-19 patients.

1.4.3.2 Tissue characterization and risk stratification using artificial intelligence in lung CT

Over the past few decades, numerous studies have been conducted ML and DL algorithms for the classification of lung CT images. These studies can be categorized into two types based on the number of risk stratification classes involved. The initial set of research focused on distinguishing COVID-19 pneumonia patients from non-COVID-19 pneumonia patients, resulting in a two-class scenario. The subsequent collection of studies explored multi-class paradigms.

In one study, the DenseNet 121 model was employed to create and segment lung masks, achieving an area under the curve (AUC) of 0.9, a sensitivity of 78.93%, and a specificity of 89.93% for categorizing COVID-19 and control patients [173]. Zhang *et al* adopted a three-class classification system that encompassed lung segmentation and categorization, including COVID-19, community pneumonia, and normal cases. They utilized the DeepLabv3 model for lung segmentation and 3D ResNet-18 for classification, achieving an accuracy of 92.49% and an AUC of 0.98 [174].

Other researchers also incorporated AI techniques in their CT lung scan studies. For instance, Li *et al* developed a DL system for CT lung analysis capable of predicting COVID-19 severity and progression [175]. Chen *et al* devised a UNet++ architecture to segment COVID-19-infected lung regions in CT scans [176, 177]. In a similar manner, Yang *et al* conducted lung segmentation on CT images by identifying pulmonary parenchyma and employing DenseNet for classification. Their approach achieved an accuracy of 92% and an AUC of 0.98 [178]. In another study, Oh *et al* employed x-ray chest images for both classification and segmentation, achieving an accuracy of 88.9% by employing a patch-based technique with the same network [179].

1.4.3.3 AI-based plaque tissue characterization and risk stratification for cardiac health

A DL-based platform is proposed for the treatment of COVID-19 patients with comorbidities. The platform utilizes preexisting facts obtained by patients suffering from COVID-19 worldwide to train the DL system. This data includes multiple ultrasound scans by patients suffering from COVID-19 with comorbidities who underwent treatment according to strict guidelines [93–97]. The AtheroEdgeTM system, which is capable of distinguishing and fragmenting plaque regions, automatically extracts tissue regions of interest (ROIs) from the ultrasound scans. The same AtheroEdgeTM technique is applied to extract ROIs from online patients' ultrasound scans. The DL model is then used to estimate the susceptibility of plaque in the online data, which is collected from testing patients after being trained with the offline data. The predictions obtained from this process are utilized to evaluate and support the clinical feasibility of the DL system.

1.5 Summary

In this review, we conducted an analysis of various imaging investigations performed on COVID-19 patients to assess the impact of the infection on key organs such as the lungs [104, 105], heart [106, 107], brain [108, 109] and liver. These imaging investigations were crucial in guiding the medical team in providing appropriate treatment for COVID-19 patients with varying degrees of symptoms. Among the available imaging modalities, ultrasonography was found to be particularly advantageous due to its portability and the ability to perform scans in isolation rooms, minimizing the risk of infection transmission across different wards. Similar mobility recommendations were also given for MRI and CT scanning as a preventive measure to curb the transmission of infections [180–182]. The availability of mass scanning for admitting patients would enable healthcare professionals to promptly design treatment plans and potentially save lives. In serious cases of patients suffering from COVID-19, the use of IVUS [98] and ventriculography (with necessary preventive measures) is recommended [42, 43, 99, 100] (figure 1.9).

The exponential growth of the COVID-19 pandemic presents challenges in evaluating and analyzing medical images in light of resource constraints and a shortage of radiologists. To address this, AI-driven medical imaging (MI) can be utilized to assist in the analysis, diagnosis, and risk stratification of patients suffering from COVID-19. AI systems have the capability to process large volumes of images simultaneously, enabling mass diagnosis to keep up with the rapidly evolving pandemic curve.

There are two main types of AI: ML and DL [54]. ML models use feature mining algorithms to make predictions, while DL models directly extract features from medical images, resulting in clearer images. An AI-based imaging-based risk evaluation model is recommended, where patients are categorized into risk levels such as zero/no-risk, low-risk, low-medium, medium-high, low-high, or high-high based on pre-test probability (PTP) tests [120, 122–126]. MI is then performed based on the patient's risk level, followed by AI utilization to assess the risk in MI. DL-driven tissue characterization systems can be particularly useful for ultrasound examinations and other imaging modalities. These DL-driven systems are trained using training data and evaluated using test data, allowing for the evaluation of tissue damage caused by COVID-19 infection.

Telemedicine, combined with AI support, can play a significant role in monitoring the well-being of patients. Telemedicine enables the management of infections by monitoring patients' health through Internet of Things (IoT) devices [183]. Social media platforms can also contribute to tracking patients' health and sharing important research findings through the application of big data analytics [184–186].

1.5.1 A note on COVID-19 precautions

In order to prevent infection, medical personnel must strictly adhere to guidelines [187–189]. This includes wearing eye protection, disposable gowns that are water-resistant, and disposable gloves, among other necessary precautions. Portable equipment should be used to avoid the need for relocating patients. Any

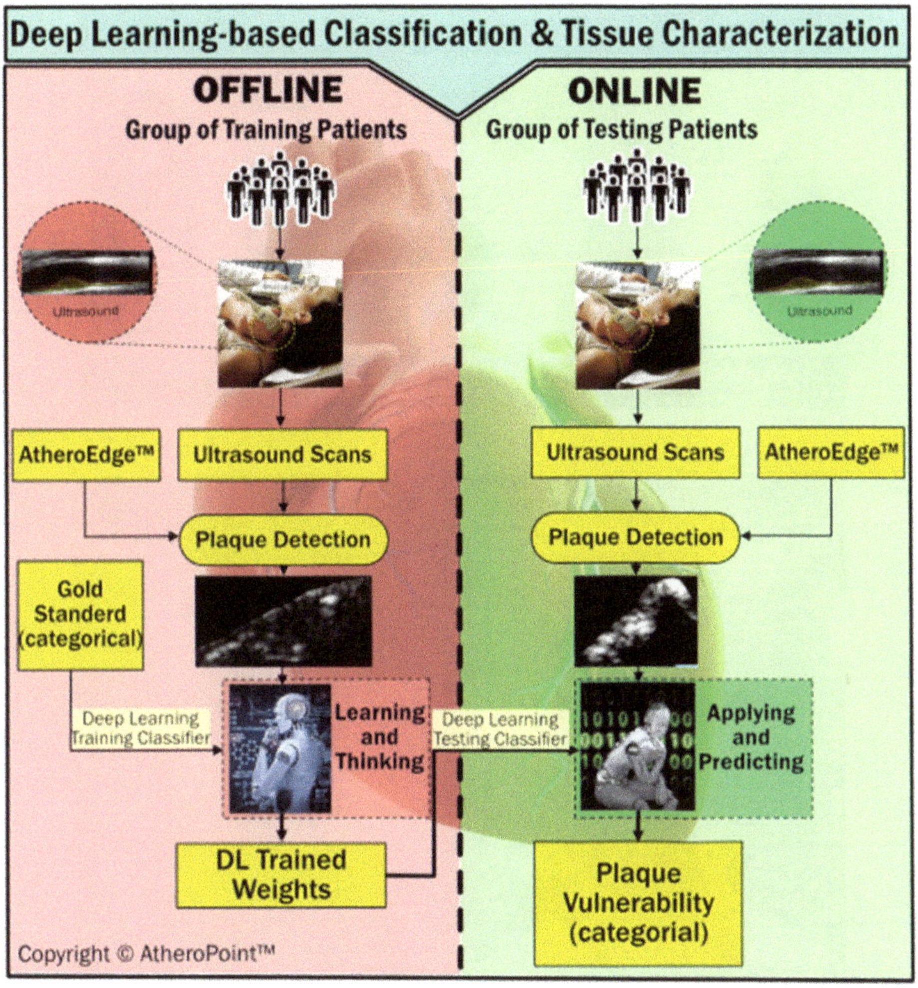

Figure 1.9. Proposed DL-based method for tissue characterization and classification of COVID-19 severity for patients with comorbidities (courtesy of AtheroPoint™, CA, USA).

medical imaging equipment that requires physical contact should be sterilized after each use, as shown in figure 1.10. Imaging equipment can be positioned outside the isolation room, allowing image acquisition through the window of the room to minimize direct interaction, as depicted in figure 1.10(b). When it is necessary to handle devices, a sterile protective disposable cover, such as an ultrasound probe cover as illustrated in figure 1.10(c), should be used.

1.6 Conclusion

COVID-19 can cause harm to the heart and blood vessels through four pathways: RAAS, neuronal, hypoxia, and immune. The severity of a patient's symptoms determines the level of risk associated with their condition, which in turn determines

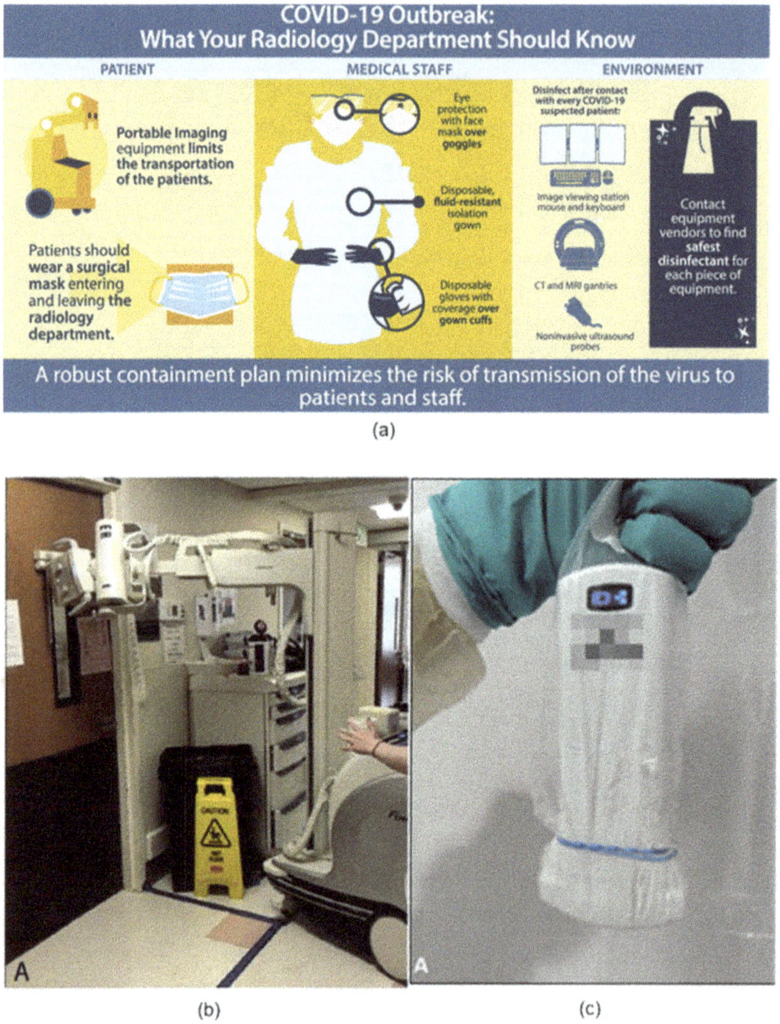

Figure 1.10. (a) Before taking scans, clinical personnel should follow these protection measures (with permission to reprint [187]); (b) Images shot through a window (with permission to reprint [188]); (c) probe covered with disposable sterile sheath (with permission to reprint [189]).

the type of imaging modality that should be used. Portable or non-portable invasive imaging modalities are recommended depending on the risk level, and appropriate safety measures must be taken during the imaging process. However, the limited availability of qualified radiologists poses a challenge to the widespread use of MI for COVID-19 diagnosis and evaluation.

To address this challenge, AI approaches such as ML and DL can be employed to expedite MI-based clinical evaluation and diagnosis. These AI methodologies have the potential to improve the efficiency and speed of diagnosing COVID-19 and assessing the risk associated with the disease. In particular, a DL-based system has been developed for COVID-19 diagnosis and risk classification, which can aid in providing timely and accurate evaluations for individuals with comorbidities who are at a higher risk of experiencing severe health complications [190].

References

[1] Yuen K-S, Ye Z-W, Fung S-Y, Chan C-P and Jin D-Y 2020 SARS-CoV-2 and COVID-19: the most important research questions *Cell Biosci.* **10** 1–5

[2] Coronavirus (COVID-19) outbreak (https://who.int/westernpacific/emergencies/covid-19)

[3] Coronavirus (https://who.int/health-topics/coronavirus#tab=tab_1)

[4] Chan J F-W, Yuan S, Kok K-H, To K K-W, Chu H, Yang J, Xing F, Liu J, Yip C C-Y and Poon R W-S 2020 A familial cluster of pneumonia associated with the 2019 novel coronavirus indicating person-to-person transmission: a study of a family cluster *Lancet* **395** 514–23

[5] Coronavirus (https://worldometers.info/coronavirus/)

[6] Wang D, Hu B, Hu C, Zhu F, Liu X, Zhang J, Wang B, Xiang H, Cheng Z and Xiong Y 2020 Clinical characteristics of 138 hospitalized patients with 2019 novel coronavirus–infected pneumonia in Wuhan, China *JAMA* **323** 1061–9

[7] Huang C, Wang Y, Li X, Ren L, Zhao J, Hu Y, Zhang L, Fan G, Xu J and Gu X 2020 Clinical features of patients infected with 2019 novel coronavirus in Wuhan, China *Lancet* **395** 497–506

[8] Shi S, Qin M, Shen B, Cai Y, Liu T, Yang F, Gong W, Liu X, Liang J and Zhao Q 2020 Association of cardiac injury with mortality in hospitalized patients with COVID-19 in Wuhan, China *JAMA Cardiol.* **5** 802–10

[9] Zhou F, Yu T, Du R, Fan G, Liu Y, Liu Z, Xiang J, Wang Y, Song B and Gu X 2020 Clinical course and risk factors for mortality of adult inpatients with COVID-19 in Wuhan, China: a retrospective cohort study *Lancet* **395** 1054–62

[10] Gao F, Zheng K I, Wang X B, Yan H D, Sun Q F, Pan K H, Wang T Y, Chen Y P, George J and Zheng M H 2020 Metabolic associated fatty liver disease increases COVID-19 disease severity in non-diabetic patients *J. Gastroenterol. Hepatol* **36** 204–7

[11] Yan Y, Yang Y, Wang F, Ren H, Zhang S, Shi X, Yu X and Dong K 2020 Clinical characteristics and outcomes of patients with severe covid-19 with diabetes *BMJ Open Diabetes Res. Care* **8** e001343

[12] Virani S S, Alonso A, Benjamin E J, Bittencourt M S, Callaway C W, Carson A P, Chamberlain A M, Chang A R, Cheng S and Delling F N 2020 Heart disease and stroke statistics—2020 update: a report from the American heart association *Circulation* **141** E139–596

[13] Zheng Y-Y, Ma Y-T, Zhang J-Y and Xie X 2020 COVID-19 and the cardiovascular system *Nat. Rev. Cardiol.* **17** 259–60

[14] Chen L, Li X, Chen M, Feng Y and Xiong C 2020 The ACE2 expression in human heart indicates new potential mechanism of heart injury among patients infected with SARS-CoV-2 *Cardiovasc. Res.* **116** 1097–100

[15] Williams V R and Scholey J W 2018 Angiotensin-converting enzyme 2 and renal disease *Curr. Opi. Nephrol. Hypertens.* **27** 35–41

[16] Wang B, Li R, Lu Z and Huang Y 2020 Does comorbidity increase the risk of patients with COVID-19: evidence from meta-analysis *Aging (Albany NY)* **12** 6049

[17] Cheng H, Wang Y and Wang G Q 2020 Organ-protective effect of angiotensin-converting enzyme 2 and its effect on the prognosis of COVID-19 *J. Med. Virol* **92** 726–30

[18] Libby P 2020 The heart in COVID19: primary target or secondary bystander? *JACC: Basic Transl. Sci.* **5** 537–42
[19] Clerkin K J, Fried J A, Raikhelkar J, Sayer G, Griffin J M, Masoumi A, Jain S S, Burkhoff D, Kumaraiah D and Rabbani L 2020 Coronavirus disease 2019 (COVID-19) and cardiovascular disease *Circulation* **141** 1648–55
[20] Libby P, Ridker P M and Maseri A 2002 Inflammation and atherosclerosis *Circulation* **105** 1135–43
[21] Suri J S, Kathuria C and Molinari F 2010 *Atherosclerosis Disease Management* (New York: Springer Science & Business Media)
[22] South A M, Diz D I and Chappell M C 2020 COVID-19, ACE2, and the cardiovascular consequences *Am. J. Physiol-Heart Circ. Physiol.* **318** H1084–90
[23] Dong B, Zhang C, Feng J B, Zhao Y X, Li S Y, Yang Y P, Dong Q L, Deng B P, Zhu L and Yu Q T 2008 Overexpression of ACE2 enhances plaque stability in a rabbit model of atherosclerosis *Arter. Thromb. Vasc. Biol.* **28** 1270–6
[24] Mossa-Basha M, Meltzer C C, Kim D C, Tuite M J, Kolli K P and Tan B S 2020 Radiology department preparedness for COVID-19: radiology scientific expert panel *Radiology* **296** 200988
[25] Kotsis V, Jamthikar A D, Araki T, Gupta D, Laird J R, Giannopoulos A A, Saba L, Suri H S, Mavrogeni S and Kitas G D 2018 Echolucency-based phenotype in carotid atherosclerosis disease for risk stratification of diabetes patients *Diabetes Res. Clin. Pract.* **143** 322–31
[26] Khanna N N, Jamthikar A D, Gupta D, Araki T, Piga M, Saba L, Carcassi C, Nicolaides A, Laird J R and Suri H S 2019 Effect of carotid image-based phenotypes on cardiovascular risk calculator: AECRS1. 0 *Med. Biol. Eng. Comput.* **57** 1553–66
[27] Khanna N N, Jamthikar A D, Araki T, Gupta D, Piga M, Saba L, Carcassi C, Nicolaides A, Laird J R and Suri H S 2019 Nonlinear model for the carotid artery disease 10-year risk prediction by fusing conventional cardiovascular factors to carotid ultrasound image phenotypes: a japanese diabetes cohort study *Echocardiography* **36** 345–61
[28] Cuadrado-Godia E, Jamthikar A D, Gupta D, Khanna N N, Araki T, Maniruzzaman M, Saba L, Nicolaides A, Sharma A and Omerzu T 2019 Ranking of stroke and cardiovascular risk factors for an optimal risk calculator design: logistic regression approach *Comput. Biol. Med.* **108** 182–95
[29] Khanna N N, Jamthikar A D, Gupta D, Piga M, Saba L, Carcassi C, Giannopoulos A A, Nicolaides A, Laird J R and Suri H S 2019 Rheumatoid arthritis: atherosclerosis imaging and cardiovascular risk assessment using machine and deep learning–based tissue characterization *Curr. Atheroscler. Rep.* **21** 7
[30] Jamthikar A, Gupta D, Khanna N N, Araki T, Saba L, Nicolaides A, Sharma A, Omerzu T, Suri H S and Gupta A 2019 A special report on changing trends in preventive stroke/cardiovascular risk assessment via B-mode ultrasonography *Curr. Atheroscler Rep.* **21** 25
[31] Schnee J M and Hsueh W A 2000 Angiotensin II, adhesion, and cardiac fibrosis *Cardiovasc. Res.* **46** 264–8
[32] Wu L L, Yang N, Roe C J, Cooper M E, Gilbert R E, Atkins R C and Lan H Y 1997 Macrophage and myofibroblast proliferation in remnant kidney: role of angiotensin II *Kidney Int. Suppl.* **63** S221–5
[33] Sun Y, Ramires F J and Weber K T 1997 Fibrosis of atria and great vessels in response to angiotensin II or aldosterone infusion *Cardiovasc. Res.* **35** 138–47

[34] Morihara K, Takai S, Takenaka H, Sakaguchi M, Okamoto Y, Morihara T, Miyazaki M and Kishimoto S 2006 Cutaneous tissue angiotensin–converting enzyme may participate in pathologic scar formation in human skin *J. Am. Acad. Dermatol.* **54** 251–7

[35] Cosyns B, Lochy S, Luchian M L, Gimelli A, Pontone G, Allard S D, de Mey J, Rosseel P, Dweck M and Petersen S E 2020 The role of cardiovascular imaging for myocardial injury in hospitalized COVID-19 patients *Eur. Heart J. Cardiovasc. Imaging* **21** 709–14

[36] Inciardi R M, Lupi L, Zaccone G, Italia L, Raffo M, Tomasoni D, Cani D S, Cerini M, Farina D and Gavazzi E 2020 Cardiac involvement in a patient with coronavirus disease 2019 (COVID-19) *JAMA Cardiol* **5** 819–24

[37] Kim I-C, Kim J Y, Kim H A and Han S 2020 COVID-19-related myocarditis in a 21-year-old female patient *Eur. Heart J.* **41** 1859–9

[38] Kiamanesh O, Harper L, Wiskar K, Luksun W, McDonald M, Ross H, Woo A and Granton J 2020 Lung ultrasound for cardiologists in the time of COVID-19 *Can. J. Cardiol* **36** 1144–7

[39] Zieleskiewicz L, Duclos G, Dransart-Rayé O, Nowobilski N and Bouhemad B 2020 Ultrasound findings in patients with COVID-19 pneumonia in early and late stages: two case-reports *Anaesth. Crit. Care Pain Med* **39** 571–3

[40] Saba L, Tiwari A, Biswas M, Gupta S K, Godia-Cuadrado E, Chaturvedi A, Turk M, Suri H S, Orru S and Sanches J M 2019 Wilson's disease: a new perspective review on its genetics, diagnosis and treatment *Front. Biosci. (Elite edition)* **11** 166–85

[41] Collaborators NASCET 1991 Beneficial effect of carotid endarterectomy in symptomatic patients with high-grade carotid stenosis *New Engl. J. Med.* **325** 445–53

[42] Sanches J M, Laine A F and Suri J S 2012 *Ultrasound Imaging* (Berlin: Springer)

[43] Suri J S, Wilson D and Laxminarayan S 2005 *Handbook of Biomedical Image Analysis* vol 2 43 (New York: Springer Science & Business Media)

[44] Suri J S and Laxminarayan S 2003 *Angiography and Plaque Imaging: Advanced Segmentation Techniques* (Boca Raton, FL: CRC Press)

[45] Acharya U R, Mookiah M R K, Sree S V, Afonso D, Sanches J, Shafique S, Nicolaides A and Pedro L M 2013 Atherosclerotic plaque tissue characterization in 2D ultrasound longitudinal carotid scans for automated classification: a paradigm for stroke risk assessment *Med. Biol. Eng. Comput.* **51** 513–23

[46] Acharya U R, Faust O, Sree S V, Alvin A P C, Krishnamurthi G, Sanches J and Suri J S 2011 Atheromatic™: symptomatic vs. asymptomatic classification of carotid ultrasound plaque using a combination of HOS, DWT & texture *2011 Annual Int. Conf. of the IEEE Engineering in Medicine and Biology Society* (Piscataway, NJ: IEEE) pp 4489–92

[47] Acharya U R, Sree S V, Kulshreshtha S, Molinari F, Koh J E W, Saba L and Suri J S 2014 GyneScan: an improved online paradigm for screening of ovarian cancer via tissue characterization *Technol. Cancer Res. Treat.* **13** 529–39

[48] Biswas M, Kuppili V, Edla D R, Suri H S, Saba L, Marinhoe R T, Sanches J M and Suri J S 2018 Symtosis: a liver ultrasound tissue characterization and risk stratification in optimized deep learning paradigm *Comput. Methods Prog. Biomed.* **155** 165–77

[49] Acharya U R, Krishnan M M R, Sree S V, Sanches J, Shafique S, Nicolaides A, Pedro L M and Suri J S 2012 Plaque tissue characterization and classification in ultrasound carotid scans: a paradigm for vascular feature amalgamation *IEEE Trans. Instrum. Meas.* **62** 392–400

[50] Molinari F, Liboni W, Pavanelli E, Giustetto P, Badalamenti S and Suri J S 2007 Accurate and automatic carotid plaque characterization in contrast enhanced 2-D ultrasound images

2007 29th Annual Int. Conf. of the IEEE Engineering in Medicine and Biology Society (Piscataway, NJ: IEEE) pp 335–8

[51] Acharya U, Vinitha Sree S, Mookiah M, Yantri R, Molinari F, Zieleźnik W, Małyszek-Tumidajewicz J, Stępień B, Bardales R and Witkowska A 2013 Diagnosis of Hashimoto's thyroiditis in ultrasound using tissue characterization and pixel classification *Proc. Inst. Mech. Eng. Part H J. Eng. Med.* **227** 788–98

[52] Sharma A M, Gupta A, Kumar P K, Rajan J, Saba L, Nobutaka I, Laird J R, Nicolades A and Suri J S 2015 A review on carotid ultrasound atherosclerotic tissue characterization and stroke risk stratification in machine learning framework *Curr. Atheroscler. Rep.* **17** 55

[53] Ravì D, Wong C, Deligianni F, Berthelot M, Andreu-Perez J, Lo B and Yang G-Z 2016 Deep learning for health informatics *IEEE J. Biomed. Health Inform.* **21** 4–21

[54] Saba L, Biswas M, Kuppili V, Godia E C, Suri H S, Edla D R, Omerzu T, Laird J R, Khanna N N and Mavrogeni S 2019 The present and future of deep learning in radiology *Eur. J. Radiol* **114** 14–24

[55] Biswas M, Kuppili V, Saba L, Edla D R, Suri H S, Cuadrado-Godia E, Laird J, Marinhoe R, Sanches J and Nicolaides A 2019 State-of-the-art review on deep learning in medical imaging *Front. Biosci. (Landmark Ed)* **24** 392–426

[56] Biswas M, Kuppili V, Araki T, Edla D R, Godia E C, Saba L, Suri H S, Omerzu T, Laird J R and Khanna N N 2018 Deep learning strategy for accurate carotid intima-media thickness measurement: an ultrasound study on Japanese diabetic cohort *Comput. Biol. Med.* **98** 100–17

[57] Jamthikar A, Gupta D, Khanna N N, Saba L, Araki T, Viskovic K, Suri H S, Gupta A, Mavrogeni S and Turk M 2019 A low-cost machine learning-based cardiovascular/stroke risk assessment system: integration of conventional factors with image phenotypes *Cardiovasc. Diagn. Ther.* **9** 420

[58] Hoffmann M, Kleine-Weber H, Schroeder S, Krüger N, Herrler T, Erichsen S, Schiergens T S, Herrler G, Wu N-H and Nitsche A 2020 SARS-CoV-2 cell entry depends on ACE2 and TMPRSS2 and is blocked by a clinically proven protease inhibitor *Cell* **181** 271–80

[59] de Wit E, van Doremalen N, Falzarano D and Munster V J 2016 SARS and MERS: recent insights into emerging coronaviruses *Nat. Rev. Microbiol.* **14** 523

[60] Wu K, Peng G, Wilken M, Geraghty R J and Li F 2012 Mechanisms of host receptor adaptation by severe acute respiratory syndrome coronavirus *J. Biol. Chem.* **287** 8904–11

[61] Patel V B, Zhong J-C, Grant M B and Oudit G Y 2016 Role of the ACE2/angiotensin 1–7 axis of the renin–angiotensin system in heart failure *Circ. Res.* **118** 1313–26

[62] Zou X, Chen K, Zou J, Han P, Hao J and Han Z 2020 Single-cell RNA-seq data analysis on the receptor ACE2 expression reveals the potential risk of different human organs vulnerable to 2019-nCoV infection *Front. Med.* **14** 185–192

[63] Hamming I, Timens W, Bulthuis M, Lely A, Navis G and van Goor H 2004 Tissue distribution of ACE2 protein, the functional receptor for SARS coronavirus. A first step in understanding SARS pathogenesis *J. Pathol.* **203** 631–7

[64] Giacomelli A, Pezzati L, Conti F, Bernacchia D, Siano M, Oreni L, Rusconi S, Gervasoni C, Ridolfo A L and Rizzardini G 2020 Self-reported olfactory and taste disorders in patients with severe acute respiratory coronavirus 2 infection: a cross-sectional study *Clin. Infect Dis.* **71** 889–90

[65] Wu A, Peng Y, Huang B, Ding X, Wang X, Niu P, Meng J, Zhu Z, Zhang Z and Wang J 2020 Genome composition and divergence of the novel coronavirus (2019-nCoV) originating in China *Cell Host Microbe* **27** 325–8
[66] Koyuncu O O, Hogue I B and Enquist L W 2013 Virus infections in the nervous system *Cell Host Microbe* **13** 379–93
[67] Desforges M, Le Coupanec A, Dubeau P, Bourgouin A, Lajoie L, Dubé M and Talbot P J 2020 Human coronaviruses and other respiratory viruses: underestimated opportunistic pathogens of the central nervous system? *Viruses* **12** 14
[68] McCray P B, Pewe L, Wohlford-Lenane C, Hickey M, Manzel L, Shi L, Netland J, Jia H P, Halabi C and Sigmund C D 2007 Lethal infection of K18-hACE2 mice infected with severe acute respiratory syndrome coronavirus *J. Virol.* **81** 813–21
[69] Li K, Wohlford-Lenane C, Perlman S, Zhao J, Jewell A K, Reznikov L R, Gibson-Corley K N, Meyerholz D K and McCray P B 2016 Middle East respiratory syndrome coronavirus causes multiple organ damage and lethal disease in mice transgenic for human dipeptidyl peptidase 4 *J. Infect. Dis.* **213** 712–22
[70] Netland J, Meyerholz D K, Moore S, Cassell M and Perlman S 2008 Severe acute respiratory syndrome coronavirus infection causes neuronal death in the absence of encephalitis in mice transgenic for human ACE2 *J. Virol.* **82** 7264–75
[71] Baig A M 2020 Neurological manifestations in COVID-19 caused by SARS-CoV-2 *CNS Neurosci. Ther.* **26** 499
[72] Ryan W 2020 There's a new symptom of coronavirus, doctors say: sudden loss of smell or taste
[73] Hopkins C and Kumar N 2020 Loss of sense of smell as marker of COVID-19 infection *ENT UK* **26** 2020 (https://www.entuk.org/sites/default/files/files/Loss_of_sense_of_smell_as_marker_of_COVID.pdf)
[74] Bohmwald K, Galvez N, Ríos M and Kalergis A M 2018 Neurologic alterations due to respiratory virus infections *Front. Cell. Neurosci.* **12** 386
[75] Zhang H and Baker A 2017 *Recombinant Human ACE2: Acing Out Angiotensin II in ARDS Therapy* (Berlin: Springer)
[76] Radermacher P, Maggiore S M and Mercat A 2017 Fifty years of research in ARDS. Gas exchange in acute respiratory distress syndrome *Am. J. Respir. Crit. Care Med.* **196** 964–84
[77] Chen N, Zhou M, Dong X, Qu J, Gong F, Han Y, Qiu Y, Wang J, Liu Y and Wei Y 2020 Epidemiological and clinical characteristics of 99 cases of 2019 novel coronavirus pneumonia in Wuhan, China: a descriptive study *Lancet* **395** 507–13
[78] Xiong T-Y, Redwood S, Prendergast B and Chen M 2020 Coronaviruses and the cardiovascular system: acute and long-term implications *Eur. Heart J.* **41** 1798–800
[79] Oudit G, Kassiri Z, Jiang C, Liu P, Poutanen S, Penninger J and Butany J 2009 SARS-coronavirus modulation of myocardial ACE2 expression and inflammation in patients with SARS *Eur. J. Clin. Investig.* **39** 618–25
[80] Abdennour L, Zeghal C, Deme M and Puybasset L 2012 Interaction brain-lungs *Ann. Fr. Anesth. Reanim.* e101–7
[81] Fountain J H and Lappin S L 2019 Physiology, Renin Angiotensin System *StatPearls* pp 81 (St Petersburg, FL: StatPearls Publishing)
[82] Rajendran P, Rengarajan T, Thangavel J, Nishigaki Y, Sakthisekaran D, Sethi G and Nishigaki I 2013 The vascular endothelium and human diseases *Int. J. Biol. Sci.* **9** 1057

[83] Lillie P J, Samson A, Li A, Adams K, Capstick R, Barlow G D, Easom N, Hamilton E, Moss P J and Evans A 2020 Novel coronavirus disease (Covid-19): the first two patients in the UK with person to person transmission *J. Infect.* **80** 578–606

[84] Bai Y, Yao L, Wei T, Tian F, Jin D-Y, Chen L and Wang M 2020 Presumed asymptomatic carrier transmission of COVID-19 *JAMA* **323** 1406–7

[85] Mehta P, McAuley D F, Brown M, Sanchez E, Tattersall R S and Manson J J 2020 COVID-19: consider cytokine storm syndromes and immunosuppression *Lancet* **395** 1033–4

[86] Siddiqi H K and Mehra M R 2020 COVID-19 illness in native and immunosuppressed states: a clinical-therapeutic staging proposal *J. Heart Lung Transp.* **39** 405–7

[87] Yin C, Wang C, Tang Z, Wen Y, Zhang S and Wang B 2004 Clinical analysis of multiple organ dysfunction syndrome in patients suffering from SARS *Zhongguo wei zhong bing ji jiu yi xue= Chin. Crit. Care Med.= Zhongguo weizhongbing jijiuyixue* **16** 646–50

[88] Schoenhagen P, Tuzcu E M and Ellis S G 2002 Plaque vulnerability, plaque rupture, and acute coronary syndromes: (multi)-focal manifestation of a systemic disease process *Circulation*

[89] Tisoncik J R, Korth M J, Simmons C P, Farrar J, Martin T R and Katze M G 2012 Into the eye of the cytokine storm *Microbiol. Mol. Biol. Rev.* **76** 16–32

[90] Tersalvi G, Vicenzi M, Calabretta D, Biasco L, Pedrazzini G and Winterton D 2020 Elevated troponin in patients with Coronavirus Disease 2019 (COVID-19): possible mechanisms *J. Cardiac Failure* **26** 470–75

[91] Gomes V A 2020 COVID-19 Cardiac repercussions *Rev. Bras. Fisiol. Exer.* **19**

[92] Zeng J H, Liu Y-X, Yuan J, Wang F-X, Wu W-B, Li J-X, Wang L-F, Gao H, Wang Y and Dong C-F 2020 *First Case of COVID-19 Infection with Fulminant Myocarditis Complication: Case Report and Insights*

[93] Cieszanowski A, Czekajska E, Giżycka B, Gruszczyńska K, Podgórska J, Oronowicz-Jaśkowiak A, Serafin Z, Szurowska E and Walecki J M 2020 Management of patients with COVID-19 in radiology departments, and indications regarding imaging studies–recommendations of the Polish Medical Society of Radiology *Pol. J. Radiol.* **85** e209

[94] Kim D J, Jelic T, Woo M Y, Heslop C and Olszynski P 2020 Just the facts: recommendations on point of care ultrasound use and machine infection control during the COVID-19 pandemic *Can. J. Emerg. Med.* **22** 1–7

[95] An X, Song Z, Gao Y, Tao J and Yang J 2020 To resume noninvasive imaging detection safely after peak period of COVID-19: experiences from Wuhan China *Dermatol. Ther.* **33** e13590

[96] Jakhar D, Kaur I and Kaul S 2020 Art of performing dermoscopy during the times of coronavirus disease (COVID-19): simple change in approach can save the day! *J. Eur. Acad. Dermatol. Venereol.*

[97] Skulstad H, Cosyns B, Popescu B A, Galderisi M, Salvo G D, Donal E, Petersen S, Gimelli A, Haugaa K H and Muraru D 2020 COVID-19 pandemic and cardiac imaging: EACVI recommendations on precautions, indications, prioritization, and protection for patients and healthcare personnel *Eur. Heart J.-Cardiovasc. Imaging* **21** 592–98

[98] Lo S, Yong A, Sinhal A, Shetty S, McCann A, Clark D, Galligan L, El-Jack S, Sader M and Tan R 2020 Consensus guidelines for interventional cardiology services delivery during COVID-19 pandemic in Australia and New Zealand *Heart Lung Circ* **29** 69–77

[99] El-Baz A, Jiang X and Suri J S 2016 *Biomedical Image Segmentation: Advances and Trends* (Boca Raton, FL: CRC Press)

[100] El-Baz A S, Acharya R, Mirmehdi M and Suri J S 2011 *Multi Modality State-of-the-Art Medical Image Segmentation and Registration Methodologies* vol 1 (New York: Springer Science & Business Media)

[101] Olusanya O 2020 Ultrasound in times of COVID-19 *ICU Management & Practice* **20** 43–50

[102] Smith M, Hayward S, Innes S and Miller A 2020 Point-of-care lung ultrasound in patients with COVID-19: a narrative review *Anaesthesia* **75** 1096–104

[103] Jacobi A, Chung M, Bernheim A and Eber C 2020 Portable chest X-ray in coronavirus disease-19 (COVID-19): a pictorial review *Clin. Imaging* **64** 35–42

[104] Chung M, Bernheim A, Mei X, Zhang N, Huang M, Zeng X, Cui J, Xu W, Yang Y and Fayad Z A 2020 CT imaging features of 2019 novel coronavirus (2019-nCoV) *Radiology* **295** 202–7

[105] Vasilev Y, Sergunova K, Bazhin A, Masri A, Vasileva Y, Suleumanov E, Semenov D, Kudryavtsev N, Panina O and Khoruzhaya A 2020 MRI of the lungs in patients with COVID-19: clinical case *JMRI* **79** 13–9

[106] Huang L, Zhao P, Tang D, Zhu T, Han R, Zhan C, Liu W, Zeng H, Tao Q and Xia L 2020 Cardiac involvement in recovered COVID-19 patients identified by magnetic resonance imaging *JACC: Cardiovasc. Imaging* **13** 2330–9

[107] Luetkens J A, Isaak A, Zimmer S, Nattermann J, Sprinkart A M, Boesecke C, Rieke G J, Zachoval C, Heine A and Velten M 2020 Diffuse myocardial inflammation in covid-19 associated myocarditis detected by multiparametric cardiac magnetic resonance imaging *Circ.: Cardiovasc. Imaging* **13** e010897

[108] Poyiadji N, Shahin G, Noujaim D, Stone M, Patel S and Griffith B 2020 COVID-19–associated acute hemorrhagic necrotizing encephalopathy: CT and MRI features *Radiology* **296** 201187

[109] Kandemirli S G, Dogan L, Sarikaya Z T, Kara S, Akinci C, Kaya D, Kaya Y, Yildirim D, Tuzuner F and Yildirim M S 2020 Brain MRI findings in patients in the intensive care unit with COVID-19 infection *Radiology* **297** 201697

[110] Bhayana R, Som A, Li M D, Carey D E, Anderson M A, Blake M A, Catalano O, Gee M S, Hahn P F and Harisinghani M 2020 Abdominal imaging findings in COVID-19: preliminary observations *Radiology* **297** 201908

[111] Eliezer M and Hautefort C 2020 MRI evaluation of the olfactory clefts in patients with SARS-CoV-2 infection revealed an unexpected mechanism for olfactory function loss *Acad. Radiol.* **27** 1191

[112] Xiao Z, Xu C, Wang D and Zeng H 2020 The experience of treating patients with acute myocardial infarction under the COVID-19 epidemic *Catheter. Cardiovasc. Interv.* **97** E244–8

[113] Meyer P, Degrauwe S, Van Delden C, Ghadri J-R and Templin C 2020 Typical takotsubo syndrome triggered by SARS-CoV-2 infection *Eur. Heart J.* **41** 1860–0

[114] Danzi G B, Loffi M, Galeazzi G and Gherbesi E 2020 Acute pulmonary embolism and COVID-19 pneumonia: a random association? *Eur. Heart J.* **41** 1858–8

[115] Zhang L, Feng X, Zhang D, Jiang C, Mei H, Wang J, Zhang C, Li H, Xia X and Kong S 2020 Deep vein thrombosis in hospitalized patients with coronavirus disease 2019 (COVID-19) in Wuhan, China: prevalence, risk factors, and outcome *Circulation* **142** 114–28

[116] Emanuel E J, Persad G, Upshur R, Thome B, Parker M, Glickman A, Zhang C, Boyle C, Smith M and Phillips J P 2020 Fair Allocation of Scarce Medical Resources in the Time of Covid-19 *N Engl. J. Med.* **382** 2049–55

[117] Rosenbaum L 2020 Facing Covid-19 in Italy—ethics, logistics, and therapeutics on the epidemic's front line *New Engl. J. Med.* **382** 1873–5

[118] Bhatt A S *et al* 2020 Declines in hospitalizations for acute cardiovascular conditions during the COVID-19 pandemic: a multicenter tertiary care experience *J. Am. Coll. Cardiol.* **76** 280–8

[119] Vaishya R, Haleem A, Vaish A and Javaid M 2020 Emerging technologies to combat COVID-19 pandemic *J. Clin. Exp. Hepatol* **10** 409–11

[120] Murphy K, Smits H, Knoops A J, Korst M B, Samson T, Scholten E T, Schalekamp S, Schaefer-Prokop C M, Philipsen R H and Meijers A 2020 COVID-19 on the chest radiograph: a multi-reader evaluation of an AI system *Radiology* **296** 201874

[121] Zheng N, Du S, Wang J, Zhang H, Cui W, Kang Z, Yang T, Lou B, Chi Y and Long H 2020 Predicting COVID-19 in China using hybrid AI model *IEEE Trans. Cybern* **50** 2891–904

[122] Chieffo A, Stefanini G G, Price S, Barbato E, Tarantini G, Karam N, Moreno R, Buchanan G L, Gilard M and Halvorsen S 2020 EAPCI position statement on invasive management of acute coronary syndromes during the COVID-19 pandemic *Eur. Heart J.* **41** 1839–51

[123] Salehi S, Abedi A, Balakrishnan S and Gholamrezanezhad A 2020 Coronavirus disease 2019 (COVID-19): a systematic review of imaging findings in 919 patients *Am. J. Roentgenol.* **215** 1–7

[124] Dangis A, Gieraerts C, Bruecker Y D, Janssen L, Valgaeren H, Obbels D, Gillis M, Ranst M V, Frans J and Demeyere A 2020 Accuracy and reproducibility of low-dose submillisievert chest CT for the diagnosis of COVID-19 *Radiol.: Cardiothorac. Imaging* **2** e200196

[125] Rubin G D, Ryerson C J, Haramati L B, Sverzellati N, Kanne J P, Raoof S, Schluger N W, Volpi A, Yim J-J and Martin I B 2020 The role of chest imaging in patient management during the COVID-19 pandemic: a multinational consensus statement from the Fleischner Society *Chest* **158** 106–16

[126] Nair A *et al* 2020 Society of Thoracic Imaging statement: considerations in designing local imaging diagnostic algorithms for the COVID-19 pandemic *Clin. Radiol.* **75** 329–34

[127] Laghi A 2020 Cautions about radiologic diagnosis of COVID-19 infection driven by artificial intelligence *Lancet Digit. Health* **2** e225

[128] Miotto R, Wang F, Wang S, Jiang X and Dudley J T 2018 Deep learning for healthcare: review, opportunities and challenges *Brief. Bioinform.* **19** 1236–46

[129] Esteva A, Robicquet A, Ramsundar B, Kuleshov V, DePristo M, Chou K, Cui C, Corrado G, Thrun S and Dean J 2019 A guide to deep learning in healthcare *Nat. Med.* **25** 24–9

[130] Hosny A, Parmar C, Quackenbush J, Schwartz L H and Aerts H J 2018 Artificial intelligence in radiology *Nat. Rev. Cancer* **18** 500–10

[131] Sinha JSS G R 2019 *Cognitive Informatics, Computer Modelling, and Cognitive Science: Volume 1: Theory, Case Studies, and Applications* (Netherlands: Elsevier)

[132] Tang X 2019 The role of artificial intelligence in medical imaging research *BJR| Open* **2** 20190031

[133] Saeian K, Rhyne T L and Sagar K B 1994 Ultrasonic tissue characterization for diagnosis of acute myocardial infarction in the coronary care unit *Am. J. Cardiol.* **74** 1211–5

[134] Mavrogeni S, Sfikakis P P, Gialafos E, Bratis K, Karabela G, Stavropoulos E, Spiliotis G, Sfendouraki E, Panopoulos S and Bournia V 2014 Cardiac tissue characterization and the

diagnostic value of cardiovascular magnetic resonance in systemic connective tissue diseases *Arthritis Care Res.* **66** 104–12
[135] Wu J, Pan J, Teng D, Xu X, Feng J and Chen Y-C 2020 Interpretation of CT signs of 2019 novel coronavirus (COVID-19) pneumonia *Eur. Radiol.* **30** 5455–62
[136] Alimadadi A, Aryal S, Manandhar I, Munroe P B, Joe B and Cheng X 2020 Artificial Intelligence and Machine Learning to Fight COVID-19 *Physiological Genomics* **52** 200–2
[137] Vaishya R, Javaid M, Khan I H and Haleem A 2020 Artificial intelligence (AI) applications for COVID-19 pandemic *Diabetes Metab. Syndr.: Clin. Res. Rev.* **14** 337–9
[138] Jamthikar A, Gupta D, Saba L, Khanna N N, Araki T, Viskovic K, Mavrogeni S, Laird J R, Pareek G and Miner M *et al* 2020 Cardiovascular/stroke risk predictive calculators: a comparison between statistical and machine learning models *Cardiovasc. Diagn. Ther* **10** 919–38
[139] Jamthikar A, Gupta D, Khanna N N, Saba L, Laird J R and Suri J S 2020 Cardiovascular/stroke risk prevention: a new machine learning framework integrating carotid ultrasound image-based phenotypes and its harmonics with conventional risk factors *Indian Heart J.* **72** 258–64
[140] Biswas M, Kuppili V, Edla D R, Suri H S, Saba L, Marinhoe R T, Sanches J M and Suri J S 2017 Symtosis: a liver ultrasound tissue characterization and risk stratification in optimized deep learning paradigm *Comput. Methods Prog. Biomed* **155** 165–77
[141] Bishop C M 2006 *Pattern Recognition and Machine Learning* (New York: Springer)
[142] LeCun Y, Bengio Y and Hinton G 2015 Deep learning *Nature* **521** 436–44
[143] Suri J S, Acharya U R, Faust O, Alvin A P C, Sree S V, Molinari F, Saba L and Nicolaides A 2011 Symptomatic vs. asymptomatic plaque classification in carotid ultrasound *J. Med. Syst.* **36** 1861–71
[144] Cortes C and Vapnik V 1995 Support-vector networks *Mach. Learn.* **20** 273–97
[145] Mirmehdi M 2008 *Handbook of Texture Analysis* (Singapore: Imperial College Press)
[146] Bharati M H, Liu J J and MacGregor J F 2004 Image texture analysis: methods and comparisons *Chemometr. Intell. Lab. Syst.* **72** 57–71
[147] Acharya U R, Chua C K, Lim T-C, Dorithy T and Suri J S 2009 Automatic identification of epileptic EEG signals using nonlinear parameters *J. Mech. Med. Biol.* **9** 539–53
[148] Acharya U R, Faust O, Sree S V, Molinari F and Suri J S 2012 ThyroScreen system: high resolution ultrasound thyroid image characterization into benign and malignant classes using novel combination of texture and discrete wavelet transform *Comput. Methods Programs Biomed.* **107** 233–41
[149] Reynolds D A 2009 Gaussian mixture models *Encyclopedia of Biometrics* (Berlin: Springer) 741
[150] Huang D-S 1999 Radial basis probabilistic neural networks: model and application *Int. J. Pattern Recognit Artif Intell.* **13** 1083–101
[151] Quinlan J R 1987 Generating production rules from decision trees *IJCAI* **87** 304–7
[152] Clark P J and Evans F C 1954 Distance to nearest neighbor as a measure of spatial relationships in populations *Ecology* **35** 445–53
[153] Rish I 2001 An empirical study of the naive Bayes classifier *IJCAI 2001 Workshop on Empirical Methods in Artificial Intelligence* pp 41–6
[154] Ross T J 2009 *Fuzzy Logic with Engineering Applications* (University of New Mexico: Wiley)

[155] Kadyrov A and Petrou M 2001 The trace transform and its applications *IEEE Trans. Pattern Anal. Mach. Intell.* **23** 811–28

[156] Jawahar C and Ray A 1996 Incorporation of gray-level imprecision in representation and processing of digital images *Pattern Recognit. Lett.* **17** 541–6

[157] Galloway M M 1974 Texture analysis using grey level run lengths *STIN* **75** 18555

[158] Boi A, Jamthikar A D, Saba L, Gupta D, Sharma A, Loi B, Laird J R, Khanna N N and Suri J S 2018 A survey on coronary atherosclerotic plaque tissue characterization in intravascular optical coherence tomography *Curr. Atheroscler. Rep.* **20** 33

[159] Jamthikar A, Gupta D, Khanna N N, Saba L, Araki T, Viskovic K, Suri H S, Gupta A, Mavrogeni S and Turk M *et al* 2019 A low-cost machine learning-based cardiovascular/stroke risk assessment system: integration of conventional factors with image phenotypes *Cardiovasc. Diagn. Ther.* **9** 420

[160] Jamthikar A, Gupta D, Khanna N N, Araki T, Saba L, Nicolaides A, Sharma A, Omerzu T, Suri H S and Gupta A *et al* 2019 A special report on changing trends in preventive stroke/cardiovascular risk assessment via B-Mode ultrasonography *Curr. Atheroscler. Rep.* 21 25

[161] Viswanathan V *et al* 2020 Does the carotid bulb offer a better 10-Year CVD/stroke risk assessment compared to the common carotid artery? A 1516 ultrasound scan study *Angiology* **71** 3319720941730

[162] Long J, Shelhamer E and Darrell T 2015 Fully convolutional networks for semantic segmentation *Proc. of the IEEE Conf. on Computer Vision and Pattern Recognition* pp 3431–40

[163] Biswas M, Kuppili V, Saba L, Edla D R, Suri H S, Sharma A, Cuadrado-Godia E, Laird J R, Nicolaides A and Suri J S 2019 Deep learning fully convolution network for lumen characterization in diabetic patients using carotid ultrasound: a tool for stroke risk *Med. Biol. Eng. Comput.* **57** 543–64

[164] Saba L, Biswas M, Suri H S, Viskovic K, Laird J R, Cuadrado-Godia E, Nicolaides A, Khanna N, Viswanathan V and Suri J S 2019 Ultrasound-based carotid stenosis measurement and risk stratification in diabetic cohort: a deep learning paradigm *Cardiovasc. Diagn. Ther.* **9** 439

[165] Dong D, Tang Z, Wang S, Hui H, Gong L, Lu Y, Xue Z, Liao H, Chen F and Yang F 2020 The role of imaging in the detection and management of COVID-19: a review *IEEE Rev. Biomed. Eng*

[166] Ito R I S and Naganawa S 2020 A review on the use of artificial intelligence for medical imaging of the lungs of patients with coronavirus disease 2019 *Diagn. Interv. Radiol*

[167] Lu W, Zhang S, Chen B, Chen J, Xian J, Lin Y, Shan H and Su Z Z 2020 A clinical study of noninvasive assessment of lung lesions in patients with coronavirus disease-19 (COVID-19) by bedside ultrasound *Ultraschall in der Medizin-Eur. J. Ultrasound* **41** 300–7

[168] Kang H, Xia L, Yan F, Wan Z, Shi F, Yuan H, Jiang H, Wu D, Sui H and Zhang C 2020 Diagnosis of coronavirus disease 2019 (covid-19) with structured latent multi-view representation learning *IEEE Trans. Med. Imaging* **39** 2606–14

[169] Xinggang Wang X D, Fu Q, Zhou Q, Zhou Q, Feng J, Ma H, Liu W and Zheng C 2020 A weakly-supervised framework for COVID-19 classification and lesion localization from chest CT *IEEE Trans. Med. Imaging* **39** 2615–25

[170] Krizhevsky A, Sutskever I and Hinton G E 2012 *Imagenet classification with deep convolutional neural networks Advances in Neural Information Processing Systems 25 (NIPS 2012)* F. Pereira *et al* (Cambridge, MA: MIT Press) 1097–105

[171] Szegedy C, Vanhoucke V, Ioffe S, Shlens J and Wojna Z 2016 Rethinking the inception architecture for computer vision *Proc. of the IEEE Conf. on Computer Vision and Pattern Recognition* pp 2818–26
[172] Wang K, Lu X, Zhou H, Gao Y, Zheng J, Tong M, Wu C, Liu C and Huang L 2019 Deep learning Radiomics of shear wave elastography significantly improved diagnostic performance for assessing liver fibrosis in chronic hepatitis B: a prospective multicentre study *Gut* **68** 729–41
[173] Wang S *et al* 2020 A fully automatic deep learning system for COVID-19 diagnostic and prognostic analysis *Eur. Respir. J.* **56** 2000775
[174] Zhang K *et al* 2020 Clinically applicable AI system for accurate diagnosis, quantitative measurements, and prognosis *Cell* **182** 1360
[175] Li Z *et al* 2020 from community acquired pneumonia to COVID-19: a deep learning based method for quantitative analysis of COVID-19 on thick-section CT scans *Medrxiv* 2020
[176] Chen J *et al* 2020 Deep learning-based model for detecting 2019 novel coronavirus pneumonia on high-resolution computed tomography: a prospective study *Sci. Rep.* **10** 19196
[177] Angel C T (http://121.40.75.149/znyx-ncov/index#/app/index) (accessed 24 July 2020)
[178] Yang S, Jiang L, Cao Z, Wang L, Cao J, Feng R, Zhang Z, Xue X, Shi Y and Shan F 2020 Deep learning for detecting corona virus disease 2019 (COVID-19) on high-resolution computed tomography: a pilot study *Ann. Transl. Med.* **8**
[179] Oh Y, Park S and Ye J C 2020 Deep learning Covid-19 features on CXR using limited training data sets *IEEE Trans. Med. Imaging* **39** 2688–700
[180] Ren Z H, Mu W C and Huang S Y 2018 Design and optimization of a ring-pair permanent magnet array for head imaging in a low-field portable MRI system *IEEE Trans. Magn.* **55** 1–8
[181] Cooley C Z, Stockmann J P, Armstrong B D, Sarracanie M, Lev M H, Rosen M S and Wald L L 2015 Two-dimensional imaging in a lightweight portable MRI scanner without gradient coils *Magn. Reson. Med.* **73** 872–83
[182] Mirvis S E 1999 Use of portable CT in the R Adams Cowley Shock Trauma Center: experiences in the admitting area, ICU, and operating room *Surg. Clin. North Am.* **79** 1317–30
[183] Wang X and Bhatt D L 2020 COVID-19: an unintended force for medical revolution *J. Invasive Cardiol.* **32** E81–2
[184] Thamman R, Gulati M, Narang A, Utengen A, Mamas M A and Bhatt D L 2020 Twitter-based learning for continuing medical education? *Eur. Heart J.*
[185] Li L, Zhang Q, Wang X, Zhang J, Wang T, Gao T-L, Duan W, Tsoi K K-F and Wang F-Y 2020 Characterizing the propagation of situational information in social media during COVID-19 epidemic: a case study on weibo *IEEE Trans. Comput. Soc. Syst.* **7** 556–62
[186] El-Baz A and Suri J S (ed) 2019 *Big Data in Multimodal Medical Imaging* (Boca Raton, FL: CRC Press)
[187] Kooraki S, Hosseiny M, Myers L and Gholamrezanezhad A 2020 Coronavirus (COVID-19) outbreak: what the department of radiology should know *J. Am. Coll. Radiol* **17** 447–51
[188] Mossa-Basha M, Medverd J, Linnau K, Lynch J B, Wener M H, Kicska G, Staiger T and Sahani D 2020 Policies and guidelines for COVID-19 preparedness: experiences from the University of Washington *Radiology* **296** 201326

[189] Buonsenso D, Piano A, Raffaelli F, Bonadia N, Donati K D G and Franceschi F 2020 novel coronavirus disease-19 pnemoniae: a case report and potential applications during COVID-19 outbreak *Eur. Rev. Med. Pharmacol. Sci.* **24** 2776–80

[190] Suri J S *et al* 2020 COVID-19 pathways for brain and heart injury in comorbidity patients: A role of medical imaging and artificial intelligence-based COVID severity classification: A review *Comput. Biol. Med.* **124** 103960

IOP Publishing

Multimodality Imaging, Volume 2
Heart, lungs and peripheral organs
Mainak Biswas and Jasjit S Suri

Chapter 2

Characterizing acute respiratory distress syndrome in COVID-19: a narrative review of artificial intelligence-based lung analysis

Ashutosh Jha, Radhakrishn Birla, Mainak Biswas and Jasjit S Suri

COVID-19 has infected millions of people and caused hundreds of thousands of deaths worldwide. One of the most serious complications of COVID-19 is acute respiratory distress syndrome (ARDS), which can be fatal. People who are at least 60 years old or have underlying health conditions are at higher risk of developing ARDS.

Medical imaging, such as computed tomography (CT) scans and chest x-rays, can be used to diagnose ARDS. However, these tests can be time-consuming and require specialised equipment. Artificial intelligence (AI) is being developed to automate the diagnosis of ARDS from medical images.

AI-based systems can be used to quickly and accurately identify patients who are at risk of developing ARDS. They can also be used to monitor patients who have already developed ARDS and to assess the severity of their condition.

AI-based systems are still under development, but they have the potential to improve the diagnosis and treatment of ARDS. By automating the diagnosis of ARDS, AI can help to reduce the time it takes to get patients the treatment they need. AI can also help to improve the accuracy of diagnosis and to identify patients who are at risk of developing ARDS.

The use of AI in the diagnosis and treatment of ARDS is still in its early stages, but it has the potential to revolutionise the way this serious condition is managed.

2.1 Introduction

In December 2019, a new coronavirus called severe acute respiratory distress syndrome coronavirus 2 (SARS-CoV-2) [1] emerged in Wuhan, Hubei province, China. Initially referred to as novel coronavirus pneumonia (NCP) by the Chinese government [2], it was later renamed coronavirus disease 2019 (COVID-19) by the

doi:10.1088/978-0-7503-2352-9ch2

World Health Organization (WHO). COVID-19 is currently a global pandemic [4] and primarily affects the respiratory system, potentially leading to acute respiratory distress syndrome (ARDS) and death. As of December 21, 2020, there have been over 77.4 million reported cases of COVID-19 worldwide, resulting in 1.7 million fatalities [5]. The virus has a high Ro value, ranging from 2.43 to 3.10, indicating its high level of contagiousness [6]. Figure 2.1 illustrates the distribution of reported cases per million population worldwide, with colors ranging from white to dark red. The countries with the highest mortality rates include France, Brazil, Mexico, Italy, the USA, India, the UK, and Spain [7]. It is worth noting that COVID-19 is considered a syndemic, as it involves both biological and social factors [3].

From a genetic standpoint, COVID-19 is found to be more closely related to SARS-CoV-1 rather than the MERS-CoV. However, it is different from SARS-CoV-1 in terms of clinical severity, incubation period, and transmissibility [8]. Despite various government measures such as social distancing, mask-wearing, quarantine, and non-pharmacological preventive treatments for overall well-being, the global spread of COVID-19 has continued to increase [9, 10].

COVID-19 exhibits distinct imaging characteristics and affects organs beyond the lungs [11, 12]. Consequently, there has been an exponential increase in research on COVID-19, with nearly 72 000 related articles published since December 2019, averaging 2000 articles per week (as seen on the PubMed website) [13]. Notably, over 900 articles focus on the intersection 2 of COVID-19 and artificial intelligence (AI), encompassing machine learning (ML), transfer learning (TL), and deep learning (DL) models (as shown in figure 2.2(d)). AI has the potential to aid in the characterization of ARDS in the lungs and the diagnosis of COVID-19's impact on other parts of the body [14]. Careful investigation of AI for ARDS

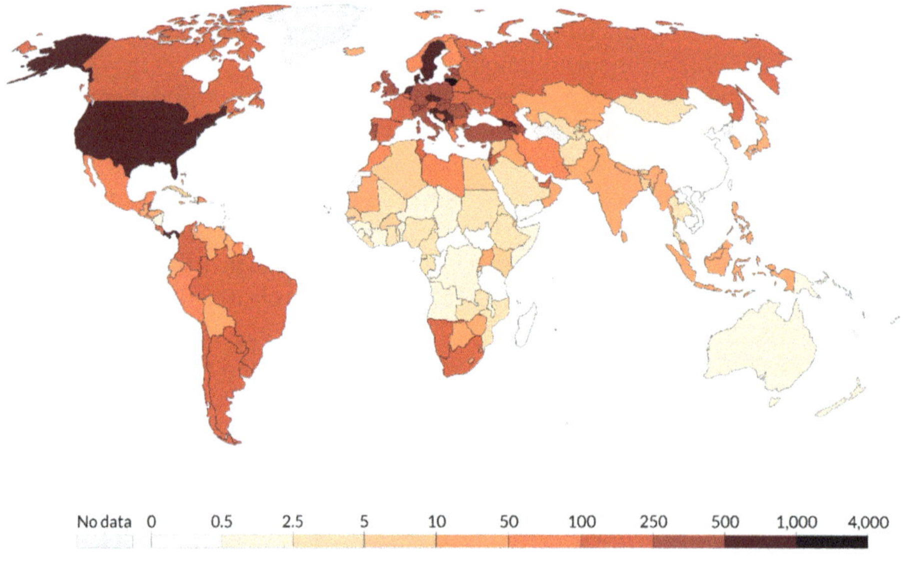

Figure 2.1. Total confirmed cases per million as of December 21, 2020 [15]. (Source: Center for Systems Science and Engineering (CSSE) at Johns Hopkins University, Maryland, USA.).

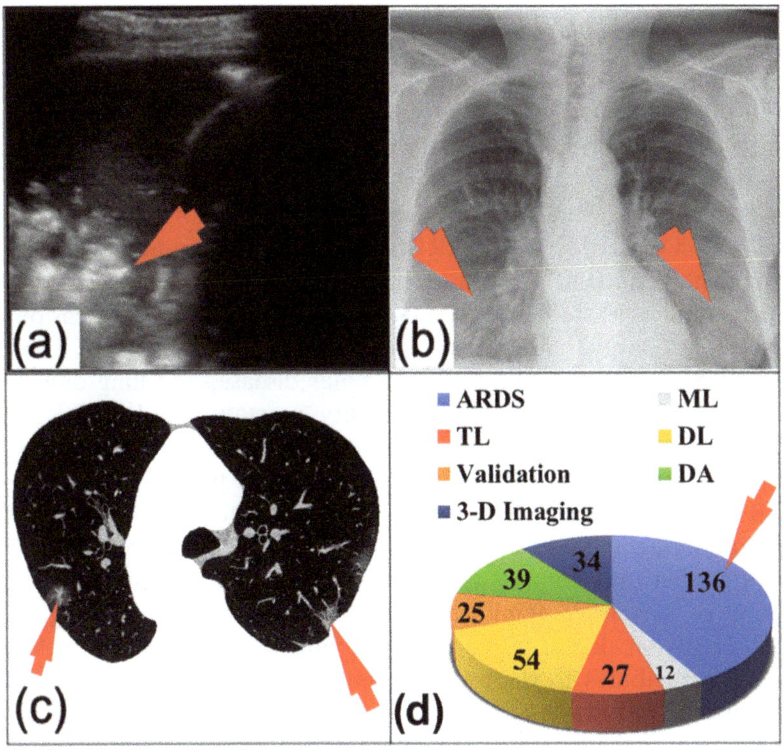

Figure 2.2. Images of COVID-19 infection: (a) lung ultrasound (hyper-echoic region of the COVID-19 lung), (b) chest x-rays (the infected region in the lung), and (c) lung CT (segmented lung region; courtesy of Luca Saba, University of Cagliari, Italy). (d) The number of COVID-19 studies involving ARDS, ML, TL, DL, validation, data acquisition (DA), and 3D imaging.

characterization is essential to assist healthcare practitioners in managing COVID-19 pneumonia and its progression to ARDS.

AI plays a significant role in the management of COVID-19, including patient follow-up, risk assessment, medical imaging, and telemedicine [16]. While AI has been applied in radiological imaging for tasks like classifying images into control, community viral pneumonia, and COVID-19 pneumonia [18–20], it remains unclear if these models align with the expectations of critical care physicians and pulmonologists. Can AI effectively incorporate information on a patient's pre-existing conditions, age, and lung scan patterns to correlate them with the severity of COVID-19? Additionally, there are important questions to address, such as the suitable imaging modalities for ARDS, optimal image-based classifiers for classifying and detecting ARDS severity considering age and comorbidity, estimating patient survival, methods for measuring severity of COVID-19 due to ARDS, assessing the impact of pre-existing conditions on ARDS mortality, and detecting and classifying early stages of ARDS. Addressing these issues is crucial for accurate diagnosis and evaluation of AI-based therapeutic applications [14].

The main imaging tools used for lung imaging include CT, x-rays, ultrasound, and a combination of PET and CT to visualise lung function [21–23]. This study examines different AI-based solutions for classifying COVID-19 lung severity, considering TL, DL, ML, and their combinations. It also highlights important issues that current AI-driven COVID-19 research should tackle in order to provide valuable contributions to the medical field.

This review makes several contributions. Firstly, it explores the pathophysiological aspects of COVID-19-induced ARDS, specifically examining the stages leading to gas-exchange disorders in the alveoli. Secondly, it links comorbid conditions derived from numerous studies to ARDS, including diabetes, hypertension, chronic kidney disease, obesity, cardiovascular diseases, hyperlipidemia, liver disease, cancer, renal dysfunction, HIV, cerebrovascular disease, and lung disease. It also establishes the connection between comorbidity, mortality, and ARDS. Thirdly, it examines the contrast between AI methods employed for characterizing lung diseases before the COVID-19 pandemic (non-ARDS periods) and those utilized during ARDS periods. Fourthly, it introduces seven distinct approaches, referred to as schools of thought (SoT), that utilize AI techniques to assess the severity of COVID-19 within the context of ARDS. Fifthly, it presents a comparative examination of various imaging modalities used for evaluating lung conditions in ARDS. Lastly, it demonstrates how the integration of AI with ARDS and comorbidity contributes to the advancement of personalized medicine, offering valuable insights into assessing ARDS conditions within the framework of COVID-19.

The study is structured as follows: section 2.2 presents the research strategy, section 2.3 describes the pathophysiology of ARDS, section 2.4 analyses the impact of comorbidity in COVID-19, section 2.5 discusses AI architectures categorised into SoT, section 2.6 explains the practical aspect of AI for COVID-19, and sections 2.7 and 2.8 provide critical discussion and conclusion, respectively.

2.2 Research strategy

Figure 2.3 depicts the flowchart of the research strategy adopted in our paper. We conducted research using four online databases: IEEE Xplore, PubMed, Web of Science and ArXiv. Initial screening used the keywords 'COVID-19,' 'coronavirus,' or 'ARDS,' with the modality terms 'lung CT,' 'x-ray,' or 'ultrasound.' The search was augmented with terms 'artificial intelligence,' 'machine learning,' or 'transfer learning,' or 'deep learning,' resulting in 1557 articles. We eliminated 115 duplicate articles and those that were not focused on COVID-19 severity, including classification, leaving us with a total of 1442 articles. From these, we further refined our selection based on relevance and novelty, resulting in 399 articles. We excluded articles that were not relevant to comorbidity, resulting in 242 articles. After removing records with insufficient data, we ended up with a final set of 230 resources. These resources were used in our narrative study, which incorporated AI-based, comorbid-based, and pathophysiology-based articles.

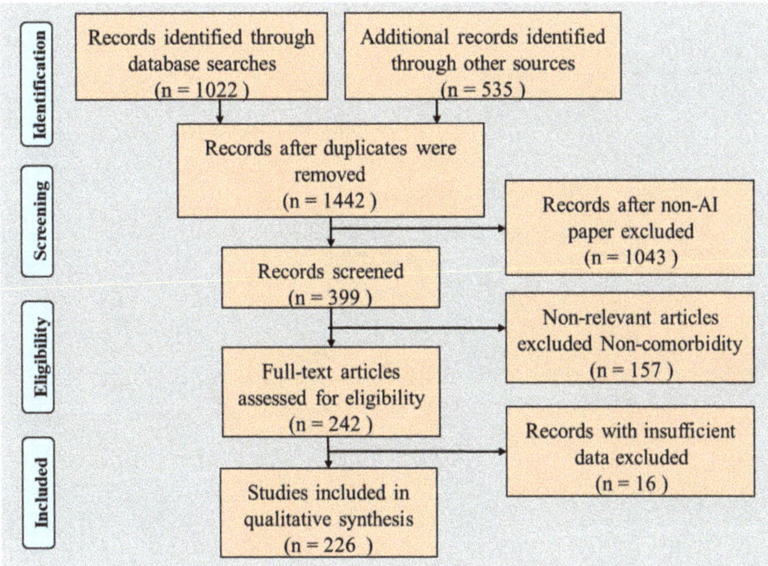

Figure 2.3. The flowchart showing the research strategy.

2.3 The pathophysiology of acute respiratory distress syndrome

The primary mode of transmission for the SARS-CoV-2 virus is through nasal droplets and saliva of an infected person [19]. Upon entry into the body, the virus targets the alveolar type 2 cells (AT2 cells) by binding its viral spike proteins (S1 and S2) to the angiotensin-converting enzyme 2 (ACE2) receptor [24], as depicted in figure 2.4. Previous research by Mossel *et al* has indicated that the SARS-CoV-1 coronavirus exhibits more aggressive replication in AT2 cells compared to alveolar type 1 cells (AT1 cells) in the lung [25]. SARS-CoV-2, with an 80% genetic similarity to SARS-CoV-1, has been shown, via molecular pathways [26], to possess a significant affinity for binding [27]. In the inflammatory phase (termed phase 3), systemic inflammatory mediators are released as a response to SARS-CoV-2 infecting alveolar type 2 (AT2) cells on the surface of the alveolar epithelium [28]. These inflammatory mediators stimulate alveolar macrophages, leading to the production of cytokines such as IL-1, IL-6, and TNF-α. This elevated cytokine production, along with the release of chemokines, can trigger a condition known as a cytokine storm. The sequential occurrence of the systemic inflammatory response, cytokine storm, and organ failure significantly impacts the pathophysiology of ARDS. Similar sequences have been observed in other coronaviruses, including MERS-CoV and SARS-CoV-1 [29, 30]. These processes contribute to the increase in trypsin.

Moreover, the inflammatory response induced by SARS-CoV-2 infection affects the integrity of zonula occludens and endothelial tight junction proteins, leading to weakening and disorganization of endothelial cells. Consequently, the vascular permeability of these cells is increased, allowing intravascular fluids to leak into the

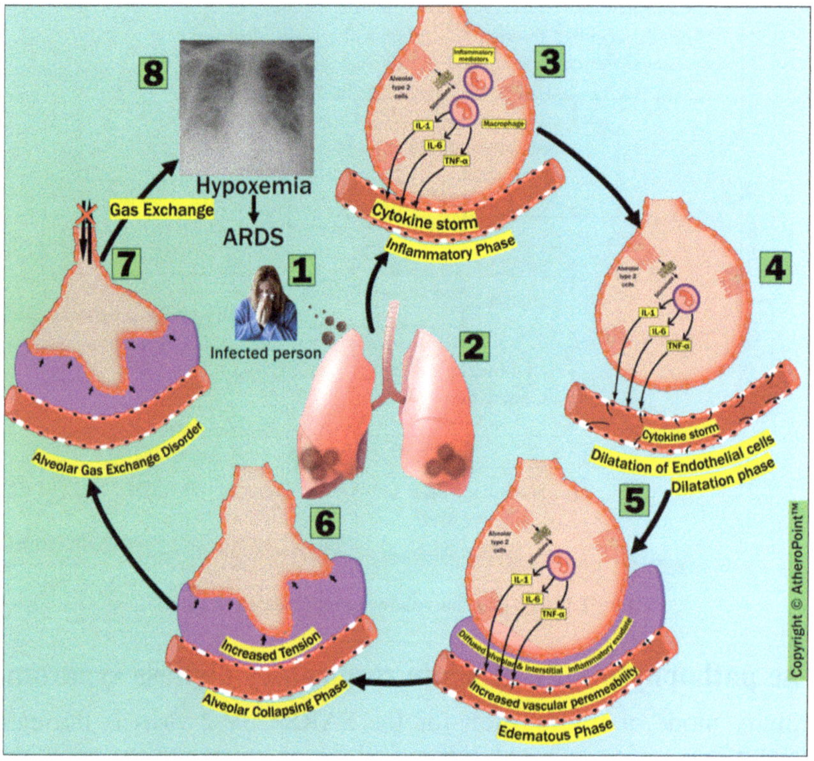

Figure 2.4. The pathophysiology of ARDS after COVID-19 infection, which consists of six phases: (i) inflammatory phase, (ii) dilatation phase, (iii) edematous phase, (iv) alveolar collapsing phase, (v) gas-exchange disorder, and (vi) hypoxemia. (Courtesy of AtheroPoint™, Roseville, CA, USA; reproduced with permission.)

surrounding tissues [31]. During the dilatation phase (referred to as phase 4), the cytokine storm disrupts the endothelial barrier function, which is a consequence of cytokine release following viral infection. This dysfunction contributes to increased vascular permeability. In the edematous phase (designated as phase 5), the elevated intravascular permeability results in the movement of fluids due to diffusion (including neutrophils, proteins, platelets, and erythrocytes) from the blood vessels into the sub-alveolar and interstitial spaces, resulting in the development of diffuse alveolar and interstitial exudates, along with alveolar edema. These manifestations can be detected through radiological consolidations observed in lung CT scans, aiding in the diagnosis and monitoring of treatment response [32]. ARDS is characterized by the presence of diffused alveolar and interstitial exudates, accompanied by elevated sub-alveolar edema [33].

In the alveolar collapsing phase (referred to as phase 6), the accumulation of fluid in the sub-alveolar and interstitial space leads to increased tension, causing the collapse of alveoli. Consequently, the disrupted alveolar structure hinders efficient gas exchange [34]. The alveolar collapse contributes to a ventilation-to-perfusion mismatch between carbon dioxide and oxygen, resulting in impaired gas exchange.

This condition, known as alveolar gas exchange disorder (designated as phase 7), leads to hypoxemia and the development of acute respiratory distress syndrome (ARDS) [35]. If left untreated, the progression of ARDS can increase mortality rates.

2.4 Comorbidity and ARDS

Comorbidities commonly associated with COVID-19 in the context of ARDS include factors such as old age, ethnicity, hypertension, diabetes mellitus (DM), elevated body mass index (BMI), cardiac diseases, and respiratory disorders. Research has indicated that ARDS tends to have more severe outcomes in older patients [36–38], African Americans (Black people in general) [39], individuals with hypertension [36, 40], diabetes [36, 38, 41], higher BMI [42–44], respiratory diseases [42], and myocarditis [45–47]. The inclusion of comorbidities is vital in classifying AI models that can be trained separately to improve diagnosis and predict COVID-19 severity. The cited comorbidity studies [48–101] were sourced from the website https://pubmed.ncbi.nlm.nih.gov/. Figure 2.5 visually presents the number of subjects with comorbidities included in the ARDS-focused studies.

During the selection process, we applied specific criteria based on keywords such as diabetes, hypertension, obesity, cardiovascular diseases, chronic kidney disease, liver disease, renal dysfunction, cancer, hyperlipidemia, human immunodeficiency virus (HIV), cerebrovascular disease, and lung disease. These keywords were utilized to identify relevant studies discussing comorbidity and age groups. 48 studies from specialized medical journals were chosen, provide a rationale and inspiration for gathering statistical data to support the development of innovative AI solutions for the monitoring and diagnosis of COVID-19 severity. Figure 2.6, presented as a pie chart, illustrates the percentage of comorbidity subjects among the selected studies. It is noteworthy that these studies primarily focused on comorbidity and age groups. Among the contributing factors, diabetes and hypertension were identified as the

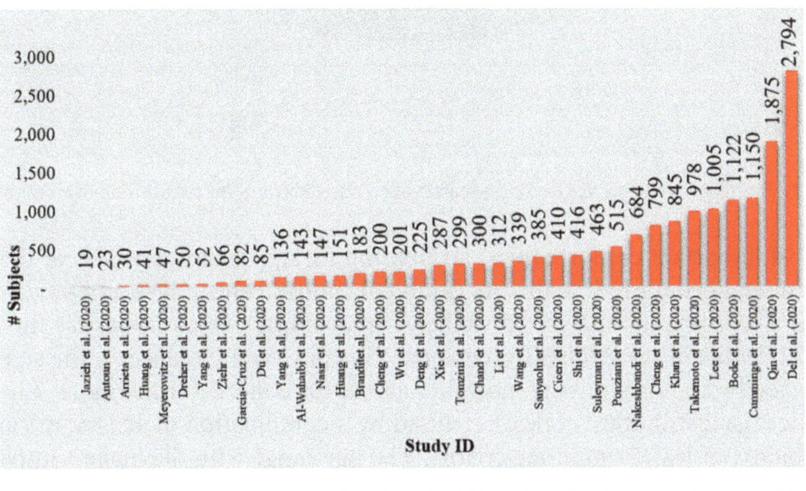

Figure 2.5. The number of subjects enrolled in the ARDS-based studies that consider comorbidities.

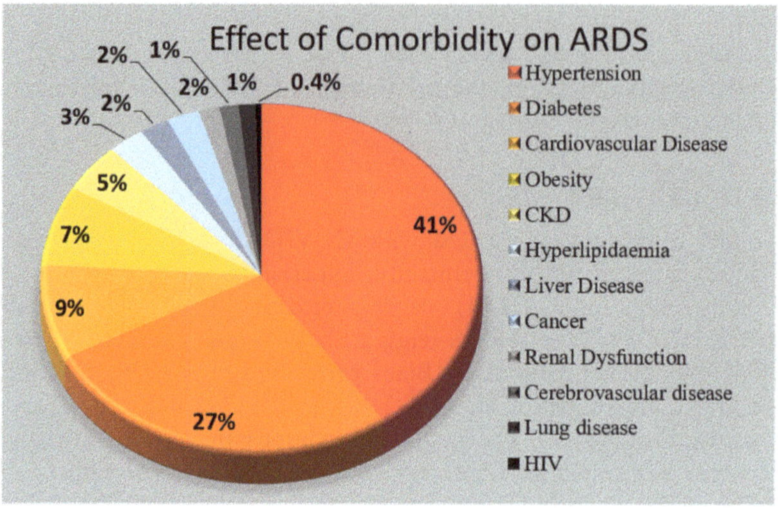

Figure 2.6. Depiction of comorbidities collected from 48 studies.

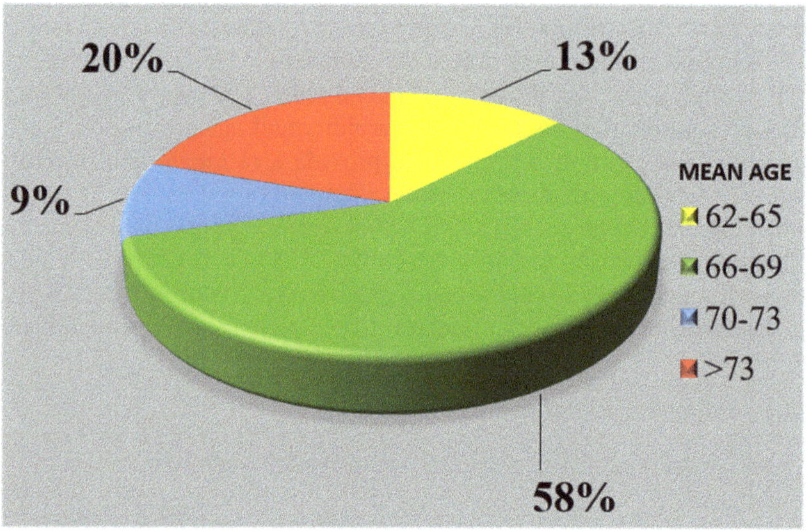

Figure 2.7. Mortality due to the age factor (in years) with comorbidities in the cohort from the selected studies.

two most important comorbidities, accounting for approximately 68% of the cases. The remaining comorbid predictors exhibited similar percentages, ranging from 4% to 13%. According to the selected studies, comorbidities were responsible for 14% of the total deaths in the ARDS framework. This pie chart emphasizes the significant role played by comorbidities in mortality across the cohort. Furthermore, Figure 2.7 illustrates the distribution of deaths caused by a combination of age factors and one comorbidity, at least, within the cohort. The age range with the highest impact was identified as 66–69, comprising 58% of the overall cohort, which is followed by the age group above 70, which constituted 20% based on the studies included in the analysis.

All the subjects included in the cohort had COVID-19 along with one or more comorbidities. These comorbidities were found to contribute to the deterioration of COVID-19. Consequently, each comorbidity depicts a distinct category within the cohort, with similar grayscale features or imaging characteristics. This observation suggests the possibility of developing multiple knowledge-based AI systems, each trained independently on cohorts specific to a particular comorbidity. By utilizing these independently trained AI systems specific to each comorbidity, it becomes possible to analyze lung scans and evaluate COVID-19's severity.

2.5 Artificial intelligence architectures for ARDS characterization by school of thought

The application of AI in disease characterization has been embraced across various medical imaging domains. This encompasses the role of AI in identifying diseases, extracting the region of interest (ROI) related to the disease, and automatically classifying the disease based on binary or multiclass events. We have selected machine learning and deep learning characterization systems that align and harmonize with ARDS frameworks. The purpose of employing this characterization system is to leverage the creative and innovative approaches developed for various modalities, organs, or applications, fostering knowledge exchange and advancement. It should be noted that certain examples mentioned are from our own research group, deliberately selected to emphasize the significance of tissue characterization and disease characterization through the utilization of AI models. These unique and exclusive solutions serve as a 'one-stop-shop' for researchers interested in understanding the relationship between ARDS paradigms and other organ paradigms. This cluster also provides direct access to our group for more detailed information on parallel characterization systems. AI-based characterization has been applied to various body and disease applications, including the brain [102–104], stroke [105–107], vascular plaque [108–110], arrhythmia [111], liver [112–114], coronary artery [115, 116], prostate [117], ovarian [118, 119], diabetes [120], thyroid cancer [121], skin cancer [122, 123], heart [124–126], rheumatoid arthritis [128], and gene expression [127]. This framework of characterization can be expanded and applied within the context of ARDS.

Regarding studies comparing non-ARDS and ARDS, research specifically focused on ARDS started after December 2019, whereas lung segmentation and classification techniques for non-ARDS cases had been in existence for several years before SARS-CoV-2 emerged. Certain AI models that were initially developed for non-ARDS lung data have been partially investigated and adapted for ARDS lung data. In table 2.1, we present studies conducted in ARDS and non-ARDS frameworks, specifically focusing on segmentation, classification, and sub-regional applications such as cancer, tumor segmentation [129–136], and nodule. Please be aware that we excluded animal studies and lung cellular images, specifically concentrating

Table 2.1. AI-based studies involved during Non-ARDS and ARDS periods. AI-based Non-ARDS: AI on ARDS lung data during pre-COVID-19. AI-based ARDS: AI on ARDS lung data postmarked December 2019 COVID-19 (post-COVID-19).

Subsystems	AI-based Non-ARDS	AI-based ARDS
Segmentation	Characteristics Watershed, region-based, contour-based, fusion-based, and model-based. References [129, 137–146]	Characteristics FC-Densenet103, Unet, DenseNet, and DenseNet121-FPN. References [147–165]
Classification	Characteristics Grayscale feature extraction and ML classifier, and model-based techniques. References [166–170]	Characteristics ResNet50, CNN, SVM, ResNet101, VGG16, and VGG19. References [148, 150–153, 157–159, 171–175]
Joint segmentation and classification	Characteristics They use the same characteristics as adapted by segmentation and classification domain for AI-based Non-ARDS. References [167, 169, 170]	Characteristics They use the same characteristics as adapted by segmentation and classification domain for AI-based ARDS. References [148, 150, 151, 154, 156–159, 162]

on human lungs. Furthermore, validation and inter- and intra-observer variability studies were not considered in table 2.1.

It is important to highlight the three key components of an AI-based ARDS system: lung region segmentation, classification, and COVID-19 severity measurement (refer to table 2.2). In this narrative review, we propose a classification of the literature based on different AI architecture categories, referred to as schools of thought (SoT), as outlined in table 2.3.

AI architecture classification has had a significant impact in numerous engineering domains, especially in medical imaging. Biswas *et al* [17] have examined a specific category of AI architectures for medical imaging, while our study demonstrates the application of various architecture classes (SoT). Biswas also delves into the evolution of architectures within the realm of deep learning and its potential for the present and future.

In the past, both manual and automated feature selection methods were utilized in AI architectures. Manual feature selection involved hand-picking specific features and feeding them into the classification model during the learning and training process. This approach belonged to an older generation of methods and represented

Table 2.2. Types of artificial intelligence architectures and severity index.

	AI components and attributes		
	Lung segmentation	AI component	Severity index
SN	Auto/semi-auto	ML/DL/TL/DL+ ML/DL+TL	Categorical/continuous/ Categorical + continuous
1	Auto	DL	Categorical + continuous
2	Auto	DL+ML	Categorical + continuous
3	Auto	DL	Categorical
4	Semi-Auto	DL+ML	Categorical + continuous
5	Auto	ML	Categorical + continuous
6	Auto	DL+TL	Categorical
7	Auto	TL	Categorical

+: both technologies are present.

a traditional school of thought. However, with advancements in AI, the concept of deep learning emerged, which aimed to mimic the brain by using a greater number of layers to filter features. DL paradigms, characterized by their ability to automatically compute millions of training parameters, represent a more recently evolved subset of machine learning. These methods belong to a newer school of thought, primarily defined by their foundational architecture.

While the AI industry has been predominantly dominated by DL and manual feature selection methods, alternative architectures are continuously being explored. DL, which has been widely used, is starting to show limitations, particularly during training. As a result, new architectures have emerged to address these challenges. One such architecture is transfer learning, which involves generating pre-trained weights to expedite and simplify DL systems. Transfer learning is considered a more recent school of thought and is regarded as a classic in its category. The classification of AI architecture has evolved to align with different schools of thought based on the components and characteristics of the architecture. This approach is similar to how image segmentation was categorized into various architectures or schools of thought in the past. For instance, segmentation was classified as region-based, contour-based, or knowledge-based. These categories were then fused together to create intermediate architectures, known as fused architectures, that combined regions with contours or regions with knowledge. This generation of fused architectures represented another architectural paradigm or school of thought. This concept can be observed in the seminal papers by Suri [176–178].

In the context of AI architecture, the term 'School of Thought' (SoT) is synonymous with architectural design. Each individual architecture, as well as fused architectures, can be considered as a distinct SoT. Fused architectures have been developed by various research groups worldwide and have been documented in the literature. Therefore, the SoT represents a more refined approach to architectural

Table 2.3. Clustering of multimodality artificial intelligence architectures and their salient features.

SoT	Reference	Modality	3D/2D imaging	Highlight/objective	Architecture description	Performance metrics
SoT-1	[147–153, 229]	CT: [147, 148, 229] X-ray: [149, 151] LUS: [150]	3D: [147, 148, 229] 2D: [149–151]	Multiview fusion [229], Multiview pyramid network with attention [147], training using human in loop [148], video-based real-time prediction [150], end- to-end DL architecture for semi quantitative prediction COVID- 19 severity [151]	ResNet50 [229], Custom CNN with attention [147], VB-Net [148], commercial deep learning system by Lunit Inc [149], Spatial Transformer Net- work [150], ensemble of multiple networks (Backbone—ResNet, VGG, DenseNet, Inception; Segmentation-UNet, UNet++; Alignment-Spatial Transformer Network; Scoring Head-Feature Pyramid Network; Custom Network) [151]	ACC: [147, 148, 150, 151, 229] AUC: [147, 229] Sensitivity: [149, 150, 229] Specificity: [149, 229] Others: [148–151]
SoT-2	[152, 153]	CT: [152, 153] X-ray: LUS:	3D: [153] 2D: [152]	Biomarker based model [152], model for severity in 3D lung abnormalities [153]	ResNet34 with logistic regression [152], Dense UNet [153]	AUC: [152] Others: [152, 153]
SoT-3	[154–156, 171, 173, 174]	CT [154–156, 171, 173] X-ray: [157, 174] LUS: NA	3D: [154, 173] 2D: [155–157, 171, 174]	3D Convolution Network [154], multi-objective differential evolution based CNN [171], comparison of ten CNNs [156], weakly supervised DL model [173], truncated InceptionNet [174], modified DarkNet CNN [157]	ResNet50 [154], Custom CNN [157, 171, 173], DenseNet [155] (AlexNet, VGG16 VGG19, SqueezeNet, GoogleNet, MobileNet-V2, ResNet18, ResNet-50, ResNet101, and Xception) [156], InceptionNetV3 [174]	ACC: [155–157, 171, 173, 174] AUC: [154–156, 173, 174] Sensitivity: [154–157, 171, 173, 174] Specificity: [154–157, 171, 173, 174] Others: [155–157, 171, 173, 174]

SoT-4	[157]	CT: [157] X-ray: NA LUS: NA	3D: [157] 2D: NA	ML and DL hybrid network for classification and prognosis [157]	ResNet18 with Gradient Boosting [157]	ACC: [157] AUC: [157]
SoT-5	[158, 159]	CT: [158] X-ray: NA LUS: [158]	3D:NA 2D: [158, 159]	Pleural line identification using ML [158], automatic severity assessment and exploration of severity related features using ML [159]	Hidden Markov Model and Viterbi Algorithm combined with SVM [158], Random forest [159]	ACC: [158, 159] AUC: [159] Sensitivity: [158] Specificity: [158] Others: [158, 159]
SoT-6	[160, 172]	CT X-ray: [160, 172] LUS:	3D:NA 2D: [160, 172]	Ensemble of DL and TL [160], multi-dilation CNN for extraction of COVID-19 features [172]	Custom CNN [160, 172], (VGG16, VGG19, Inception-V3, Xception, InceptionResNet- V2, MobileNet- V2, DenseNet201, NasNet-mobile) [160]	ACC: [160, 172] AUC: [160, 172] Sensitivity: [160, 172] Specificity: [160, 172] Others: [160, 172]
SoT-7	[161–165, 175]	CT:[165] X-ray: [163–165, 175] LUS: [161]	3D:NA 2D: [161–165, 175]	Explainable DL to provide explainability about the prediction [175], real-time internet based COVID-19 detection service [161], TL model trained on ensemble of two publicly available datasets [163], interpretable AI framework for COVID-19 classification [164], applying TL on comprehensive custom COVID- 19 CT and x-ray datasets [165]	VGG-16 [161, 175], Alexnet [165] DenseNet201 [162], Xception [163], InceptionNetV3 [164]	ACC: [161–165, 175] AUC: [161, 162, 164] Sensitivity: [161–165, 175] Specificity: [161–165, 175] Others: [161–164, 175]

design, allowing for a more nuanced classification. In our previous studies, we have also employed the concept of SoT. It is important to note that while the SoT aligns with the generations of architecture, it provides a more nuanced understanding of the differences between them.

The AI models used in different SoT along with their salient features are listed below.

SoT-1 represents an effective approach for quantifying the severity of COVID-19 lung damage and classifying the lung based on binary or multiclass frameworks. In this approach, lung segmentation is automated without the need for human intervention. This can be achieved using commercial software like XMedCon [179] or AI techniques such as threshold segmentation [180] or UNet-based segmentation [146]. The AI component in SoT-1 utilizes state-of-the-art or custom DL architectures.

SoT-2 serves a similar purpose as SoT-1, aiming to provide quantitative and categorical assessments of COVID-19 severity. However, it utilizes a more complex hybrid architecture compared to SoT-1. By integrating more than one AI architecture, including both DL and ML techniques, SoT-2 has exhibited comparable accuracy and the capability to accomplish supplementary sub-goals, including prognosis analysis. One drawback of SoT-2 is the increased effort and time needed to combine and fine-tune the different components of the AI.

SoT-3 is specifically designed to classify patients into different categories, such as non-COVID-19, COVID-19, and other types of pneumonia. Many COVID-19 studies utilizing AI have adopted SoT-3 for this purpose. Like SoT-1, SoT-3 automates lung segmentation. It employs an end-to-end automated pipeline that focuses on classifying pneumonia cases, including COVID-19. This approach enables faster and more straightforward implementation of research. However, one drawback of SoT-3 is its limited capacity to provide actionable insights for healthcare professionals in the treatment of patients suffering from COVID-19. For instance, SoT-3 does not facilitate the measurement of biomarkers that could indicate disease severity.

SoT-4 is centred around semi-automatic lung segmentation conducted by experts before the AI-driven processing of radiography images. It incorporates a hybrid AI component that combines different AI architectures to calculate multiple metrics. SoT-4 provides researchers with the flexibility to construct a pipeline that attains the desired balance between accuracy, complexity, and speed, depending on the specific metrics of interest. Nevertheless, a drawback of SoT-4 is the significant time required for designing the semi-automatic lung segmentation process and developing appropriate hybrid models.

SoT-5 involves the use of automated lung segmentation along with feature selection methods, manual feature engineering, and traditional ML models. The computed features are employed to calculate biomarkers, which are subsequently utilized for predicting the severity of COVID-19 and classifying patients. SoT-6 research utilizes a hybrid model for automated segmentation that combines DL and TL to generate categorical metrics for patient classification [181]. Similarly, SoT-7 research focuses on TL and comprises both an online and an offline system. In the offline system, researchers manually extract relevant features from radiography images, which are then used to train a classifier. The trained coefficients are subsequently transferred to the online classifier, enabling real-time classification and prediction of COVID-19 severity. Effective lung segmentation using ML techniques has been demonstrated in previous studies [146, 167]. Figure 2.8 showcases an online ML-based COVID-19 risk prediction system, which is consistent with the ML systems previously published by the research group [165, 182, 183].

Figure 2.9 displays a DL architecture reference, established using [184]. Among various DL architectures, the convolutional neural network (CNN) is extensively employed. CNNs employ multiple convolutional filters to extract basic visual features from input image data. These filters are stacked in layers to capture more complex visual features. Pooling layers are frequently integrated with CNN layers to reduce spatial information in the intermediate representation, while padding layers help preserve the correct data dimensions.

TL is an advanced technique that enhances the capabilities of DL architectures. It involves pre-training these architectures on large datasets, enabling them to achieve

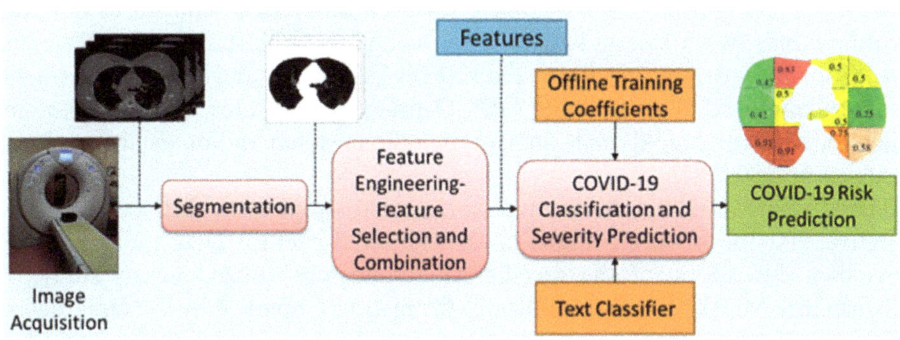

Figure 2.8. An online ML-based COVID-19 risk prediction system. (Courtesy of AtheroPoint™, Roseville, CA, USA; reproduced with permission).

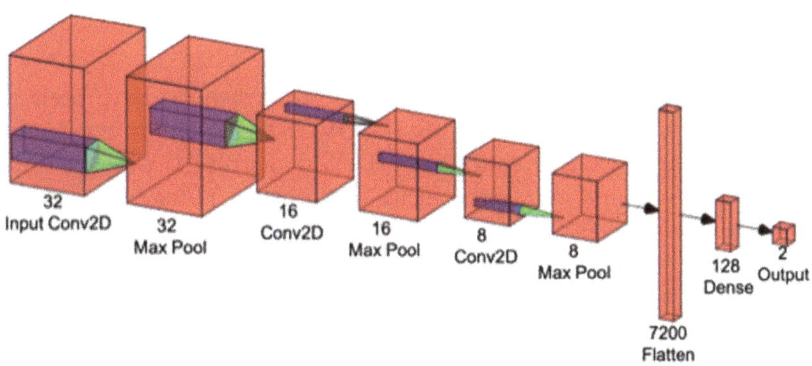

Figure 2.9. A custom CNN-based DL architecture comprising different layers.

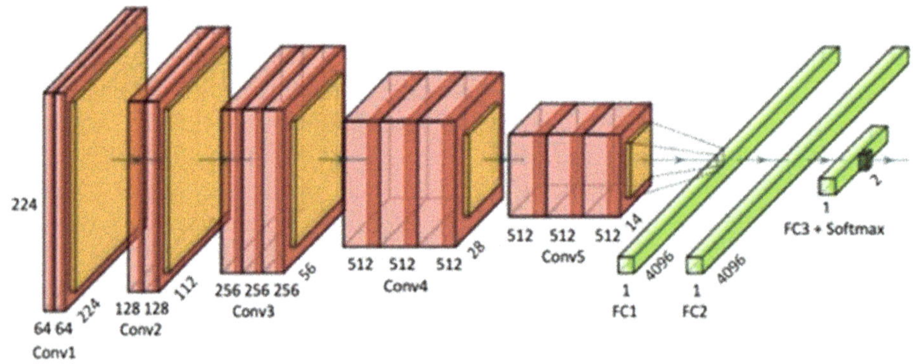

Figure 2.10. An example of transfer learning (TL) architecture using VGG16.

higher accuracy even with limited training data, less powerful hardware, or restricted training time. The reference architecture for TL, called VGG16, is depicted in figure 2.10, as described in [185].

Table 2.3 gives us a comparative analysis of multi-modal AI architectures that are utilized for the diagnosis of COVID-19, highlighting their main characteristics. The 'Arch Type' column provides reference to seven different architectures developed according to the workflow and goals of various AI research studies. The 'Reference' and 'Modality' attributes indicate the sources referenced and the specific imaging modality employed. The attribute '3D/2D Imaging' indicates that the research is either focused on three-dimensional or two-dimensional radiological imaging, as discussed in [186, 187].

The attribute 'Highlight/Objective' in the table signifies the distinctive aspect or objective that the research might have. The 'Architecture Description' attribute provides a description of the specific AI architecture utilized in the study. The 'Performance Metrics' attribute indicates the metrics employed by the researchers to assess the effectiveness of their work. For a more in-depth understanding of AI and the applications, readers are advised to explore specialized reviews that focus on AI-based methodologies.

2.6 Workflow considerations for COVID-19 lung characterization: CT vs. x-ray

The three chest-imaging modalities x-ray, CT, and ultrasound have been strongly recommended by WHO for the diagnosis of COVID-19 [90]. In their guidelines, the following observations were given for COVID-19 diagnosis. (i) x-rays was found to be lower in sensitivity and higher specificity than chest CT imaging [188, 189]. (ii) Chest x-rays are less expensive, have lower radiation, take less acquisition time, and are less expensive to use for monitoring than CT. (iii) Chest CT was found to have higher sensitivity but lower specificity and emits more radiation than x-ray. (iv) Lung ultrasound was found to have very low diagnostic accuracy but offers an alternative for several other applications, such as the abdomen, carotid, urology, obstetrics, and gynaecology imaging. On the other hand, ultrasound has a high risk of COVID-19 infection transmission due to close contact with the patient compared to other imaging modalities. In accordance with the guidelines published by the American College of Radiology on April 8, 2020 [190], MRI is not recommended for COVID-19 patients due to the elevated risk of infection [91]. While there are studies available on MRI and COVID-19 [92–94], no specific literature was found that directly addresses the combination of COVID-19, MRI, and AI. The main reason is the long scanning time, risk of infection, and high cost associated with MRI. On the contrary, FDG-PET/CT imaging, which is a more advanced technique compared to CT alone [191], has been used in many studies for diagnosing COVID-19 [95, 96]. However, no relevant studies combining FDG-PET/CT imaging, COVID-19, and AI were found. This could be attributed to the higher cost associated with PET/CT imaging [23, 192].

X-ray and CT imaging modalities are considered appropriate for COVID-19 diagnosis according to the recommendations of the World Health Organization, as indicated in table 2.4. Figure 2.11(c) demonstrates the percentages of studies utilizing either CT, x-ray, or both modalities, with CT being regarded as the gold standard. Several significant studies have been conducted using AI for automated COVID-19 diagnosis with x-rays and CT [21–23, 192–202]. Open-source COVID-19 datasets for x-ray and CT imaging, such as the one provided by RSNA, have gained popularity in the scientific community.

To compare studies involving x-ray and CT imaging, several attributes were considered, including the risk classes, number of subjects, 2D vs. 3D imaging, AI models, automated vs. non-automated ROI segmentation, augmented vs. non-augmented techniques, hardware and software used, K-fold cross-validation, performance, and optimal models. Several notable studies [203–208] included cohorts of over 1000 subjects. The number of classes in the studies was categorized into multiclass and binary classifications, with a majority focusing on binary classification [203, 204, 209–216]. While most studies utilized 2D x-ray images, two studies employed 3D CT volumes as input [210, 212]. ROI segmentation, an essential step in chest image analysis, was mostly automated, with five studies

Table 2.4. Compatibility of imaging modality for COVID-19 and adaptability for AI [230]

Imaging modality	Suitable for COVID-19 as per WHO guidelines	Cost	Risk of radiation	Risk of infection due to close contact	Compatible with AI for COVID-19 diagnosis
PET/CT	High	Very high	Very high	Low	Low
CT	High	High	Very high	Low	Very high
X-ray	Medium	Low	High	Low	Medium
Ultrasound	Low	Low	No	High	Low
MRI	Low	High	No	Low	Low

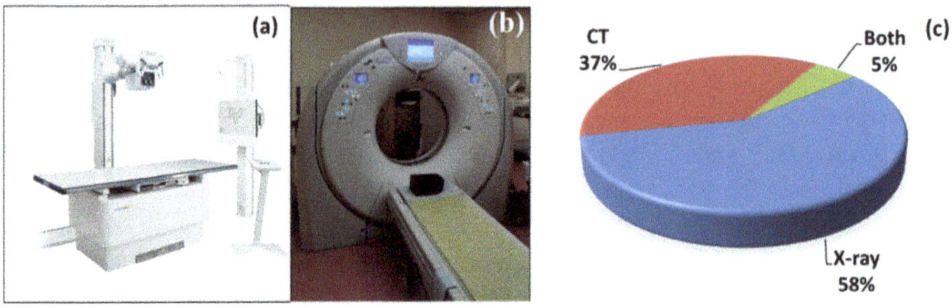

Figure 2.11. (a) An x-ray scanner. (b) A CT-scanner (Courtesy of Luca Saba, University of Cagliari, Italy). (c) Studies using CT vs. x-ray.

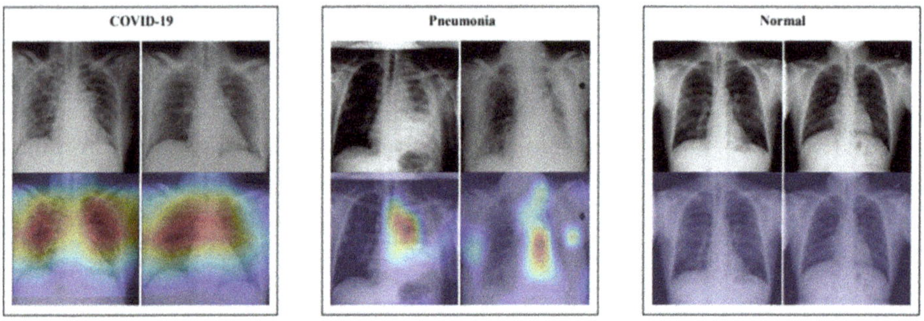

Figure 2.12. X-ray scans of COVID-19, pneumonia, and normal lungs [219] (permission pending).

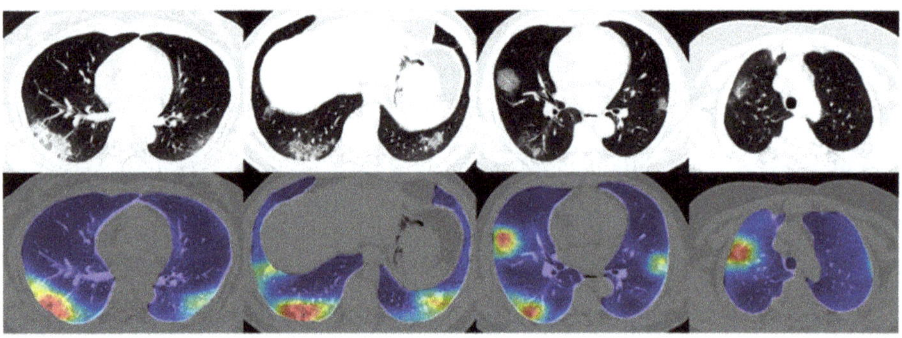

Figure 2.13. CT scans classified as positive for coronavirus abnormalities and their corresponding color heatmaps [220] (permission pending).

employing automated methods [203, 211, 212, 217, 218], and one study using manual segmentation [209]. Full-sized images were predominantly utilized in the classification process.

Regarding AI models, the studies primarily fell into two categories: ML and DL. In one study, a network based on DL achieved a classification accuracy of 90.13% when distinguishing pneumonia, normal, other diseases, and COVID-19 cases in chest x-rays [219]. Figure 2.12 visualizes sample images and their corresponding colour maps from this study. Another study utilized a network based on DL to detect COVID-19 abnormalities in lung CT scans and generate a heatmap indicating severity levels [220]. Figure 2.13 provides an illustration of the resulting images.

To address data limitations in medical studies, image augmentation techniques were commonly employed. Transfer learning with DL models was particularly utilized when datasets were small, often consisting of a few thousand samples [210–212, 215, 218]. These augmentation techniques involved various operations such as image transformation, blurring, and colour manipulation. Some studies also adopted a patch-based framework with a small number of trainable parameters, incorporating multi-scaling for spatial features in diagnosis of COVID-19 [221, 222]. Cross-validation, specifically the K-fold strategy, was frequently used to evaluate AI performance. Authors employed five-fold [203, 205, 215, 218], and ten-fold [207] cross-validation approaches to comprehensively assess model performance.

To prevent overfitting during AI model optimization, researchers utilized techniques like cross-validation and transfer learning in supervised learning. Overfitting occurs when a model performs well on known data but struggles with unknown data. Increasing the training dataset size was among the approaches used to mitigate overfitting. The performance of the classifier was influenced by various factors, including the training iterations, number of subjects, training time, and sample size. At present, there is no definitive best-performing model or method that can be identified. However, as further trials and studies are conducted, it is anticipated that superior AI models will be emerging.

2.7 Critical discussion

Clinical requirements for COVID-19 AI systems. To develop an optimal AI system for COVID-19, it is crucial to meet several key clinical requirements. Firstly, the system should demonstrate robustness and stability, ensuring that its output remains reliable and consistent even in the face of variations in patient demographics or other related characteristics. Secondly, it should demonstrate reliability and reprehensibility, consistently producing results that are similar in multiple trials. In addition, when deployed in an operating room, the AI system should be cost-effective and have reasonable speed compared to traditional techniques for diagnosing COVID-19, like RT-PCR tests. To ensure broad applicability, the AI system should be trained and tested on a diverse and representative dataset, ensuring its generalization. It is essential for the AI system to provide an accurate assessment of COVID-19 severity, as this metric plays a critical role in guiding treatment decisions for patients. Incorporating this capability into an AI diagnostic system is highly desirable. Overall, an ideal system embedded with AI for COVID-19 must meet these clinical requirements to effectively assist healthcare professionals in diagnosing and managing the disease.

Validation of the AI system's results by radiologists, pulmonologists, and doctors is necessary to assess its effectiveness. Furthermore, the AI system should utilize 3D imaging to segment and analyze the impact of COVID-19 on various organs, providing additional support to clinicians. By integrating these clinical requirements, there is significant potential for enhancing the current state of AI-assisted diagnostics for patients suffering from COVID-19. This has the potential to greatly improve the current standard of care by incorporating AI technology into the diagnostic process.

System optimization. DL techniques are extensively utilized in DL-based AI systems for the detection and classification of COVID-19. Researchers have employed data augmentation techniques to increase the size of the training dataset for COVID-19. The choice of the number of convolution layers in DL models is typically based on researchers' intuition. However, it is crucial to optimize DL-based AI systems for COVID-19 by considering the relationship between augmentation, the resulting classification accuracy, and the number of convolution layers. This optimization process can lead to enhanced performance and accuracy in COVID-19 classification tasks. Please refer to figure 2.14 for a visual representation of these concepts.

Scientific validation. To validate and evaluate an AI-based COVID-19 diagnostics system, it is essential to consider a wide range of comorbidity conditions. Various scenarios should be taken into account where assumptions may differ. For instance, modifying the thickness of CT scans or providing different views (coronal, sagittal, axial) to the system for the purpose of diagnosing can impact its performance. Moreover, the validation of system stability must be done by altering the combination of data used and employing partition protocols like K10, K5, and K2. A AI system which is stable will demonstrate minimal standard deviation across different data combinations [165]. Additionally, the system should undergo validation using patients from diverse age groups and with different comorbidities to ensure its robustness and reliability in real-world applications. These validations have great importance in establishing effectiveness and generalizability of the AI-based diagnostics systems that are used for COVID-19 [165].

Clinical validation. To gain approval from regulatory authorities, it is vital to conduct gold standard validation for the research. This involves comparing the predictions made by the AI system with a physical examination of the body's organs. One method to accomplish this is to validate the body tissues

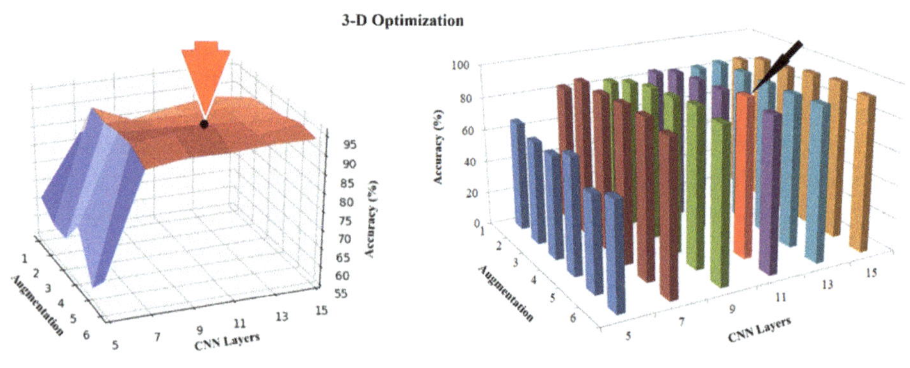

Figure 2.14. A 3D graph representing the relationship between CNN layers, data augmentation, and accuracy. (Courtesy of AtheroPoint™, Roseville, CA, USA; reproduced with permission [14].)

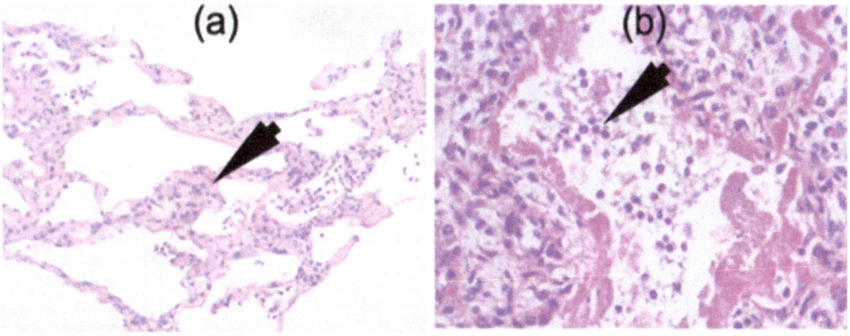

Figure 2.15. Microscopic views of (a) interstitial pneumonia and (b) COVID-19 pneumonia. (Courtesy of Luca Saba, A.O.U., Cagliari, Italy.)

microscopically to evaluate how severe the ARDS is which has resulted from COVID-19, as shown in figure 2.15. Furthermore, it is crucial to perform an analysis of intra-observer and inter-observer variability [223] to address any potential human bias in the system. These validations play a critical role in ensuring the accuracy and reliability of the AI system's predictions, thereby increasing its credibility and acceptance by regulatory authorities [223].

AI for COVID-19 comorbidity and age-group frameworks. The AI system's design should be flexible enough to consider patients' comorbidities, including conditions such as cardiovascular diseases, diabetes, obesity, pancreatic diseases, retinopathy, angiography, and blood vessel diseases [224], with their corresponding age groups. In [225], a study was conducted to develop an AI system for predicting cardiovascular diseases in a diverse patient population. Similarly, the AI systems discussed in [226, 227] can be extended to incorporate comorbidities and different age groups to enhance the accuracy of COVID-19 diagnosis.

To address this challenge, the primary AI system can be divided into multiple subsystems, each tailored for specific comorbidity categories and further categorized based on different age groups. For example, if there are five comorbidities and three age groups, a total of 15 subsystems would be developed. Each subsystem should be trained independently using a distinct gold standard database [228]. By incorporating additional input data related to the patient's comorbidities and age group, an appropriate AI subsystem can be identified for each new patient.

3D image acquisition and processing. The majority of AI research on COVID-19 has predominantly focused on evaluating the severity of lung infection using 2D imaging, which may not fully capture the disease's progression. To overcome this limitation, there is a need to explore COVID-19 data in 3D, as depicted in figure 2.16. One potential approach is to utilize scans acquired during the contrast enhancement of a lung ultrasound, which can provide valuable insights for early management and prognosis of COVID-19 [231]. In a study cited in [232], researchers employed 3D CT scans to diagnose and classify lung lesions affected by COVID-19, offering a more comprehensive understanding of the disease's characteristics. Furthermore, in another study mentioned in [233], the authors utilized the

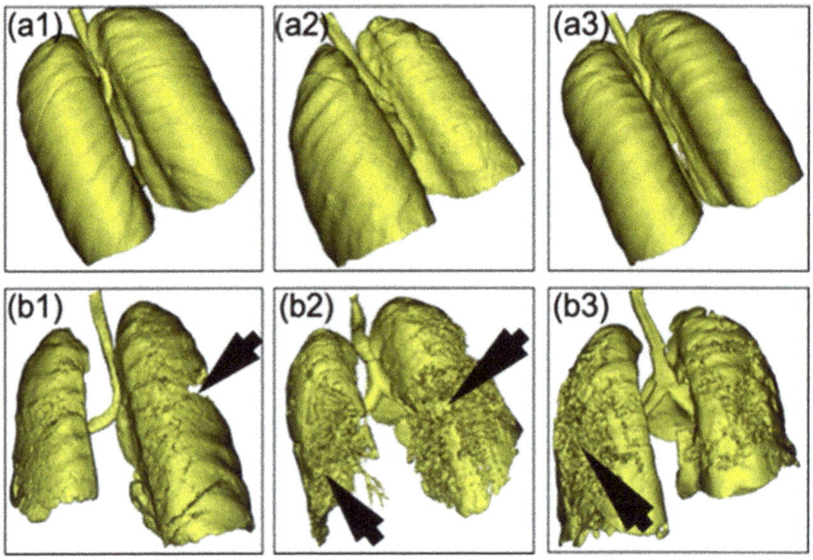

Figure 2.16. Three lungs with non-COVID-19 pneumonia (a1, a2, and a3). Three lungs with COVID-19 pneumonia with different COVID-19 severities (b1, b2, and b3). (Courtesy of AtheroPoint™, Roseville, CA, USA; reproduced with permission.)

DensNet121-FPN architecture and a deep learning strategy for lung segmentation and classification, leading to improved accuracy in the analysis.

Multi-modality data. Radiographic techniques provide several diagnostic options for COVID-19, including lung chest x-rays, CT, PET/CT scans, and lung ultrasound. The choice of the appropriate imaging modality depends on factors such as the patient's age, pre-test probabilities (PTPs), and comorbidities [18]. To evaluate the severity of a COVID-19 infection, troponin levels can be used as a PTP test, serving as an indirect indicator of hypoxia. In the early stages of COVID-19, x-rays and lung ultrasounds are cost-effective and easily accessible diagnostic tools, given their compact size. However, for patients with comorbidities, advanced age, or COVID-19-induced acute respiratory distress syndrome (ARDS) leading to hypoxia, lung CT is often the preferred choice. Lung CT provides robust diagnostic capabilities and higher resolution in such scenarios. When developing systems embedded with AI for COVID-19, it is crucial to ensure their adaptability and generalizability across multiple imaging modalities to cater to the specific needs of individual patients. By considering these factors and the insights presented in the cited references, AI systems can be effectively designed to integrate and analyze data from different imaging modalities, thereby optimizing the diagnostic process for COVID-19 patients.

Tissue characterization of lung scans. In one of the last investigations [3], an analysis was conducted using bispectrum (B) on lung tissues affected by COVID-19. The study employed a higher-order spectrum (HoS) approach which validated the

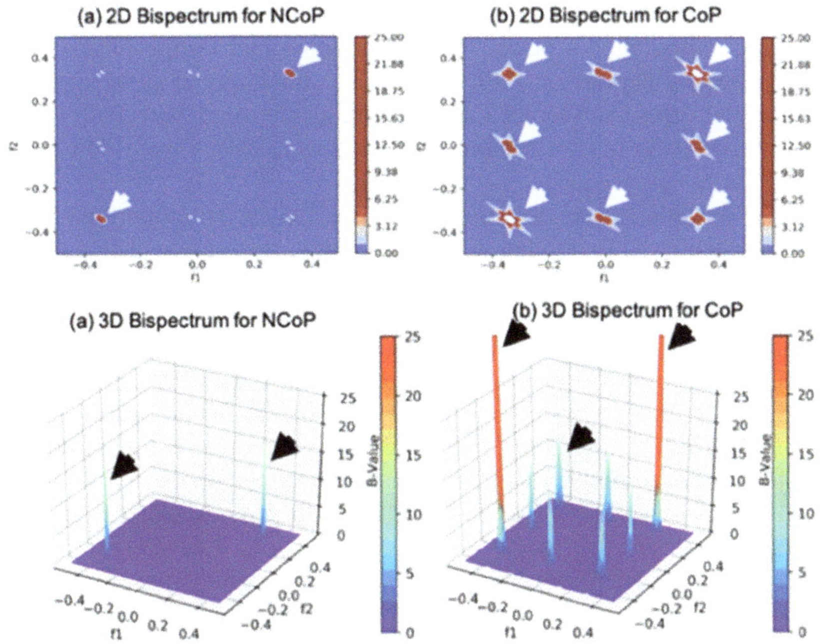

Figure 2.17. Bispectrum analysis of non-COVID-19 pneumonia (NCoP) and COVID-19 pneumonia (CoP). (Courtesy of AtheroPoint™, Roseville, CA, USA; reproduced with permission.)

findings and was not dependent upon the AI for any validation. The results, illustrated in figure 2.17, revealed that samples infected by COVID-19 exhibited significantly higher B-values compared to non-infected samples. This observation suggests that AI systems can effectively utilize these patterns to differentiate between non-COVID-19 and COVID-19 samples. By incorporating such insights and methodologies, AI systems can enhance their capability to accurately identify and distinguish cases of COVID-19.

Data collection. To mitigate biases related to ethnicity, comorbidity, age, and other factors among COVID-19 patients, it is highly advantageous to collect data over an extended period and from diverse geographical locations [234]. This approach ensures the inclusion of a more representative and comprehensive dataset. Furthermore, the patient data should encompass multiple classes, including a control group and various types of pneumonia classes such as HIV, viral, COVID-19, MERS, and bacterial [235]. By adhering to these criteria, the development of a more robust and generalized AI system for COVID-19 diagnostics becomes feasible, eliminating the need for localized validations or retraining. Such an approach facilitates the creation of a versatile solution that accounts for the inherent variations and complexities in different patient populations. By considering these aspects, AI systems can be designed to deliver accurate and reliable COVID-19 diagnostics, benefiting healthcare providers worldwide.

AI and hardware constraints. GPUs have a vital role in DL investigations, providing the necessary computational power for both training and testing phases [234, 235]. Researchers heavily depend on GPUs for their computational capabilities. Popular open-source platforms like Google Colab are frequently used, with Python being the programming language of choice, along with frameworks like TensorFlow or PyTorch [14, 102].

Strengths, limitations, and extensions. This research offers important findings on different aspects of ARDS in relation to comorbidity, medical imaging modalities for COVID-19 patients with ARDS, workflow considerations for imaging tools, AI architecture design for diagnosing the severity of ARDS-related lung conditions, and the role of AI in managing comorbidities. The research emphasizes the essential elements required for a dependable and secure AI-driven approach to evaluate the severity of COVID-19.

Although this study introduces novel approaches by incorporating AI designs based on comorbidity, it acknowledges that there are other aspects that could be included for a more comprehensive analysis. For example, further exploration of the biological processes underlying comorbidities could provide valuable insights, but due to limitations in the length of the manuscript, they were not included. Additionally, more rigorous comparisons of deep learning paradigms could be incorporated, and the study references previous dedicated reviews that offer opportunities for further exploration in these areas (table 2.5 and 2.6).

Future research should focus on developing methods that utilize GPS within targeted lung regions in 3D and 2D medical images for rapid assessment of COVID-19 severity. These methods should take into account factors such as comorbidity and time constraints to enable accurate and timely determination of severity [236–238]. Additionally, this foundational study creates opportunities for further exploration in various medical fields, including cardiology, neurology [239], ophthalmology [241], and diabetology [240]. The application of statistical tools, such as systematic review and meta-analysis (SRMA), can provide comprehensive and evidence-based insights in these areas. However, it is important to note that the inclusion of SRMA goes beyond the current scope of this narrative review.

2.8 Conclusion

This recent study focuses on the investigation of various comorbidities, their impact on an ARDS-based framework, and their association with mortality rates. The study proposes solutions based on AI that incorporate comorbidity as a factor that is independent in the design. It presents the key features and architectural characteristics of seven different SoT approaches. The study highlights the need for a thorough investigation of AI system's important components before its clinical application for diagnosing severity of COVID-19. These components include clinical and scientific validation, optimizing the architecture design by balancing layers and augmentation, and selecting the appropriate imaging modality based on the troponin release and severity of COVID-19 symptoms. The study concludes that the incorporation of comorbidity and age group as novel features in AI design holds great promise for characterizing ARDS in the future [242–245].

Table 2.5. Artificial Intelligence-based studies for automatic COVID-19 detection using lung CT.

SN	References	#Subj.	Risk Class	2D vs. 3D	*ROI	AI Model	Augm.	CV	H/W-S/W	Optimal model	Performances
1	Ardakani *et al* (2020) [209]	194	2	2D	×	CNN+TL	×	NA	SPSS software (version 24, IBM)	ResNet101	ACC:99.51% SE:100% SP:99.02% AUC:0.994
2	Wu *et al* (2020) [217]	495	3	2D	✓	CNN+TL	×	NA	GPU Python	ResNet50	ACC:76.0% SE:81.1% SP:61.5% AUC:0.819
3	Zheng *et al* (2020) [210]	499	2	3D	×	3D CNN+TL	✓	NA	GPU PyTorch	UNet	ACC: 90.1% SE: 90.7% SP: 911% AUC: 0.976
4	Yang *et al* (2020) [211]	679	2	2D	✓	CNN	✓	NA	GPU PyTorch	DenseNet169	ACC:89% FS :90% AUC:0.98
5	Gozes *et al* (2020) [212]	270	2	2D and 3D	✓	3D + 2D CNN	✓	NA	NA	ResNet50	SE: 98.2% SP: 92.2% AUC: 0.996

(Continued)

Table 2.5. (*Continued*)

SN	References	#Subj.	Risk Class	2D vs. 3D	*ROI	AI Model	Augm.	CV	H/W-S/W	Optimal model	Performances
6	Shi *et al* (2020) [203]	2685	2	2D	✓	ML (RF)	×	K5	NA	RF	ACC:87.9% SE: 90.7% SP: 83.3% AUC: 0.942
7	Liu *et al* (2020) [213]	746	2	2D	×	LA-DNN+ TL	×	NA	NA	LA-DNN	ACC:88.8% F1S: 94.7% AUC: 0.88
8	Panwar *et al* (2020) [204]	2482	2	2D	×	CNN+TL	×	NA	NA	VGG19	ACC:87.9% SE: 90.7% SP: 83.3% AUC: 0.942

#Subj.: Number of subjects in the study; Augm.: Augmentation; NA: Not available; ACC: Accuracy, SE: Sensitivity; SP: Specificity; AUC: Area under the curve; F-M: F-measure; KP: Kappa statistics; TL: Transfer Learning; FS: F1-Score LA-DNN: lesion-attention deep neural network; RF: Random forest; ML: machine learning; H/W: Hardware; S/W: Software; CV: Cross-Validation; K5: five-fold; ROI: Automated Region of Interest.

Table 2.6. Artificial Intelligence-based studies for automatic COVID-19 detection using lung x-ray.

SN	Reference	#Subj.	Risk Class	2D vs. 3D	Auto ROI	AI Model	Augm.	CV	H/W-S/W	Optimal Model	Performances
1	Narayan Das *et al* (2020) [216]	NA	3	2D	✗	CNN+TL	✗	NA	NA	Proposed CNN	ACC: 97% FM: 96% SE: 97% SP:97% KP:0.97
2	Ouchicha *et al* (2020) [205]	2905	3	2D	✗	CNN	✗	K5	NA	CVDNet (proposed CNN)	ACC: 90% PR: 96.72% RC: 96.84% FS: 96.68%
3	Hemdan *et al* (2020) [214]	50	2	2D	✗	CNN	✗	NA	GPU Python	VGG19 DenseNet201	ACC: 96.69% PR: 83% RC: 100% FS: 91%
4	Zhang *et al* (2020) [215]	43 370	1, 2	2D	✗	CNN	✓	K5	NA	CAAD (proposed CNN)	ACC: 78.57% SE: 71.70% SP:79.40% AUC: 0.83
5	Togacar *et al* (2020) [218]	458	3	2D	✓	CNN+TL	✓	K5	CPU MATLAB	MobileNet-V2	ACC: 99.24% SE: 100% SP: 97.72% FS: 99.43%

(Continued)

Table 2.6. (*Continued*)

SN	Reference	#Subj.	Risk Class	2D vs. 3D	Auto ROI	AI Model	Augm.	CV	H/W-S/W	Optimal Model	Performances
6	Farooq *et al* (2020) [206]	2839	4	2D	×	CNN+TL	✓	NA	NA	ResNet50	ACC: 96.23% RC: 100% PR: 100% FS: 100%
7	Cozzi *et al* (2020) [207]	1427	3	2D	×	CNN+TL	×	K10	NA	VGG19	ACC: 96.78% SE: 98.66% SP: 96.46%
8	Pereira *et al* (2020) [208]	1144	7	2D	×	ML+DL+TL	×	NA	NA	Multilayer Perceptron	FS: 89%

#Subj.: Number of subjects in the study; Augm: Augmentation; NA: Not available; ACC: Accuracy; SE: Sensitivity; SP: Specificity; AUC: Area under the curve; F-M: F-measure; KP: Kappa statistics; PR: precision; RC: recall; FS:F-Score; TL: Transfer Learning; LA-DNN: lesion-attention deep neural network; ML: machine learning; H/W: Hardware; S/W: Software; CAAD: confidence-aware anomaly detection model; ROI: Region of Interest.

Symbol table

SN	Acronyms	Description
1	2D	Two dimensions
2	3D	Three dimensions
3	ACC	Accuracy
4	AI	Artificial intelligence
5	ARDS	Acute respiratory distress syndrome
6	AUC	Area-under-the-curve
7	CNN	Convolution neural network
8	CT	Computed tomography
9	DL	Deep learning
10	FS	F1-score
11	GPU	Graphics Processing Unit
12	H/W and S/W	Hardware and software
13	LUS	Lung ultrasound
14	ML	Machine learning
15	MRI	Magnetic resonance imaging
16	RF	Random forest
17	RSNA	Radiological Society of North America
18	SE	Sensitivity
19	SP	Specificity
20	TL	Transfer learning
21	US	Ultrasound
22	X-ray	Röntgen radiation

References

[1] Shereen M A, Khan S, Kazmi A, Bashir N and Siddique R 2020 COVID-19 infection: origin, transmission, and characteristics of human coronaviruses *J. Adv. Res.* **24** 91–8

[2] Guo Y-R, Cao Q-D, Hong Z-S, Tan Y-Y, Chen S-D, Jin H-J, Tan K-S, Wang D-Y and Yan Y 2020 The origin, transmission and clinical therapies on coronavirus disease 2019 (COVID-19) outbreak - an update on the status *Mil Med Res.* **7** 11–1

[3] Horton R J L 2020 Offline: COVID-19 is not a pandemic *Lancet* **396** 874

[4] Cucinotta D and Vanelli M 2020 WHO declares COVID-19 a pandemic *Acta Biomed* **91** 157–60

[5] https://worldometers.info/coronavirus/

[6] D'Arienzo M and Coniglio A 2020 Assessment of the SARS-CoV-2 basic reproduction number, R (0), based on the early phase of COVID-19 outbreak in Italy *Biosaf Health* **2** 57–9

[7] Ravalli S and Musumeci G 2020 *Coronavirus Outbreak in Italy: Physiological Benefits of Home-Based Exercise During Pandemic* (Multidisciplinary Digital Publishing Institute)

[8] Wilder-Smith A, Chiew C J and Lee V J 2020 Can we contain the COVID-19 outbreak with the same measures as for SARS? *Lancet Infect Dis* **20** e102–7
[9] Maugeri G, Castrogiovanni P, Battaglia G, Pippi R, D'Agata V, Palma A, Di Rosa M and Musumeci G J H 2020 The impact of physical activity on psychological health during Covid-19 pandemic in Italy **6** e04315
[10] Lesser I A, Nienhuis C P J I J O E R and Health P 2020 The impact of COVID-19 on physical activity behavior and well-being of Canadians *Int. J. Environ. Res. Public Health* **17** 3899
[11] Viswanathan V, Puvvula A and Jamthikar A D A pathophysiological bidirectional association between diabetes mellitus and COVID-19 leading to heart and brain injury: a mini-review
[12] Saba L, Gerosa C, Wintermark M, Hedin U, Fanni D, Suri J S, Balestrieri A and Faa G 2020 Can COVID19 trigger the plaque vulnerability—a Kounis syndrome warning for 'asymptomatic subjects' *Cardiovasc. Diagn. Ther.* **10** 1352–5
[13] Pubmed COVID-19 publicatons
[14] Skandha S S *et al* 2020 3-D optimized classification and characterization artificial intelligence paradigm for cardiovascular/stroke risk stratification using carotid ultrasound-based delineated plaque: Atheromatic™ 2.0 *Comput. Biol. Med.* **125** 103958
[15] https://ourworldindata.org/grapher/total-confirmed-cases-of-covid-19-per-million-people
[16] Saba L *et al* 2019 The present and future of deep learning in radiology *Eur. J. Radiol.* **114** 14–24
[17] Biswas M *et al* 2019 State-of-the-art review on deep learning in medical imaging *Front. Biosci.* **24** 392–426
[18] Suri J S *et al* 2020 COVID-19 pathways for brain and heart injury in comorbidity patients: a role of medical imaging and artificial intelligence-based COVID severity classification: a review *Comput. Biol. Med.* **124** 103960
[19] Gatto M, Bertuzzo E, Mari L, Miccoli S, Carraro L, Casagrandi R and Rinaldo A 2020 Spread and dynamics of the COVID-19 epidemic in Italy: effects of emergency containment measures *Proc. Natl Acad. Sci. U S A.* **117** 10484–91
[20] Rekha Hanumanthu S J C 2020 Solitons, fractals, role of intelligent computing in COVID-19 prognosis: a state-of-the-art review *Chaos Solit. Fractals.* **138** 109947
[21] Deng Y, Lei L, Chen Y and Zhang W 2020 The potential added value of FDG PET/CT for COVID-19 pneumonia *Eur. J. Nucl. Med. Mol. Imaging* **47** 1634–5
[22] Liu C, Zhou J, Xia L, Cheng X and Lu D 2020 18F-FDG PET/CT and serial chest CT findings in a COVID-19 patient With dynamic clinical characteristics in different period *Clin. Nucl. Med.* **45** 495–6
[23] Maurea S, Mainolfi C G, Bombace C, Annunziata A, Attanasio L, Petretta M, Del Vecchio S and Cuocolo A 2020 FDG-PET/CT imaging during the Covid-19 emergency: a southern Italian perspective *Eur. J. Nucl. Med. Mol. Imaging* **47** 2691–7
[24] Verdecchia P, Cavallini C, Spanevello A and Angeli F 2020 The pivotal link between ACE2 deficiency and SARS-CoV-2 infection *Eur. J. Intern. Med.* **76** 14–20
[25] Mossel E C *et al* 2008 SARS-CoV replicates in primary human alveolar type II cell cultures but not in type I-like cells *Virology* **372** 127–35
[26] Saba L *et al* 2020 Molecular pathways triggered by COVID19 in different organs: ACE2 receptor-expressing cells under attack? A review *Eur Rev Med Pharmacol Sci.* **24** 12609–22

[27] Zhou P *et al* 2020 A pneumonia outbreak associated with a new coronavirus of probable bat origin *Nature* **579** 270–3

[28] Qian Z, Travanty E A, Oko L, Edeen K, Berglund A, Wang J, Ito Y, Holmes K V and Mason R J 2013 Innate immune response of human alveolar type II cells infected with severe acute respiratory syndrome-coronavirus *Am. J. Respir. Cell Mol. Biol.* **48** 742–8

[29] Ding Y *et al* 2003 The clinical pathology of severe acute respiratory syndrome (SARS): a report from China *J. Pathol.* **200** 282–9

[30] Liu J, Zheng X, Tong Q, Li W, Wang B, Sutter K, Trilling M, Lu M, Dittmer U and Yang D 2020 Overlapping and discrete aspects of the pathology and pathogenesis of the emerging human pathogenic coronaviruses SARS-CoV, MERS-CoV, and 2019-nCoV *J. Med. Virol* **92** 491–4

[31] Wang S, Le T Q, Kurihara N, Chida J, Cisse Y, Yano M and Kido H 2010 Influenza virus-cytokine-protease cycle in the pathogenesis of vascular hyperpermeability in severe influenza *J. Infect. Dis.* **202** 991–1001

[32] Huang C *et al* 2020 Clinical features of patients infected with 2019 novel coronavirus in Wuhan, China *Lancet* **395** 497–506

[33] Matthay M A, Ware L B and Zimmerman G A 2012 The acute respiratory distress syndrome *J. Clin. Invest.* **122** 2731–40

[34] Katzenstein A L, Bloor C M and Leibow A A 1976 Diffuse alveolar damage–the role of oxygen, shock, and related factors. A review *Am. J. Pathol.* **85** 209–28

[35] Nuckton T J, Alonso J A, Kallet R H, Daniel B M, Pittet J-F, Eisner M D and Matthay M A 2002 Pulmonary dead-space fraction as a risk factor for death in the acute respiratory distress syndrome *New Engl. J. Med.* **346** 1281–6

[36] Wu C *et al* 2020 Risk factors associated with acute respiratory distress syndrome and death in patients with coronavirus disease 2019 pneumonia in Wuhan, China *JAMA Intern. Med.* **180** 934–43

[37] Lian J *et al* 2020 Analysis of epidemiological and clinical features in older patients with coronavirus disease 2019 (COVID-19) outside Wuhan *Clin. Infect. Dis* **71** 740–7

[38] Liu Y, Sun W, Li J, Chen L, Wang Y, Zhang L and Yu L 2020 *Clinical Features And Progression of Acute Respiratory Distress Syndrome in Coronavirus Disease 2019* (Cold Spring Harbor Laboratory)

[39] Khan A, Chatterjee A and Singh S 2020 *Comorbidities and Disparities in Outcomes of COVID-19 Among African American and White Patients* (Cold Spring Harbor Laboratory)

[40] Zhang P *et al* 2020 Association of inpatient use of angiotensin-converting enzyme inhibitors and angiotensin II receptor blockers with mortality among patients with hypertension hospitalized with COVID-19 *Circ. Res.* **126** 1671–81

[41] Maniruzzaman M, Kumar N, Menhazul Abedin M, Shaykhul Islam M, Suri H S, El-Baz A S and Suri J S 2017 Comparative approaches for classification of diabetes mellitus data: machine learning paradigm *Comput. Methods Programs Biomed.* **152** 23–34

[42] Dreher M *et al* 2020 The characteristics of 50 hospitalized COVID-19 patients with and without ARDS *Dtsch. Arztebl Int* **117** 271–8

[43] Palaiodimos L, Kokkinidis D G, Li W, Karamanis D, Ognibene J, Arora S, Southern W N and Mantzoros C S 2020 Severe obesity, increasing age and male sex are independently associated with worse in-hospital outcomes, and higher in-hospital mortality, in a cohort of patients with COVID-19 in the Bronx, New York *Metabolism* **108** 154262–2

[44] Yu T, Cai S, Zheng Z, Cai X, Liu Y, Yin S, Peng J and Xu X 2020 Association between clinical manifestations and prognosis in patients with COVID-19 *Clin Ther* **42** 964–72
[45] Bandyopadhyay D *et al* 2020 COVID-19 Pandemic: cardiovascular complications and future implications *Am. J. Cardiovasc. Drugs : Drugs, Devices, and Other Interventions* **20** 311–24
[46] Doyen D, Moceri P, Ducreux D and Dellamonica J 2020 Myocarditis in a patient with COVID-19: a cause of raised troponin and ECG changes *Lancet* **395** 1516
[47] Madjid M, Safavi-Naeini P, Solomon S D and Vardeny O 2020 Potential effects of coronaviruses on the cardiovascular system: a review *JAMA Cardiol* **5** 831–40
[48] Suleyman G *et al* 2020 Clinical characteristics and morbidity associated with coronavirus disease 2019 in a series of patients in metropolitan detroit *JAMA Netw. Open* **3** e2012270
[49] Meyerowitz E A *et al* 2020 Disproportionate burden of coronavirus disease 2019 among racial minorities and those in congregate settings among a large cohort of people with *HIV, AIDS* **34** 1781–7
[50] Chandran M, Chan Maung A, Mithal A and Parameswaran R 2020 Vitamin D in COVID - 19: dousing the fire or averting the storm?—A perspective from the Asia-Pacific *Osteoporosis and Sarcopenia* **6** 97–105
[51] Chang T S *et al* 2020 Prior diagnoses and medications as risk factors for COVID-19 in a Los Angeles health system *Medrxiv*
[52] Sanyaolu A, Okorie C, Marinkovic A, Patidar R, Younis K, Desai P, Hosein Z, Padda I, Mangat J and Altaf M 2020 Comorbidity and its impact on patients with COVID-19 *SN. Compr. Clin. Med.* **2** 1069–76
[53] Takemoto M *et al* 2020 C. Brazilian group for studies of, pregnancy, clinical characteristics and risk factors for mortality in obstetric patients with severe COVID-19 in Brazil: a surveillance database analysis *BJOG* **127** 1618–26
[54] Gudipati S, Brar I, Murray S, McKinnon J E, Yared N and Markowitz N J J O A I D S 2020 Descriptive analysis of patients living with HIV affected by COVID-19 *JAIDS* **85** 123–6
[55] Bornstein S R *et al* 2020 Practical recommendations for the management of diabetes in patients with COVID-19 *Lancet Diabetes Endocrinol.* **8** 546–50
[56] Guzik T J *et al* 2020 COVID-19 and the cardiovascular system: implications for risk assessment, diagnosis, and treatment options *Cardiovasc Res.* **116** 1666–87
[57] Bassendine M F, Bridge S H, McCaughan G W and Gorrell M D 2020 COVID-19 and comorbidities: a role for dipeptidyl peptidase 4 (DPP4) in disease severity? *J. Diabetes* **12** 649–58
[58] Bode B, Garrett V, Messler J, McFarland R, Crowe J, Booth R and Klonoff D C 2020 Glycemic characteristics and clinical outcomes of COVID-19 patients hospitalized in the United States *J. Diabetes Sci. Technol.* **14** 813–21
[59] Akram J, Azhar S, Shahzad M, Latif W and Khan K S 2020 Pakistan randomized and observational trial to evaluate coronavirus treatment (PROTECT) of hydroxychloroquine, oseltamivir and azithromycin to treat newly diagnosed patients with COVID-19 infection who have no comorbidities like diabetes mellitus: a structured summary of a study protocol for a randomized controlled trial *Trials* **21** 702
[60] Yang Q *et al* 2020 Analysis of the clinical characteristics, drug treatments and prognoses of 136 patients with coronavirus disease 2019 *J. Clin. Pharm. Ther.* **45** 609–16

[61] Halaji M, Farahani A, Ranjbar R, Heiat M and Dehkordi F J L I I M 2020 Emerging coronaviruses: first SARS, second MERS and third SARS-CoV-2: epidemiological updates of COVID-19 *Infez. Med.* **28** 6–17

[62] Grimaldi D *et al* 2020 Characteristics and outcomes of acute respiratory distress syndrome related to COVID-19 in Belgian and French intensive care units according to antiviral strategies: the COVADIS multicentre observational study *Ann. Intensive Care.* **10** 131

[63] Zaim S, Chong J H, Sankaranarayanan V and Harky A 2020 COVID-19 and multiorgan response *Curr. Probl. Cardiol.* **45** 100618

[64] Ponziani F R, Del Zompo F, Nesci A, Santopaolo F, Ianiro G, Pompili M and Gasbarrini A'Gemelli against COVID-19' group 2020 Liver involvement is not associated with mortality: results from a large cohort of SARS-CoV-2 positive patients *Aliment Pharmacol Ther* **52** 1060–8

[65] Brandt J S, Hill J, Reddy A, Schuster M, Patrick H S, Rosen T, Sauer M V, Boyle C and Ananth C V 2020 Epidemiology of coronavirus disease 2019 in pregnancy: risk factors and associations with adverse maternal and neonatal outcomes *Am. J. Obstet. Gynecol.* **224** 389. E1–9

[66] Cheng L L *et al* 2020 Effect of recombinant human granulocyte colony-stimulating factor for patients with coronavirus disease 2019 (COVID-19) and lymphopenia: a randomized clinical trial *JAMA Intern. Med.* **181** 71–8

[67] Cummings M J *et al* 2020 Epidemiology, clinical course, and outcomes of critically ill adults with COVID-19 in New York City: a prospective cohort study *Lancet* **395** 1763–70

[68] Nakeshbandi M *et al* 2020 The impact of obesity on COVID-19 complications: a retrospective cohort study *Int. J. Obes. (Lond)* **44** 1832–7

[69] Zhao J, Li X, Gao Y and Huang W J I J O M S 2020 Risk factors for the exacerbation of patients with 2019 Novel Coronavirus: a meta-analysis *Int. J. Med. Sci.* **17** 1744

[70] Chen T *et al* 2020 Clinical characteristics of 113 deceased patients with coronavirus disease 2019: retrospective study *Brit. Med. J.* **368** m1091

[71] Qin C, Zhou L, Hu Z, Yang S, Zhang S, Chen M, Yu H, Tian D S and Wang W 2020 Clinical characteristics and outcomes of COVID-19 patients with a history of stroke in Wuhan, China *Stroke* **51** 2219–23

[72] Shi S *et al* 2020 Association of cardiac injury with mortality in hospitalized patients with COVID-19 in Wuhan, China *JAMA Cardiol* **5** 802–10

[73] Deng Y *et al* 2020 Clinical characteristics of fatal and recovered cases of coronavirus disease 2019 in Wuhan, China: a retrospective study *Chin. Med. J. (Engl)* **133** 1261–7

[74] Yang F, Shi S, Zhu J, Shi J, Dai K and Chen X 2020 Clinical characteristics and outcomes of cancer patients with COVID-19 *J. Med. Virol* **92** 2067–73

[75] Del Sole F, Farcomeni A, Loffredo L, Carnevale R, Menichelli D, Vicario T, Pignatelli P and Pastori D 2020 Features of severe COVID-19: a systematic review and meta-analysis *Eur. J. Clin. Invest.* **50** e13378–8

[76] Ciceri F *et al* 2020 Early predictors of clinical outcomes of COVID-19 outbreak in Milan, Italy *Clin. Immunol.* **217** 108509

[77] Kutluhan M A, Taş A, Şahin A, Ürkmez A, Topaktas R, Ataç Ö and Verit A J I J O C P 2020 Assessment of clinical features and renal functions in Coronavirus disease-19: a retrospective analysis of 96 patients *Int. J. Clin. Pract.* **74** e13636

[78] Derespina K R, Kaushik S, Plichta A, Conway E E, Bercow A, Choi J, Eisenberg R, Gillen J, Sen A I and Hennigan C M J T J O P 2020 Clinical manifestations and outcomes of

critically ill children and adolescents with coronavirus disease 2019 in New York City *J. Pediatr.* **226** 55–63
[79] Blumfield E and Levin T L J P R 2020 COVID-19 in pediatric patients: a case series from the Bronx, NY *Pediatr. Radiol.* **50** 1369–74
[80] Jazieh A-R, Alenazi T H, Alhejazi A, Al Safi F and Al Olayan A J J G O 2020 Outcome of oncology patients infected with coronavirus *LCO Glob. Oncol.* **6** 471–5
[81] Al-Wahaibi K, Al-Wahshi Y and Mohamed O 2020 Elfadil, myocardial injury is associated with higher morbidity and mortality in patients with 2019 Novel Coronavirus Disease (COVID-19) *SN. Compr. Clin. Med.* **2** 1–7
[82] Chand S, Kapoor S, Orsi D, Fazzari M J, Tanner T G, Umeh G C, Islam M and Dicpinigaitis P V 2020 COVID-19-associated critical illness-report of the first 300 patients admitted to intensive care units at a New York City medical center *J. Intensive Care Med.* **35** 963–70
[83] Lee Y R *et al* 2020 Clinical outcomes of coronavirus disease 2019 in patients with pre-existing liver diseases: a multicenter study in South Korea *Clin. Mol. Hepatol.* **26** 562–76
[84] Wang L, He W, Yu X, Hu D, Bao M, Liu H, Zhou J and Jiang H 2020 Coronavirus disease 2019 in elderly patients: characteristics and prognostic factors based on 4-week follow-up *J. Infect.* **80** 639–45
[85] Du Y *et al* 2020 Clinical features of 85 fatal cases of COVID-19 from Wuhan. A retrospective observational study *Am. J. Respir. Crit. Care Med.* **201** 1372–9
[86] Ziehr D R, Alladina J, Petri C R, Maley J H, Moskowitz A, Medoff B D, Hibbert K A, Thompson B T and Hardin C C 2020 Respiratory pathophysiology of mechanically ventilated patients with COVID-19: a cohort study *Am. J. Respir. Crit. Care Med.* **201** 1560–4
[87] Garcia-Cruz E *et al* 2020 Critical care ultrasonography during COVID-19 pandemic: the ORACLE protocol *Echocardiography* **37** 1353–61
[88] Huang Y, Guo H, Zhou Y, Guo J, Wang T, Zhao X, Li H, Sun Y, Bian X and Fang C 2020 The associations between fasting plasma glucose levels and mortality of COVID-19 in patients without diabetes *Diabetes Res. Clin. Pract.* **169** 108448
[89] Arrieta J, Galwankar S, Lattanzio N, Ray D and Agrawal A 2020 Studying the clinical data of COVID positive patients admitted to a tertiary care academic hospital *J. Emerg. Trauma Shock.* **13** 131–4
[90] Oltean M, Søfteland J M, Bagge J, Ekelund J, Felldin M, Schult A, Magnusson J, Friman V and Karason K J I D 2020 Covid-19 in kidney transplant recipients: a systematic review of the case series available three months into the pandemic *Infect. Dis.* **52** 830–7
[91] Marinaki S, Tsiakas S, Korogiannou M, Grigorakos K, Papalois V and Boletis I J J O C M 2020 A systematic review of COVID-19 infection in kidney transplant recipients: a universal effort to preserve patients' lives and allografts *J. Clin. Med.* **9** 2986
[92] Rajpal A, Rahimi L and Ismail-Beigi F J J O D 2020 Factors leading to high morbidity and mortality of COVID-19 in patients with type 2 diabetes *J. Diabetes* **12** 895–908
[93] Huang I, Lim M A and Pranata R 2020 Diabetes mellitus is associated with increased mortality and severity of disease in COVID-19 pneumonia—a systematic review, meta-analysis, and meta-regression *Diabetes Metab. Syndr.* **14** 395–403
[94] Salerno M, Sessa F, Piscopo A, Montana A, Torrisi M, Patanè F, Murabito P, Volti G L and Pomara C 2020 No autopsies on COVID-19 deaths: a missed opportunity and the lockdown of science *J. Clin. Med.* **9** 1472

[95] Arrieta J, Galwankar S, Lattanzio N, Ray D and Agrawal A 2020 Common clinical characteristics and complications determining the outcome in a COVID-positive predominantly geriatric population *J. Emerg. Trauma Shock.* **13**

[96] Tomazini B M *et al* 2020 Effect of dexamethasone on days alive and ventilator-free in patients with moderate or severe acute respiratory distress syndrome and COVID-19: the codex randomized clinical trial *JAMA* **324** 1307–16

[97] Nasir N, Farooqi J, Mahmood S F and Jabeen K 2020 COVID-19-associated pulmonary aspergillosis (CAPA) in patients admitted with severe COVID-19 pneumonia: an observational study from pakistan *Mycoses* **63** 766–70

[98] Antoun L, Taweel N E, Ahmed I, Patni S and Honest H 2020 Maternal COVID-19 infection, clinical characteristics, pregnancy, and neonatal outcome: a prospective cohort study *Eur. J. Obstet. Gynecol. Reprod. Biol.* **252** 559–62

[99] Khan M, Khan H, Khan S and Nawaz M 2020 Epidemiological and clinical characteristics of coronavirus disease (COVID-19) cases at a screening clinic during the early outbreak period: a single-centre study *J. Med. Microbiol.* **69** 1114–23

[100] Xie J *et al* 2020 Metabolic syndrome and COVID-19 mortality among adult black patients in New Orleans *Diabetes Care*

[101] Li T, Lu L, Zhang W, Tao Y, Wang L, Bao J, Liu B and Duan J 2020 Clinical characteristics of 312 hospitalized older patients with COVID-19 in Wuhan, China *Arch. Gerontol. Geriatr.* **91** 104185

[102] Saba L, Agarwal M, Sanagala S, Gupta S, Sinha G, Johri A, Khanna N, Mavrogeni S, Laird J and Pareek G J E L 2020 Brain MRI-based Wilson disease tissue classification: an optimised deep transfer learning approach *Electron. Lett.* **56** 1395–8

[103] Tandel G S, Biswas M, Kakde O G, Tiwari A, Suri H S, Turk M, Laird J R, Asare C K, Ankrah A A and Khanna N J C 2019 A review on a deep learning perspective in brain cancer classification *Cancers* **11** 111

[104] Saba L, Tiwari A, Biswas M, Gupta S K, Godia-Cuadrado E, Chaturvedi A, Turk M, Suri H S, Orru S and Sanches J M J F I B 2019 Wilson's disease: a new perspective review on its genetics, diagnosis and treatment *FBE* **11** 166–85

[105] Acharya U R, Mookiah M R K, Sree S V, Afonso D, Sanches J, Shafique S, Nicolaides A, Pedro L M, Fernandes J F E and Suri J S J M 2013 Atherosclerotic plaque tissue characterization in 2D ultrasound longitudinal carotid scans for automated classification: a paradigm for stroke risk assessment *Med. Biol. Eng. Comput.* **51** 513–23

[106] Sharma A M, Gupta A, Kumar P K, Rajan J, Saba L, Nobutaka I, Laird J R, Nicolades A and Suri J S J C A R 2015 A review on carotid ultrasound atherosclerotic tissue characterization and stroke risk stratification in machine learning framework *Curr. Atheroscler. Rep.* **17** 55

[107] Biswas M, Kuppili V, Saba L, Edla D R, Suri H S, Sharma A, Cuadrado-Godia E, Laird J R, Nicolaides A and Suri J S J M 2019 Deep learning fully convolution network for lumen characterization in diabetic patients using carotid ultrasound: a tool for stroke risk *Med. Biol. Eng. Comput.* **57** 543–64

[108] Saba L, Jain P K, Suri H S, Ikeda N, Araki T, Singh B K, Nicolaides A, Shafique S, Gupta A and Laird J R J J O M S 2017 Plaque tissue morphology-based stroke risk stratification using carotid ultrasound: a polling-based PCA learning paradigm *J. Med. Syst.* **41** 98

[109] Acharya U, Sree S V, Mookiah M, Saba L, Gao H, Mallarini G and Suri J S 2013 Computed tomography carotid wall plaque characterization using a combination of discrete wavelet transform and texture features: a pilot study *J. Med. Eng.* **227** 643–54

[110] Acharya U R, Molinari F, Saba L, Nicolaides A, Shafique S and Suri J S 2012 Carotid ultrasound symptomatology using atherosclerotic plaque characterization: a class of Atheromatic systems *2012 Annual Int. Conf. of the IEEE Engineering in Medicine and Biology Society* (Piscataway, NJ: IEEE) pp 3199–202

[111] Prasad H, Martis R J, Acharya U R, Min L C and Suri J S 2013 Application of higher order spectra for accurate delineation of atrial arrhythmia *2013 35th Annual Int. Conf. of the IEEE Engineering in Medicine and Biology Society (EMBC)* (Piscataway, NJ: IEEE) pp 57–60

[112] Saba L *et al* 2016 Automated stratification of liver disease in ultrasound: an online accurate feature classification paradigm *Comput. Methods Programs Biomed.* **130** 118–134

[113] Acharya U R, Sree S V, Ribeiro R, Krishnamurthi G, Marinho R T, Sanches J and Suri J S J M P 2012 Data mining framework for fatty liver disease classification in ultrasound: a hybrid feature extraction paradigm *Med. Phys.* **39** 4255–64

[114] Biswas M, Kuppili V, Edla D R, Suri H S, Saba L, Marinhoe R T, Sanches J M, Suri J S J C M and biomedicine p i 2018 Symtosis: a liver ultrasound tissue characterization and risk stratification in optimized deep learning paradigm *Comput. Methods Programs Biomed.* **155** 165–77

[115] Boi A, Jamthikar A D, Saba L, Gupta D, Sharma A, Loi B, Laird J R, Khanna N N and Suri J S J C A R 2018 A survey on coronary atherosclerotic plaque tissue characterization in intravascular optical coherence tomography *Curr. Atheroscler. Rep.* **20** 33

[116] Acharya U R, Faust O, Kadri N A, Suri J S and Yu W J C I B 2013 Medicine, Automated identification of normal and diabetes heart rate signals using nonlinear measures *Comput. Biol. Med.* **43** 1523–9

[117] Pareek G, Acharya U R, Sree S V, Swapna G, Yantri R, Martis R J, Saba L, Krishnamurthi G, Mallarini G and El-Baz A J T I C R 2013 Treatment, prostate tissue characterization/classification in 144 patient population using wavelet and higher order spectra features from transrectal ultrasound images *Technol. Cancer Res. Treat.* **12** 545–57

[118] Acharya U R, Molinari F, Sree S V, Swapna G, Saba L, Guerriero S and Suri J S J T I C R 2015 Treatment, ovarian tissue characterization in ultrasound: a review *Technol. Cancer Res. Treat.* **14** 251–61

[119] Acharya U R, Sree S V, Kulshreshtha S, Molinari F, Koh J E W, Saba L and Suri J S J T I C R 2014 Treatment, GyneScan: an improved online paradigm for screening of ovarian cancer via tissue characterization**13** 529–39

[120] Viswanathan V, Jamthikar A D, Gupta D, Shanu N, Puvvula A, Khanna N N, Saba L, Omerzum T, Viskovic K and Mavrogeni S J F I B 2020 Low-cost preventive screening using carotid ultrasound in patients with diabetes *Front. Biosci.* **25** 1132–71

[121] Acharya U R, Swapna G, Sree S V, Molinari F, Gupta S, Bardales R H, Witkowska A and Suri J S J T I C R 2014 Treatment, a review on ultrasound-based thyroid cancer tissue characterization and automated classification *Technol. Cancer Res. Treat.* **13** 289–301

[122] Shrivastava V K, Londhe N D, Sonawane R S and Suri J S J C M 2016 Computer-aided diagnosis of psoriasis skin images with HOS, texture and color features: a first comparative study of its kind *Comput. Methods Programs Biomed.* **126** 98–109

[123] Shrivastava V K, Londhe N D, Sonawane R S, Suri J S J C M and biomedicine p i 2017 A novel and robust Bayesian approach for segmentation of psoriasis lesions and its risk stratification *Comput. Methods Programs Biomed.* **150** 9–22

[124] Acharya U R, Joseph K P, Kannathal N, Lim C M, Suri J S J M and Engineering B 2006 Computing, heart rate variability: a review *Med. Bio. Eng. Comput.* **44** 1031–51

[125] Corrias G, Cocco D, Suri J S, Meloni L, Cademartiri F and Saba L J C D 2020 Therapy, Heart applications of 4D flow**10** 1140

[126] Acharya U R, Sree S V, Krishnan M M R, Krishnananda N, Ranjan S, Umesh P and Suri J S 2013 Automated classification of patients with coronary artery disease using grayscale features from left ventricle echocardiographic images *Comput. Methods Programs Biomed.* **112** 624–32

[127] Maniruzzaman M *et al* 2019 Statistical characterization and classification of colon microarray gene expression data using multiple machine learning paradigms**176** 173–93

[128] Khanna N N, Jamthikar A D, Gupta D, Piga M, Saba L, Carcassi C, Giannopoulos A A, Nicolaides A, Laird J R and Suri H S J C A R 2019 Rheumatoid arthritis: atherosclerosis imaging and cardiovascular risk assessment using machine and deep learning–based tissue characterization**21** 7

[129] Jiang J, Hu Y C, Tyagi N, Zhang P, Rimner A, Deasy J O and Veeraraghavan H 2019 Cross-modality (CT-MRI) prior augmented deep learning for robust lung tumor segmentation from small MR datasets *Med. Phys.* **46** 4392–404

[130] Aresta G, Jacobs C, Araujo T, Cunha A, Ramos I, van Ginneken B and Campilho A 2019 iW-Net: an automatic and minimalistic interactive lung nodule segmentation deep network *Sci Rep.* **9** 11591

[131] Anthimopoulos M, Christodoulidis S, Ebner L, Geiser T, Christe A and Mougiakakou S 2019 Semantic segmentation of pathological lung tissue with dilated fully convolutional networks *IEEE J. Biomed. Health Inform.* **23** 714–22

[132] Weikert T, Akinci D'Antonoli T, Bremerich J, Stieltjes B, Sommer G and Sauter A W 2019 Evaluation of an AI-powered lung nodule algorithm for detection and 3d segmentation of primary lung tumors *Contrast Media Mol. Imaging.* **2019** 1545747

[133] Liu H, Cao H, Song E, Ma G, Xu X, Jin R, Jin Y and Hung C C 2019 A cascaded dual-pathway residual network for lung nodule segmentation in CT images *Phys. Med.* **63** 112–21

[134] Wong Yuzhen N and Barrett S 2019 A review of automatic lung tumour segmentation in the era of 4DCT *Rep. Pract. Oncol. Radiother.* **24** 208–20

[135] Park B, Park H, Lee S M, Seo J B and Kim N 2019 Lung segmentation on HRCT and volumetric CT for diffuse interstitial lung disease using deep convolutional neural networks *J. Digit. Imaging.* **32** 1019–26

[136] Nasrullah N, Sang J, Alam M S, Mateen M, Cai B and Hu H 2019 Automated lung nodule detection and classification using deep learning combined with multiple strategies *Sensors (Basel)* **19**

[137] Xu M, Qi S, Yue Y, Teng Y, Xu L, Yao Y and Qian W 2019 Segmentation of lung parenchyma in CT images using CNN trained with the clustering algorithm generated dataset *Biomed. Eng. Online.* **18** 2

[138] Baek S *et al* 2019 Deep segmentation networks predict survival of non-small cell lung cancer *Sci. Rep.* **9** 17286

[139] Pang T, Guo S, Zhang X and Zhao L 2019 Automatic lung segmentation based on texture and deep features of HRCT images with interstitial lung disease *BioMed. Res. Int.* **2019** 2045432
[140] Chen G, Xiang D, Zhang B, Tian H, Yang X, Shi F, Zhu W, Tian B and Chen X 2019 Automatic pathological lung segmentation in low-dose CT image using eigenspace sparse shape composition *IEEE Trans. Med. Imaging* **38** 1736–49
[141] Senthil Kumar K, Venkatalakshmi K and Karthikeyan K 2019 Lung cancer detection using image segmentation by means of various evolutionary algorithms *Comput. Math. Methods Med.* **2019** 4909846
[142] Liu C, Zhao R and Pang M 2020 A fully automatic segmentation algorithm for CT lung images based on random forest *Med. Phys.* **47** 518–29
[143] Geng L, Zhang S, Tong J and Xiao Z 2019 Lung segmentation method with dilated convolution based on VGG-16 network *Comput. Assist. Surg. (Abingdon)* **24** 27–33
[144] Sousa A M, Martins S B, Falcao A X, Reis F, Bagatin E and Irion K 2019 ALTIS: a fast and automatic lung and trachea CT-image segmentation method *Med. Phys.* **46** 4970–82
[145] Souza J C, Bandeira Diniz J O, Ferreira J L, Franca da Silva G L, Correa Silva A and de Paiva A C 2019 An automatic method for lung segmentation and reconstruction in chest x-ray using deep neural networks *Comput. Methods Programs Biomed.* **177** 285–96
[146] Noor N M, Than J C M, Rijal O M, Kassim R M, Yunus A, Zeki A A, Anzidei M, Saba L and Suri J S 2015 Automatic lung segmentation using control feedback system: morphology and texture paradigm *J. Med. Syst.* **39**
[147] Ni Q *et al* 2020 A deep learning approach to characterize 2019 coronavirus disease (COVID-19) pneumonia in chest CT images *Eur. Radiol* **30** 6517–27
[148] Shan F, Gao Y, Wang J, Shi W, Shi N, Han M, Xue Z and Shi Y J A P A 2020 Lung infection quantification of covid-19 in CT images with deep learning
[149] Hwang E J, Kim H, Yoon S H, Goo J M and Park C M 2020 Implementation of a deep learning-based computer-aided detection system for the interpretation of chest radiographs in patients suspected for COVID-19 *Korean J. Radiol* **21** 1150–60
[150] Roy S *et al* 2020 Deep learning for classification and localization of COVID-19 markers in point-of-care lung ultrasound *IEEE Trans. Med. Imaging.* **39** 2676–87
[151] Signoroni A, Savardi M, Benini S, Adami N, Leonardi R, Gibellini P, Vaccher F, Ravanelli M, Borghesi A and Maroldi R J A P A 2020 End-to-end learning for semiquantitative rating of COVID-19 severity on chest x-rays
[152] Li Z *et al* 2020 From community-acquired pneumonia to COVID-19: a deep learning-based method for quantitative analysis of COVID-19 on thick-section 2.CT scans *Eur Radiol* **30** 6828–37
[153] Chaganti S *et al* 2020 Automated quantification of CT patterns associated with COVID-19 from chest CT *Radiol.: Artif. Intell.* **2** e200048
[154] Li L *et al* 2020 Using artificial intelligence to detect COVID-19 and community-acquired pneumonia based on pulmonary CT: evaluation of the diagnostic accuracy *Radiology* **296** E65–71
[155] Yang S, Jiang L, Cao Z, Wang L, Cao J, Feng R, Zhang Z, Xue X, Shi Y and Shan F 2020 Deep learning for detecting corona virus disease 2019 (COVID-19) on high-resolution computed tomography: a pilot study *Ann. Transl. Med.* **8** 450–0
[156] Hu S *et al* 2020 Weakly supervised deep learning for COVID-19 infection detection and classification from CT images *IEEE Access.* **8** 118869–83

[157] Zhang K *et al* 2020 Clinically applicable AI system for accurate diagnosis, quantitative measurements, and prognosis of COVID-19 pneumonia using computed tomography *Cell* **181** 1423–33 e1411

[158] Carrer L *et al* 2020 Automatic pleural line extraction and COVID-19 scoring from lung ultrasound data *IEEE Trans. Ultrason. Ferroelectr. Freq. Control* **67** 2207–17

[159] Tang Z, Zhao W, Xie X, Zhong Z, Shi F, Liu J and Shen D J A P A 2020 Severity assessment of coronavirus disease 2019 (COVID-19) using quantitative features from chest CT images *Med Image Anal.* arXiv:2003.11988

[160] Rajaraman S, Siegelman J, Alderson P O, Folio L S, Folio L R and Antani S K 2020 Iteratively pruned deep learning ensembles for COVID-19 detection in chest x-rays *IEEE Access* **8** 115041–50

[161] Born J, Brändle G, Cossio M, Disdier M, Goulet J, Roulin J and Wiedemann N J A P A 2020 POCOVID-Net: automatic detection of COVID-19 from a new lung ultrasound imaging dataset (POCUS) arXiv:2004.12084

[162] Jaiswal A, Gianchandani N, Singh D, Kumar V and Kaur M 2020 Classification of the COVID-19 infected patients using DenseNet201 based deep transfer learning *J. Biomol. Struct. Dyn.* 1–8

[163] Tsiknakis N *et al* 2020 Interpretable artificial intelligence framework for COVID-19 screening on chest x-rays *Exp. Ther. Med.* **20** 727–35

[164] Maghdid H S, Asaad A T, Ghafoor K Z, Sadiq A S and Khan M K J A P A 2020 Diagnosing COVID-19 pneumonia from x-ray and CT images using deep learning and transfer learning algorithms *Proc. IEEE* **11734** 117340E

[165] Shrivastava V K, Londhe N D, Sonawane R S and Suri J S 2016 Computer-aided diagnosis of psoriasis skin images with HOS, texture and color features: a first comparative study of its kind *Comput. Methods Programs Biomed.* **126** 98–109

[166] Chen A, Karwoski R A, Gierada D S, Bartholmai B J and Koo C W 2020 Quantitative CT analysis of diffuse lung disease *Radiographics* **40** 28–43

[167] Than J C M, Saba L, Noor N M, Rijal O M, Kassim R M, Yunus A, Suri H S, Porcu M and Suri J S 2017 Lung disease stratification using amalgamation of Riesz and Gabor transforms in machine learning framework *Comput. Biol. Med.* **89** 197–211

[168] Hattori A, Takamochi K, Oh S and Suzuki K 2019 New revisions and current issues in the eighth edition of the TNM classification for non-small cell lung cancer *Jpn. J. Clin. Oncol.* **49** 3–11

[169] Saba T 2019 Automated lung nodule detection and classification based on multiple classifiers voting *Microsc. Res. Tech.* **82** 1601–9

[170] Zhang G, Yang Z, Gong L, Jiang S and Wang L 2019 Classification of benign and malignant lung nodules from CT images based on hybrid features *Phys. Med. Biol.* **64** 125011

[171] Singh D, Kumar V, Vaishali and Kaur M 2020 Classification of COVID-19 patients from chest CT images using multi-objective differential evolution-based convolutional neural networks *Eur. J. Clin. Microbiol. Infect. Dis.* **39** 1379–89

[172] Mahmud T, Rahman M A and Fattah S A 2020 CovXNet: a multi-dilation convolutional neural network for automatic COVID-19 and other pneumonia detection from chest x-ray images with transferable multi-receptive feature optimization *Comput. Biol. Med.* **122** 103869–9

[173] Das D, Santosh K C and Pal U 2020 Truncated inception net: COVID-19 outbreak screening using chest x-rays, research Square *Phys. Eng. Sci. Med.* **43** 915–25

[174] Ozturk T, Talo M, Yildirim E A, Baloglu U B, Yildirim O and Rajendra U 2020 Automated detection of COVID-19 cases using deep neural networks with x-ray images *Comput. Biol. Med.* **121** 103792–2
[175] Brunese L, Mercaldo F, Reginelli A and Santone A 2020 Explainable deep learning for pulmonary disease and coronavirus COVID-19 detection from x-rays *Comput. Methods Programs Biomed.* **196** 105608
[176] Suri J S, Singh S and Reden L J P A 2002 Applications, fusion of region and boundary/ surface-based computer vision and pattern recognition techniques for 2-D and 3-D MR cerebral cortical segmentation (part-II): a state-of-the-art review **5** 77–98
[177] Suri J S, Singh S and Reden L J P A 2002 Applications, computer vision and pattern recognition techniques for 2-D and 3-D MR cerebral cortical segmentation (Part I): a state-of-the-art review**5** 46–76
[178] Suri J S, Liu K, Reden L and Laxminarayan S 2002 A review on MR vascular image processing: skeleton versus nonskeleton approaches: part II *IEEE Trans. Inf. Technol. Biomed.* **6** 338
[179] https://xmedcon.sourceforge.io/
[180] Sahu S P, Kamble B and Doriya R 2020 3D lung segmentation using thresholding and active contour method *Advances in Intelligent Systems and Computing* (Singapore: Springer) pp 369–80
[181] Taylor M E 2009 Transfer between different reinforcement learning methods *Transfer in Reinforcement Learning Domains* (Berlin: Springer) pp 139–79
[182] Acharya U R, Swapna G, Sree S V, Molinari F, Gupta S, Bardales R H, Witkowska A and Suri J S 2014 A review on ultrasound-based thyroid cancer tissue characterization and automated classification *Technol. Cancer Res. Treat.* **13** 289–301
[183] Shih A R *et al* 2019 Problems in the reproducibility of classification of small lung adenocarcinoma: an international interobserver study *Histopathology* **75** 649–59
[184] LeNail A 2019 NN-SVG: publication-ready neural network architecture schematics *J. Open Source Softw.* **4** 747
[185] Iqbal H 2018 HarisIqbal88/PlotNeuralNet v1.0.0
[186] Ye F, Pu J, Wang J, Li Y and Zha H 2017 Glioma grading based on 3D multimodal convolutional neural network and privileged learning *2017 IEEE Int. Conf. on Bioinformatics and Biomedicine (BIBM)* (Piscataway, NJ: IEEE)
[187] Wang S *et al* 2020 A fully automatic deep learning system for COVID-19 diagnostic and prognostic analysis *Eur. Res. J.* **56**
[188] Yoo S H *et al* 2020 Deep learning-based decision-tree classifier for COVID-19 diagnosis from chest x-ray imaging *Front. Med. (Lausanne)* **7** 427
[189] Zhu J, Shen B, Abbasi A, Hoshmand-Kochi M, Li H and Duong T Q 2020 Deep transfer learning artificial intelligence accurately stages COVID-19 lung disease severity on portable chest radiographs *PLoS One* **15** e0236621
[190] Joseph J. C and Howard P. F The Economic Impact of the COVID-19 Pandemic on Radiology Practices
[191] Yasar H and Ceylan M 2020 A novel comparative study for detection of Covid-19 on CT lung images using texture analysis, machine learning, and deep learning methods *Multimed. Tools Appl.* **80** 1–25
[192] Setti L, Kirienko M, Dalto S C, Bonacina M and Bombardieri E 2020 FDG-PET/CT findings highly suspicious for COVID-19 in an Italian case series of asymptomatic patients *Eur. J. Nucl. Med. Mol. Imaging* **47** 1649–56

[193] Alonso Sanchez J, García Prieto J, Galiana Morón A and Pilkington-Woll J P 2020 PET/CT of COVID-19 as an organizing pneumonia *Clin. Nucl. Med.* **45** 642–3

[194] Castanheira J, Mascarenhas Gaivão A, Mairos Teixeira S, Pereira P J and Costa D C 2020 Asymptomatic COVID-19 positive patient suspected on FDG-PET/CT *Nucl. Med. Commun.* **41** 598–9

[195] Cohen J P *et al* 2020 Predicting COVID-19 pneumonia severity on chest x-ray with deep learning *Cureus* **12** e9448

[196] Galougahi M K, Ghorbani J, Bakhshayeshkaram M, Naeini A S and Haseli S J A R 2020 Olfactory bulb magnetic resonance imaging in SARS-CoV-2-induced anosmia: the first report *Acad. Radiol.* **27** 892–3

[197] Gunraj H, Wang L and Wong A J A P A 2020 Covidnet-CT: a tailored deep convolutional neural network design for detection of covid-19 cases from chest CT images *Front. Med.* **7** 608525

[198] Ismael A M and Sengur A 2021 Deep learning approaches for COVID-19 detection based on chest x-ray images *Expert Syst. Appl.* **164** 114054

[199] Kandemirli S G *et al* 2020 Brain MRI findings in patients in the intensive care unit with COVID-19 infection *Radiology* **297** E232–5

[200] Kay F and Abbara S 2020 The many faces of COVID-19: spectrum of imaging manifestations *Radiol. Soc. North Am.* **2** 1–2

[201] Litjens G, Kooi T, Bejnordi B E, Setio A A A, Ciompi F, Ghafoorian M, van der Laak J A W M, van Ginneken B and Sánchez C I 2017 A survey on deep learning in medical image analysis *Med. Image Anal.* **42** 60–88

[202] Minaee S, Kafieh R, Sonka M, Yazdani S and Jamalipour G 2020 Deep-COVID: predicting COVID-19 from chest x-ray images using deep transfer learning *Med. Image Anal.* **65** 101794

[203] Shi F, Xia L, Shan F, Wu D, Wei Y, Yuan H, Jiang H, Gao Y, Sui H and Shen D J A P A 2020 Large-scale screening of Covid-19 from community acquired pneumonia using infection size-aware classification *Phys. Med. Biol.* **66** 065031

[204] Panwar H, Gupta P K, Siddiqui M K, Morales-Menendez R, Bhardwaj P and Singh V 2020 A deep learning and grad-CAM based color visualization approach for fast detection of COVID-19 cases using chest x-ray and CT-Scan images *Chaos Solit. Fractals.* **140** 110190–0

[205] Ouchicha C, Ammor O and Meknassi M 2020 CVDNet: a novel deep learning architecture for detection of coronavirus (Covid-19) from chest x-ray images *Chaos Solit. Fractals* **140** 110245–5

[206] Farooq M and Hafeez A J A P A 2020 Covid-resnet: a deep learning framework for screening of covid19 from radiographs arXiv:2003.14395

[207] Cozzi D, Albanesi M, Cavigli E, Moroni C, Bindi A, Luvarà S, Lucarini S, Busoni S, Mazzoni L N and Miele V 2020 Chest x-ray in new Coronavirus Disease 2019 (COVID-19) infection: findings and correlation with clinical outcome *Radiol. Med.* **125** 730–7

[208] Pereira R M, Bertolini D, Teixeira L O, Silla C N and Costa Y M G 2020 COVID-19 identification in chest x-ray images on flat and hierarchical classification scenarios *Comput. Methods Programs Biomed.* **194** 105532–2

[209] Ardakani A A, Kanafi A R, Acharya U R, Khadem N and Mohammadi A 2020 Application of deep learning technique to manage COVID-19 in routine clinical practice using CT images: results of 10 convolutional neural networks *Comput. Biol. Med.* **121** 103795–5

[210] Zheng C, Deng X, Fu Q, Zhou Q, Feng J, Ma H, Liu W and Wang X 2020 *Deep Learning-Based Detection for COVID-19 from Chest CT using Weak Label* (Cold Spring Harbor Laboratory)

[211] Yang X, He X, Zhao J, Zhang Y, Zhang S and Xie P J A E-P 2020 Covid-CT-dataset: a CT scan dataset about covid-19 arXiv:2003.13865

[212] Gozes O, Frid-Adar M, Sagie N, Kabakovitch A, Amran D, Amer R and Greenspan H 2020 A weakly supervised deep learning framework for COVID-19 CT detection and analysis *Thoracic Image Analysis* (Springer International Publishing) pp 84–93

[213] Liu B, Gao X, He M, Liu L and Yin G 2020 A fast online COVID-19 diagnostic system with chest CT scans *Proc. of KDD*

[214] Hemdan E E-D, Shouman M A and Karar M E J A P A 2020 Covidx-net: a framework of deep learning classifiers to diagnose covid-19 in x-ray images

[215] Zhang J, Xie Y, Liao Z, Pang G, Verjans J, Li W, Sun Z, He J and Yi Li C S J A P A 2020 Viral pneumonia screening on chest x-ray images using confidence-aware anomaly detection *IEEE Trans. on Med. Imag.* **40** 879–90

[216] Narayan Das N, Kumar N, Kaur M, Kumar V and Singh D 2020 Automated deep transfer learning-based approach for detection of COVID-19 infection in chest x-rays *Ing Rech Biomed.* **43** 114–9

[217] Wu Y-H, Gao S-H, Mei J, Xu J, Fan D-P, Zhao C-W and Cheng M-M 2020 JCS: an explainable COVID-19 diagnosis system by joint classification and segmentation *IEEE Trans. on Imag. Proc.* **30** 3113–26

[218] Toğaçar M, Ergen B and Cömert Z 2020 COVID-19 detection using deep learning models to exploit social mimic optimization and structured chest x-ray images using fuzzy color and stacking approaches *Comput. Biol. Med.* **121** 103805–5

[219] Basu S and Mitra S J A P A 2020 Deep learning for screening COVID-19 using chest x-ray images *MedRxiv* IEEE - PMC COVID-19 Collection

[220] Gozes O, Frid-Adar M, Greenspan H, Browning P D, Zhang H, Ji W, Bernheim A and Siegel E J A P A 2020 Rapid AI development cycle for the coronavirus (covid-19) pandemic: initial results for automated detection & patient monitoring using deep learning CT image analysis

[221] Yan T, Wong P K, Ren H, Wang H, Wang J and Li Y J C 2020 Solitons, fractals, automatic distinction between Covid-19 and common pneumonia using multi-scale convolutional neural network on chest CT scans *Chaos Solit. Fractals* **140** 110153

[222] Oh Y, Park S and Ye J C 2020 Deep learning COVID-19 features on CXR using limited training data sets *IEEE Trans. on Med. Imag.* **39** 2688–700

[223] Saba L, Than J C M, Noor N M, Rijal O M, Kassim R M, Yunus A, Ng C R and Suri J S 2016 Inter-observer variability analysis of automatic lung delineation in normal and disease patients *J. Med. Syst.* **40**

[224] Liu K and Suri J S 2005 Automatic vessel indentification for angiographic screening, Google Patents

[225] Ambale-Venkatesh B *et al* 2017 Cardiovascular event prediction by machine learning: the multi-ethnic study of atherosclerosis *Circ. Res.* **121** 1092–101

[226] Tandel G S, Balestrieri A, Jujaray T, Khanna N N, Saba L and Suri J S 2020 Multiclass magnetic resonance imaging brain tumor classification using artificial intelligence paradigm *Comput. Biol. Med.* **122** 103804

[227] Shrivastava V K, Londhe N D, Sonawane R S and Suri J S 2017 A novel and robust Bayesian approach for segmentation of psoriasis lesions and its risk stratification *Comput. Methods Programs Biomed.* **150** 9–22

[228] Jamthikar A, Gupta D, Saba L, Khanna N N, Viskovic K, Mavrogeni S, Laird J R, Sattar N, Johri A M and Pareek G J C I B 2020 Medicine, artificial intelligence framework for predictive cardiovascular and stroke risk assessment models: a narrative review of integrated approaches using carotid ultrasound *Comput. Biol. Med.* **126** 104043

[229] Wu X *et al* 2020 Deep learning-based multi-view fusion model for screening 2019 novel coronavirus pneumonia: a multicentre study *Eur. J. Radiol.* **128** 109041–1

[230] Organization W H 2020 *Use of Chest Imaging in COVID-19* 1–56

[231] Yusuf G T, Wong A, Rao D, Tee A, Fang C and Sidhu P S 2020 The use of contrast-enhanced ultrasound in COVID-19 lung imaging *J. Ultrasound* **25** 319–23

[232] Ni Q, Sun Z Y, Qi L, Chen W, Yang Y, Wang L, Zhang X, Yang L, Fang Y and Xing Z J E R 2020 A deep learning approach to characterize 2019 coronavirus disease (COVID-19) pneumonia in chest CT images**30** 6517–27

[233] Wang S, Zha Y, Li W, Wu Q, Li X, Niu M, Wang M, Qiu X, Li H and Yu H J E R J 2020 A fully automatic deep learning system for COVID-19 diagnostic and prognostic analysis *Eur. Respir. J.* **56** 2000775

[234] Narayanan R, Werahera P N, Barqawi A, Crawford E D, Shinohara K, Simoneau A R and Suri J S 2008 Adaptation of a 3D prostate cancer atlas for transrectal ultrasound guided target-specific biopsy *Phys. Med. Biol.* **53** N397–406

[235] Shen F, Narayanan R and Suri J S 2008 Rapid motion compensation for prostate biopsy using GPU *Annual Int. Conf. of the IEEE Engineering in Medicine and Biology Society. IEEE Engineering in Medicine and Biology Society* 2008 3257–60

[236] State of the Art in Neural Networks and Their Applications 1st edn

[237] Acharya U R *et al* 2013 Diagnosis of Hashimoto's thyroiditis in ultrasound using tissue characterization and pixel classification *Proc. Inst. Mech. Eng. Part H J. Eng. Med.* **227** 788–98

[238] Kandemirli S G, Altundag A, Yildirim D, Tekcan Sanli D E and Saatci O 2020 Olfactory bulb MRI and paranasal sinus CT findings in persistent COVID-19 Anosmia *Acad. Radiol.* **28** 28–35

[239] Cuadrado-Godia E, Dwivedi P, Sharma S, Santiago A O, Gonzalez J R, Balcells M, Laird J, Turk M, Suri H S and Nicolaides A J J O S 2018 Cerebral small vessel disease: a review focusing on pathophysiology, biomarkers, and machine learning strategies *J. Stroke* **20** 302

[240] Maniruzzaman M, Kumar N, Abedin M M, Islam M S, Suri H S, El-Baz A S, Suri J S J C M and biomedicine p i 2017 Comparative approaches for classification of diabetes mellitus data: machine learning paradigm *Comput. Methods Programs Biomed.* **152** 23–34

[241] Acharya R, Ng Y E and Suri J S 2008 *Image Modeling of the Human Eye* (Norwood, MA: Artech House)

[242] El-Baz A and Suri J 2019 *Lung Imaging and CADx* (Boca Raton, FL: CRC Press)

[243] El-Baz A and Suri J S 2011 *Lung Imaging and Computer Aided Diagnosis* (Boca Raton, FL: CRC Press)

[244] 2020 *Online COVID-19 Diagnosis with Chest CT Images: Lesion-Attention Deep Neural Networks* (Boston, MA: Rescognito, Inc.)

[245] Suri J S, Agarwal S and Gupta S K *et al* 2021 A narrative review on characterization of acute respiratory distress syndrome in COVID-19-infected lungs using artificial intelligence *Comput. Biol. Med.* **130** 104210

IOP Publishing

Multimodality Imaging, Volume 2
Heart, lungs and peripheral organs
Mainak Biswas and Jasjit S Suri

Chapter 3

A multicenter study using COVLIAS 2.0: eight pruned deep learning models for efficient COVID-19 CT lung segmentation and lesion localization

Venkateshh Moningi, Mohit Agarwal, Mainak Biswas and Jasjit S Suri

The study proposes COVLIAS 2.0, an improved version of an automated lung segmentation system for COVID-19 diagnosis. The previous version, COVLIAS 1.0, had issues with storage space and speed. COVLIAS 2.0 addresses these problems by using pruned AI (PAI) networks. The goal is to reduce the size of AI models while maintaining optimal performance. The study collected multicenter CT slices from two different nations: CroMed from Croatia (experimental data) and NovMed from Italy (validation data). The researchers hypothesize that by using pruning and evolutionary optimization algorithms, they can significantly reduce the size of AI models without sacrificing performance. They designed eight different pruning techniques using two deep learning frameworks: fully connected network (FCN) and SegNet. COVLIAS 2.0 was validated using the 'Unseen NovMed' dataset and compared against MedSeg. Statistical tests were conducted to evaluate the stability and reliability of the system. The results showed that all eight PAI networks achieved an area under the curve (AUC) greater than 0.94 on the CroMed dataset and greater than 0.86 on the NovMed dataset ($P < 0.0001$). The pruning algorithms resulted in significant improvements in storage efficiency, with reductions ranging from 92.4% to 99.8% compared to the solo FCN or SegNet models. Additionally, the PAI networks demonstrated a processing time of less than 0.25 seconds per image. DenseNet-121-based Grad-CAM heatmaps were used to validate the segmentation of glass ground opacity lesions. In conclusion, the study successfully validated eight PAI networks that are five times faster and more storage-efficient than previous models. These improvements make them suitable for clinical settings in COVID-19 diagnosis.

3.1 Introduction

The declaration of COVID-19 as a global pandemic was made by the World Health Organization (WHO) on March 11, 2020 [1]. This disease is characterized by its rapid and exponential spread, posing significant challenges to healthcare systems worldwide, which often have limited resources [2]. As of February 15, 2022, COVID-19 had resulted in the loss of over 5.8 million lives and infected more than 410 million people across the globe [2]. Individuals with pre-existing conditions such as diabetes [7], atherosclerosis [8], coronary artery disease [3, 5, 6], and fetal programming [9] experience more severe molecular effects when infected with COVID-19 [3, 4]. Furthermore, the disease can adversely affect the vasa vasorum of the aorta, leading to increased vulnerability of atherosclerotic plaques [10, 11] and microthrombosis. These interactions between alveoli and blood vessels can result in architectural distortion [12], impacting daily activities, including nutrition [13]. Pathological studies have shown that vaccine-induced immune thrombotic thrombocytopenia (VITT) can occur even after receiving the ChAdOx1 nCoV-19 vaccine [14]. Additionally, adults with limited physical growth, referred as intrauterine growth restriction (IUGR), face an elevated risk of COVID-19 infection [9]. Quick and accurate identification of COVID-19 is crucial, as there is currently no widely available immunization or specific treatment. While reverse transcription-polymerase chain reaction (RT-PCR) tests [15–17] are commonly used, their specificity is relatively lower compared to other diagnostic methods. Consequently, research has increasingly focused on image-based analysis for the identification and diagnosis of COVID-19.

Categorization and pneumonia detection have greatly benefited from the advancements in AI technology, particularly in the fields of DL and ML [18, 21]. However, while these approaches offer real-time analysis capabilities, they also come with inherent risks [22]. Hybrid DL (HDL) represents an advanced stage of DL that combines the strengths of DL and ML [23–30]. DL/HDL models exhibit improved accuracy and performance, but they require significant training time and anticipation of results, leading to increased costs. Addressing the processing issue can be achieved through the utilization of supercomputers or GPUs [29, 31–33], but the long-term maintenance of these solutions poses challenges and can be costly. The concept of pruning in deep learning was introduced in the seminal paper titled 'Optimal Brain Damage' by LeCun *et al* published in 1989 [34]. Pruning involves the removal of excess or irrelevant components from a model or search area, similar to trimming [35]. This approach gained traction in deep learning research, particularly with the selection of appropriate hyperparameters [38, 39]. By implementing pruning, improvements were achieved in storage efficiency [36, 37], as well as speeding up model training. Pruning techniques have been predominantly applied in the domain of COVID-19 and non-COVID imaging using x-rays [40]. The proposed research assumes that leveraging these pruning techniques can significantly accelerate lung segmentation in computed tomography (CT) and lesion localization, without the need for additional hardware or administration. Our proposed work represents the initial publication of eight state-of-the-art evolutionary algorithms

(EA)-based pruning models, which are DL-based, for CT lung segmentation in the context of COVID-19. This contribution serves as our primary innovation. The research yielded a total of eight pruned models, namely: (i) FCN-DE, (ii) FCN-GA, (iii) FCN-PSO, and (iv) FCN-WO, which were developed using the solo FCN architecture. Additionally, (v) SegNet-DE, (vi) SegNet-GA, (vii) SegNet-PSO, and (viii) SegNet-WO were created by employing the solo SegNet architecture. To scientifically validate our contributions and innovations, we employed a multicenter paradigm in an unexplored setting. The EA-based DL models were trained using the Croatia's high-GGO cohort (CroMed) and used to predict segmented lung data from Novara, Italy (NovMed), which represented the 'Unseen NovMed' data. This data was collected from diverse regions and ethnic groups using multiple CT scanners, highlighting the Unseen AI paradigm as our second innovation. To showcase the efficacy of our innovation, we devised a specialized process for precise identification of COVID-19 lesions within segmented lungs. This involved leveraging PAI as the lesion source and integrating DenseNet-121 and GRAD-cam to generate visually appealing color heatmaps. Consequently, our third noteworthy invention emerged—an evolutionary-based lung segmentation technique with lesion localization. To validate our contribution, we compared our evolutionary-based COVLIAS 2.0 with MedSwg [41], a web-based lung segmentation program, in terms of infrastructure speed and storage capabilities. This marked our fourth significant achievement. We conducted a comprehensive 360-degree evaluation of our entire data analysis, encompassing various metrics such as the Jaccard Index (JI) [42], Dice Similarity (DS), lung area error (AE), Bland-Altman plots (BA) [43, 44], correlation coefficient (CC) plot [45, 46], and receiver operator characteristic (ROC) curve. This evaluation forms our fifth contribution. Lastly, our sixth contribution involved performing clinical and statistical tests on COVLIAS 2.0 to demonstrate the accuracy and stability of our approach. Figure 3.1 illustrates the COVLIAS 2.0 system pipeline, which showcases a versatile AI system for COVID-19-based lung segmentation. The pipeline utilizes pruned and optimized AI (POAI) models to minimize storage requirements and expedite processing. It encompasses volume capture, online lung segmentation, benchmarking over MedSeg, and performance assessment using the 'Unseen NovMed' dataset. Our study proposal is structured

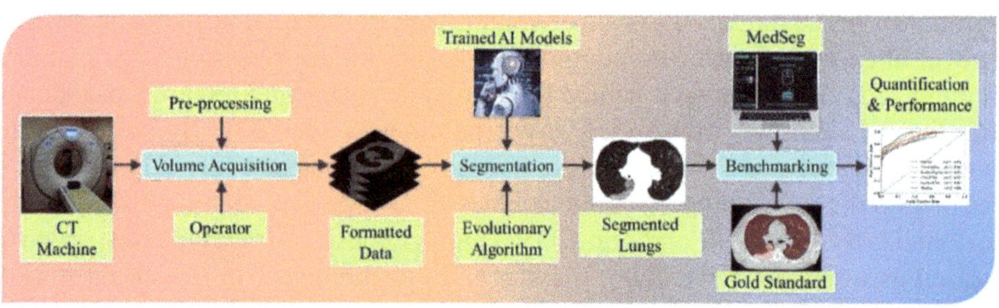

Figure 3.1. COVLIAS 2.0 using pruned AI system for segmentation of CT lungs. The benchmarking is conducted using MedSeg. The performance is evaluated using borders of lungs manually delineated by a radiologist. The scientific validation was conducted using the Unseen NovMed data set.

into seven main sections. Section 3.1 introduces the study and provides an overview of its objectives. Section 3.2 focuses on the background literature, presenting a comprehensive review of relevant studies and research in the field. Section 3.3 outlines the methods employed in our study, including details about the demographics of the participants, the image capture process, and a short description of the AI models used. Section 3.4 presents the results obtained from the AI models and provides an analysis of their performance in the context of CT lung segmentation [54]. This section also includes relevant visualizations and metrics to evaluate the effectiveness of the models. In section 3.5, we performed a statistical analysis and validation of the obtained results using a different dataset, specifically MedSeg. This section aims to benchmark our models and assess their performance against an established reference. Section 3.6 comprises the discussion, where we interpret and analyze the findings of our study in relation to existing literature. We explore the implications of our results, highlight any limitations, and propose potential areas for future research. Finally, section 3.7 concludes the study by summarizing the main findings, reiterating their significance, and providing a closing remark. This section also offers a concise summary of the overall study proposal and may include recommendations or suggestions for further investigation.

3.2 Literature survey

Segmentation has been a well-established practice in medical imaging, but its integration with AI frameworks is a relatively recent development [47–49]. The incorporation of segmentation within AI has enabled improved categorization and characterization of tissues for early disease diagnosis [19, 20]. Initially, machine learning (ML) approaches were utilized, followed by the introduction of point-based models, addressing various issues such as infant mortality [52], diabetes [50, 51], neonatal analysis, and gene categorization [53]. As the field transitioned from point-based to image-based machine learning for classification, significant advancements were made with applications in thyroid [60, 61], carotid plaques [55–59], coronary [64, 65], liver [62, 63], prostate [65], ovarian [64], Wilson disease [32], ophthalmology [69] and skin cancer [66–68].

In the context of COVID-19 diagnosis, two crucial types of medical imaging are widely used: CT scans and x-rays. CT imaging has demonstrated repeatability and high sensitivity in the general diagnosis of COVID-19, particularly in identifying different variants of opacities such as consolidation, crazy paving, ground-glass opacity (GGO) and other opacities frequently observed in COVID-19 patients [73–75]. Body imaging [62, 70–72] plays a significant role in COVID-19 diagnosis [76–81]. Jiang *et al* introduced a COVID-19-based CT lung classification approach that employed CycleGAN to generate 1186 artificial CT images [82]. The network used raw input lung cancer CT scans from LUAN16 [83] and processed them, with each image patch measuring 512 × 512 pixels. Five AI models, namely ResNet-50, VGG16, Inception_v3, Inception ResNet_v2, and DenseNet-169, were trained on the generated data. The study found that DenseNet-169 achieved the highest accuracy of 98.92%, outperforming ResNet-50, VGG16, Inception ResNet_v2, and Inception_v3, having efficiency values of 94.19%, 94.83%, 96.55%, and 95.91%, respectively. Notably, the study did not utilize PAI in their research. Kogilavani *et al* [84] conducted a study where they focused on the categorization of CT scan images into COVID-19 and non-COVID-19 cases using deep learning architectures. The authors employed six different deep learning models, namely MobileNet [87], VGG16 [85], DenseNet [86], NASNet [89], Xception [88], and EfficientNet [90]. A dataset consisting of 3873 CT scans with an image resolution of 224 × 224 pixels was used to train all of the models. According to the performance assessment measures employed by the authors, VGG16 achieved the highest accuracy of 97.68%. The assessment measures also included precision, recall, and F1-score. It is worth noting that the study by Kogilavani *et al* did not incorporate PAI in their system design or evaluation. In the study by Paluru *et al* [91], a novel architecture called AnamNet, which is a combination of ENet and UNet, was proposed for the purpose of detecting COVID-19 lesions in segmented CT lung images. The input to AnamNet was the segmented lung images, and the model was trained using a dataset comprising data from 69 patients [92]. The performance of AnamNet was compared to other models including UNet++ [94], ENet [93], LEDNet [95], and SegNet. The evaluation metric used for lesion identification was the Dice similarity coefficient, which achieved a value of 0.77. Notably, the study did not employ data augmentation during the preprocessing stage. Instead of using PAI, the authors deployed AnamNet on an edge device using an Android application. It is important to mention that the study by Paluru *et al* did not include the production of Jaccard Index (JI) or Bland-Altman (BA) plots, nor did it compare lung area errors as part of the evaluation process. The study by Cai *et al* [96] utilized a library of 250 images collected from 99 patients to implement the UNet model for lesion and lung segmentation. The achieved Dice similarity coefficient for lesion segmentation was 0.77. Additionally, based on the outcomes of the lesion segmentation, the authors proposed a technique for estimating the length of an intensive care unit (ICU) stay. However, the study did not include the production of BA or JI plots, nor did it compare lung area errors. On the other hand, Saood *et al* [97] described a COVID-19-based CT lung image segmentation method using 100 downscaled COVID-19 lung CT images with a resolution of 256 × 256. The researchers compared the results of two models, SegNet and UNet, and found that their Dice similarity scores were comparable, with values of 0.73 and 0.74, respectively. Similar to the previous study, the authors did not plot JI or BA errors, compare lung area errors, or apply the concept of PAI. In contrast to these studies, the proposed study incorporates several components: (i) the use of eight AI pruning models that offer low storage and high speed, (ii) an analysis using the unseen NovMed dataset and training of the pruning models using the CroMed dataset, (iii) the development of a classification framework to identify COVID-19 lesions, (iv) benchmarking COVLIAS 2.0 with MedSeg to authenticate prior contributions, and (v) the inclusion of statistical analysis and performance evaluation to avoid hypothesis and provide clinical evidence.

3.3 Methodology

3.3.1 Patient demographics

3.3.1.1 CroMed data set

In the proposed study, two distinct cohorts from different countries are utilized. The initial dataset, referred to as the experimental dataset, consists of 80 COVID-19-positive individuals from the CroMed cohort. Among these individuals, 57 are male, and the remaining are female. The presence of COVID-19 in the chosen cohort was confirmed through an RT-PCR test. The average value of consolidation, ground-glass opacity (GGO), and other opacities in the CroMed cohort was close to 4. Additionally, the following observations were made regarding the characteristics of the CroMed patients who participated in the study: 83% had a cough, 60% had dyspnea (difficulty breathing), 50% had hypertension (high blood pressure), 8% were smokers, 12% had a sore throat, 15% had diabetes, and 3% had chronic obstructive pulmonary disease (COPD). Out of the total 17 patients who were hospitalized in the Intensive Care Unit (ICU), three patients unfortunately succumbed to COVID-19 infection. These demographic and clinical details provide important insights into the composition of the CroMed cohort and the impact of COVID-19 on their health.

3.3.1.2 NovMed data set

The second dataset used in the study is the NovMed cohort, which consists of 72 COVID-19-positive individuals and serves as the validation dataset. Among these individuals, 47 are male, and the remaining are female. Similar to the CroMed cohort, the presence of COVID-19 in the NovMed cohort was confirmed through an RT-PCR test. The average value of GGO, consolidation, and other opacities in the NovMed cohort was close to 2.4. Furthermore, the following observations were made regarding the characteristics of the NovMed patients who participated in the study: 61% had a cough, 9% had a sore throat, 54% had dyspnea, 42% had hypertension, 12% had diabetes, 11% had COPD, and 11% were smokers. Unfortunately, a total of 10 individuals in the NovMed cohort passed away due to COVID-19 infection. These demographic and clinical details provide important insights into the composition of the NovMed cohort and the impact of COVID-19 on their health outcomes.

3.3.2 Image acquisition and data preparation

3.3.2.1 CroMed and NovMed dataset

In the proposed experiment, a total of 80 COVID-19-positive individuals from the CroMed dataset were included. The retrospective cohort study was conducted at the University Hospital for Infectious Diseases in Zagreb, Croatia, from March 1 to December 31, 2020. The study participants were adults aged 18 and above who provided their consent and had thoracic multidetector computed tomography (MDCT) during their hospital stay. They also tested positive for the SARS-CoV-2 virus through RT-PCR testing. To be included in the study, the patients had to exhibit at least one of the following conditions: hypoxia (oxygen saturation of 92%

or lower), tachypnea (respiration rate of 22 breaths per minute or higher), tachycardia (heart rate of 100 beats per minute or higher), or hypotension (systolic blood pressure of 100 mmHg or lower). The proposal for the study was approved by the University Hospital for Infectious Diseases (UHID) Ethics Committee. The CT scans of the CroMed dataset were acquired using a 64-detector FCT Speedia HD scanner from Fujifilm Corporation, Tokyo, Japan, which allowed for a complete inspiratory breath-hold in the craniocaudal direction. The scanner was capable of capturing images with high resolution and detail (figure 3.2).

For the NovMed dataset, which consisted of 72 patients from Italy, chest CT images were obtained using a 128-slice multidetector-row CT scanner (Philips Ingenuity Core, by Philips Healthcare). The patients underwent CT scans while lying flat. No oral or intravenous contrast agent was administered during the scanning process. The CT images were reconstructed with a one-millimeter slice thickness using a soft tissue kernel and a lung kernel of a 768 × 768 matrix (lung window). The CT scans of the NovMed dataset were performed using the Philips automated tube current modulation (Z-DOM) technique. The scanning parameters involves a tube voltage of 120 kV, a tube current of 226 mAs per slice, a spiral pitch factor of 1.08, and a gantry rotation time of 0.5 seconds. The CT scanner had a 64 × 0.625 detector configuration, which contributed to the image acquisition process. These details provide information about the CT scanning equipment and protocols used for collecting the imaging data from both the CroMed and NovMed datasets in the proposed study (as shown in figures 3.2 and 3.3).

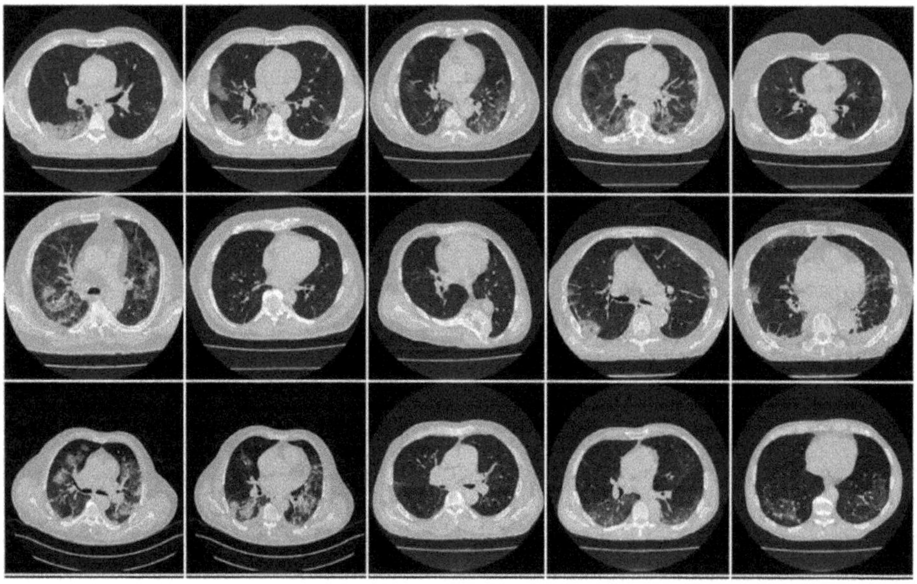

Figure 3.2. Sample CT scans taken from raw CroMed data set.

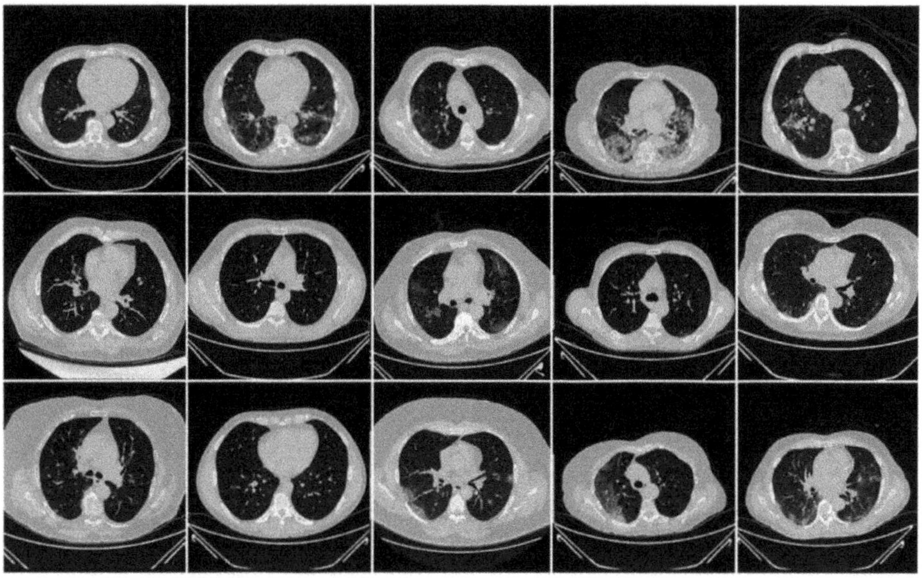

Figure 3.3. Sample CT scans taken from raw unseen NovMed data set.

3.3.3 Artificial intelligence models

3.3.3.1 Overall system for pruned AI and unpruned AI

Figure 3.4 displays the CT lung segmentation using a local trimmed AI system. The system consists of two components: the testing component and the training component. Within the training system, there are two parts referred as Part A and Part B. Part A focuses on training the unpruned AI (UnPAI) model, which is an AI model without evolutionary algorithms (EAs). The traditional system of base models such as SegNet or FCN is utilized in this part. The training process involves using gold standard and unprocessed, raw, grayscale images. The goal is to train the UnPAI model to perform lung segmentation accurately. In Part B, the training system incorporates a PAI system that utilizes evolutionary algorithms, including Differential Evolution (DE), Genetic Algorithm (GA), Particle Swarm Optimization (PSO), and Whale Optimization (WO) algorithms. These algorithms are used to create pruned AI models with improved efficiency and performance. The PAI models undergo training using the same base models as in Part A. The fundamental DL models, which are well-established and widely recognized in the DL field, are the subject of the subsequent discussion. These DL models serve as the backbone for the AI systems in the proposed study, facilitating the segmentation of CT lung images.

3.3.3.2 Base model 1: fully convolutional networks

Figure 3.5 illustrates the design of fully convolutional networks (FCNs), which are primarily used for semantic segmentation tasks. FCNs rely on locally connected layers for operations such as upsampling, pooling, and convolution. By avoiding fully connected layers, FCNs reduce the number of parameters, resulting in faster

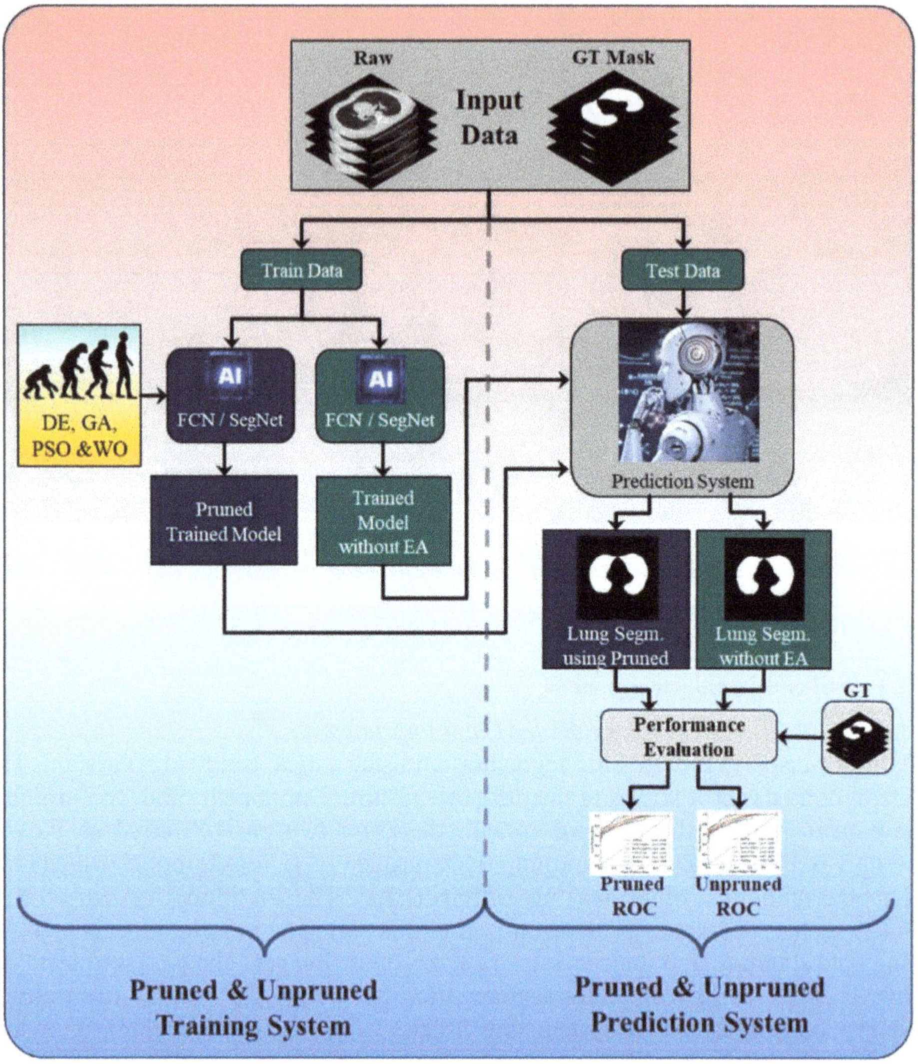

Figure 3.4. Local COVLIAS 2.0 system for pruned and unpruned AI models.

network training. Furthermore, FCNs have the advantage of being able to handle images of different dimensions if all connections are local. The model consists of an encoder, which is a convolutional neural network (CNN) feature extractor. The role of the encoder is to learn the essential features of an input image. As the image passes through the convolutional layers of the encoder, it undergoes downsampling to extract high-level representations. The decoder section of the model comprises additional convolutional layers. It takes the output from the encoder and progressively upsamples it to the original dimensions of the image. This upsampling process is performed incrementally, and as the decoder layers reconstruct the image, they generate pixelwise labels or segmentation masks. These masks indicate the class

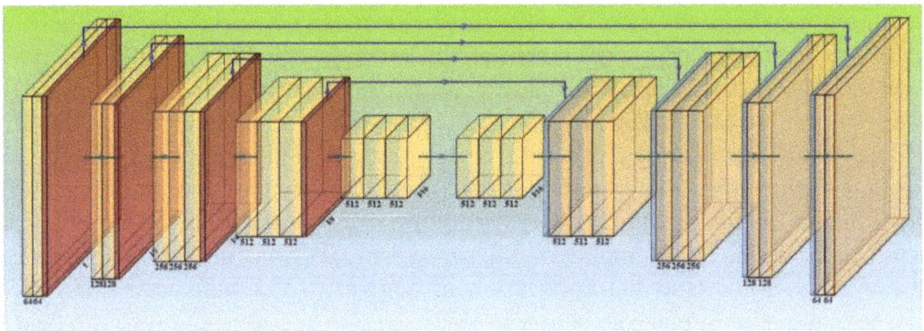

Figure 3.5. Architecture for base model one—FCN.

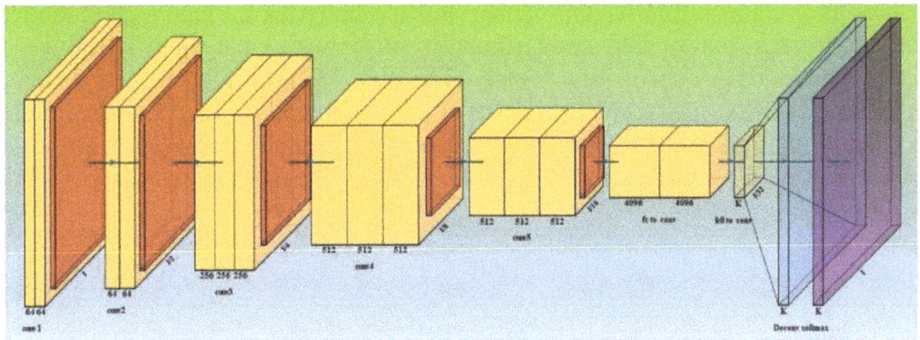

Figure 3.6. Architecture for base model two—SegNet.

or category of each pixel in the original image. FCN-16, FCN-8, and FCN-32 are examples of FCNs in their standard forms. These variants differ in the level of downsampling and the corresponding upsampling factor, which affects the resolution of the final segmentation output [98]. Overall, FCNs have proven to be effective in semantic segmentation tasks by leveraging the strengths of convolutional operations and enabling pixelwise labeling of images.

3.3.3.3 Base model 2: segmentation network

Figure 3.6 presents SegNet, which is a semantic segmentation model that has been utilized in previous works [99]. However, in our proposed study, we introduce a novel aspect by combining evolutionary techniques with the SegNet base model, applying it for the first time in CT lung segmentation. SegNet consists of three main components: an encoder network, a pixelwise classification layer, and a corresponding decoder network. The encoder network mirrors the topology of 13 convolutional layers. The design of the encoder involves storing the indices of max-pooling (with a filter size of 2×2) operations, which are used for downsampling. At the end of each block in the encoder, a max-pooling layer is added, effectively doubling the depth of the subsequent layer (64-128-256-512). In the decoder network, which is the right half of figure 3.6, the layer depth is reduced by two (512-256-128-64) and upsampling is performed. During the upsampling process, the max-pooling indices from the corresponding encoder layers are utilized to retrieve the spatial information. Finally, a K-class SoftMax classifier is employed to predict the class for each pixel in the output, enabling semantic segmentation. This design enables acceptable performance while conserving memory space. Examples of SegNet segmentation in various applications can be found in [23, 25, 28], showcasing its versatility and effectiveness. In our study, we extend the application of SegNet to CT lung segmentation by incorporating evolutionary techniques, which is a novel contribution to the field.

3.3.3.4 Design of eight PAI models

Figure 3.7 illustrates the general framework of the interaction between the four evolutionary algorithm (EA) approaches and the underlying FCN/SegNet models. The system takes binary gold standard masks and raw grayscale lung CT images as inputs. The four EA approaches, namely GA, WO, DE, and PSO, are employed to optimize the FCN and SegNet models. By combining these four EA approaches with the FCN and SegNet models, a total of eight pruning strategies are formed. Each approach is multiplied by two, representing the use of SegNet and FCN as the

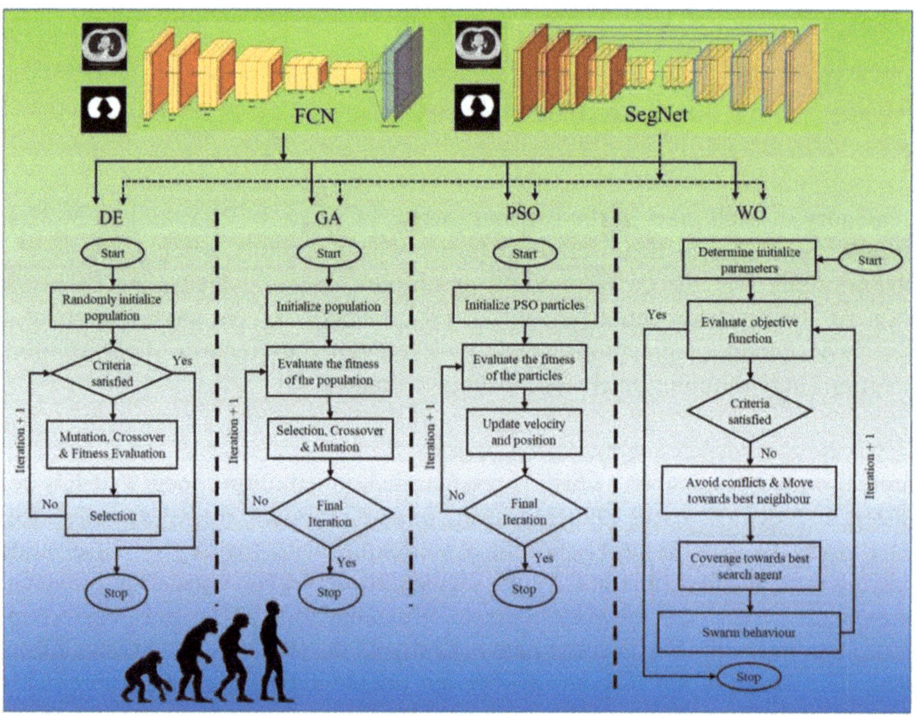

Figure 3.7. Four pruning techniques (DE, GA, PSO, and WO) leading to eight systems: FCN-DE, FCN-GA, FCN–PSO, FCN-WO, SegNet-DE, SegNet-GA, SegNet-PSO, and SegNet-WO.

base models. The resulting eight innovative AI model pruning approaches are as follows:

1. FCN-GA: FCN model optimized using genetic algorithm (GA)
2. FCN-PSO: FCN model optimized using particle swarm optimization (PSO)
3. FCN-DE: FCN model optimized using differential evolution (DE)
4. FCN-WO: FCN model optimized using the weight optimization (WO) technique
5. SegNet-WO: SegNet model optimized using weight optimization (WO)
6. SegNet-DE: SegNet model optimized using differential evolution (DE)
7. SegNet-GA: SegNet model optimized using genetic algorithm (GA)
8. SegNet-PSO: SegNet model optimized using particle swarm optimization (PSO)

These eight pruning models represent the combination of the respective EA approach with either FCN or SegNet as the base model. Each model is trained and evaluated to achieve improved performance in CT lung segmentation.

The process described outlines the steps involved in compressing the base networks (SegNet/FCN) using the evolutionary algorithm (EA) approaches. Here is a breakdown of the steps:

1. Compression Initialization: The compression process starts by generating a binary vector, consisting of random 0s and 1s, with the same length as the number of hidden neurons in the first convolutional layer. The hidden neurons corresponding to the zero positions in the vector are removed temporarily, and the value of fitness is calculated. The fitness value is determined based on two objectives: compressed nodes ratio and accuracy. The goal is to find the best vector that achieves the highest compression ratio while maintaining high accuracy. This process is repeated for a certain number of EA iterations (e.g., 20–30).
2. Compression of Hidden Layers: The compression procedure described in step 1 is repeated for each subsequent hidden convolutional layer. Redundant filters/neurons are removed based on the optimal binary vector obtained from the EA. This step is performed iteratively for all hidden layers.
3. Construction of Compressed Model: Once the compression of all hidden layers is completed, a new compressed model is made. The number of neurons in each hidden layer is estimated by summing up the 1s in the corresponding EA vector. The weights from the original trained FCN/SegNet model are then transferred to the newly constructed compressed model.
4. Fine-tuning of Compressed Model: To recover any accuracy lost during the compression process, the compressed model is further trained for a certain number of epochs (e.g., 10–50). This fine-tuning step helps to restore the model's accuracy while reducing its size.
5. Evaluation and Iterative Compression: The compressed model size and accuracy are recorded. If the efficiency of the compressed model is inside an acceptable range (e.g., within 1% of the original efficiency), the

compressed model is considered as the new uncompressed model. The process returns to step 1, and further compression iterations are performed. This iterative process continues for a specific number of compression iterations (e.g., 10) to achieve significant compression.

Each of the four evolutionary algorithms (GA, WO, DE, and PSO) generates binary vectors using an initial population of approximately 50 individuals. The intermediary processes may vary slightly for each algorithm, but the overall objective is to find the best fitness value and optimal vector. The trained FCN/SegNet model can discard unnecessary neurons based on the optimal vector obtained through the EA process, resulting in a compressed model.

DE, which stands for Differential Evolution, is an evolutionary algorithm that utilizes unit vectors to represent distance and orientation information. It improves solutions through evolutionary processes such as mutation and recombination. DE is known for its ability to explore large design spaces without making strong assumptions about the optimization problem [100–103]. GA, or Genetic Algorithm, is another evolutionary approach inspired by Darwin's Theory of Evolution. It maintains a population of individuals with distinct characteristics and operates based on the principle of 'survival of the fittest.' Individuals that are better adapted to their environment have a higher chance of reproducing and passing on their traits to subsequent generations. GA generates optimized solutions through selection, crossover, and mutation operations [104–106]. PSO, or Particle Swarm Optimization, was originally proposed in 1995 by Eberhart and Kennedy [107]. It draws inspiration from the behavior of a flock of birds or a school of fish, where individuals learn from one another to determine the optimal location, such as finding food. In PSO, position vectors are created with arbitrary values of 0 and 1, with the vector representing the highest fitness considered as the position of the food. Through a series of equations, new position vectors are discovered in each iteration [108, 109]. WO, or Whale Optimization, is an EA approach inspired by the hunting behavior of humpback whales that spiral around their prey [110–112].

The algorithm assumes that the position of the highest fitness, or the optimal vector, represents the prey's location. By employing specific equations, WO converges towards the position of the prey and identifies new position vectors in subsequent iterations. Figures 3.7(a)–(d) provides a diagrammatic representation of these four EA approaches, with (a), (b), (c), and (d) representing DA, GA, PSO, and WO, respectively.

3.3.4 Loss function for AI base models

During the model generation process, the cross-entropy (CE)-loss functions is used in the new models. If α_{CE} stood for the CE-loss function, p_i for the classifier's probability in the AI model, x_i for the input gold standard label 1, and $(1-x_i)$ for the gold standard label 0.

The mathematical expression for the loss function is given in equation (3.1).

$$\alpha_{CE} = -[(x_i \times \log p_i) + (1 - x_i) \times \log(1 - p_i)] \quad (3.1)$$

Here × represents the product of the two terms.

3.3.5 Fitness function for AI EA models

The primary tenet of EA is fitness function, which enables us to maximize or minimize the objective fitness criteria by guiding us towards the optimum solution set. The fitness function value is compared to each iteration's previous best, and the solution space is modified in accordance with its result. A dual objective fitness function that depends on the test set's mean intersection over union (mIoU) accuracy (represented as β_{ACC}) and the nodes' compression ratio (C_r) has been utilized in all of the evolutionary algorithms in this work [113, 114]. The optimized fitness function F_{Opt} may be mathematically stated by equation (3.2) below when $HN_{Original}$ represents the originally hidden neuron and $HN_{Compress}$ is the compressed hidden neuron.

$$\begin{cases} F_{Opt} = (\mu_1 \times \beta_{ACC}) + (\mu_2 \times C_r) \\ \mu_1 + \mu_2 = 1 \\ C_r = \dfrac{HN_{Original}}{HN_{Compress}} \end{cases} \quad (3.2)$$

where $\mu 1$ and $\mu 2$ are weight variables that help in emphasizing one of the goals. Our implementations use these weights at a value of 0.5. More compression can be accomplished, but accuracy may suffer if we chose combinations like 0.2 and 0.8, and vice versa.

3.3.6 Experimental protocol

In the AI training process, a standardized cross-validation (CV) method was utilized, specifically the K5 method. This approach involves dividing the available data into five folds. Among these folds, one fold is used as the test set while the remaining four folds are used for training. This process is repeated five times, each time using a different fold as the test set. In total, 4000 CT scans were used for training the AI models, while 1000 CT images were reserved for testing. The K5 protocol ensures that each fold serves as the test set once, allowing for a comprehensive evaluation of the models' performance across different subsets of the data. Additionally, within each fold, 10% of the data was further reserved for internal validation. This validation subset helps monitor the model's performance and assess its generalization ability on unseen data. By incorporating internal validation, the models' accuracy and reliability can be assessed during the training process. Overall, the K5 CV protocol with internal validation ensures robust training and evaluation of the AI models using the available data.

To assess the accuracy of the AI system, the predicted output, in the form of a lung mask, is compared with the pixel values from the ground truth. Since the lung mask is binary, with black or white readings, these values are translated into binary integers, where white corresponds to 1 and black corresponds to 0. The accuracy of the AI system can be calculated using equation (3.3), which takes into account the number of false negatives (FN), true negatives (TN), true positives (TP), and false positives (FP). These terms represent the following:

- False Positive (FP): The number of pixels that are incorrectly predicted as lung when they are not.
- True Positive (TP): The number of pixels that are correctly predicted as lung.
- True Negative (TN): The number of pixels that are correctly predicted as non-lung.
- False Negative (FN): The number of pixels that are incorrectly predicted as non-lung when they are lung.

The accuracy is determined by dividing the sum of true positive and true negative values (TP + TN) with the total number of pixels in the image. Equation (3.3): Accuracy = (TP + TN)/(TP + TN + FP + FN). By calculating the accuracy using this equation, we can evaluate how well the AI system performs in terms of accurately predicting lung pixels compared to the ground truth.

$$\text{Accuracy}(\%) = \left(\frac{\text{TP} + \text{TN}}{\text{TP} + \text{FN} + \text{TN} + \text{FP}}\right) \times 100 \tag{3.3}$$

3.4 Results and performance evaluation

The findings of the study were categorized into eight sets, each consisting of different pruning and optimization strategies recommended by COVLIAS 2.0. These sets were used to evaluate the performance of the AI models in terms of speed and storage utilization. The first set focused on the first base model, FCN, and employed four FCN-based optimization approaches: FCN-WO, FCN-DE, FCN-GA, and FCN-PSO. These approaches were used to optimize and prune the FCN model, resulting in AI models that are faster and require less storage. Similarly, the second set utilized SegNet as the base model and employed four SegNet-based optimization algorithms: SegNet-WO, SegNet-PSO, SegNet-GA, and SegNet-DE. These approaches aimed to optimize and prune the SegNet model, leading to AI models that exhibit improved speed and reduced storage requirements. By exploring these different combinations of pruning and optimization strategies, the study aimed to identify the most effective approaches for improving the performance of AI models in terms of speed and storage utilization.

3.4.1 Results

(A) Visual results from the CroMed experimental data set using pruning models

With a senior radiologist's singleujs set of ground truth annotation, the CroMed data set of 5000 COVID-19 CT lung images was used to train the AI modes. The segmented lungs' overlay results are shown in grayscale in figures 3.8 and 3.9. Figures 3.8 and 3.9 show the combinations of SegNet +

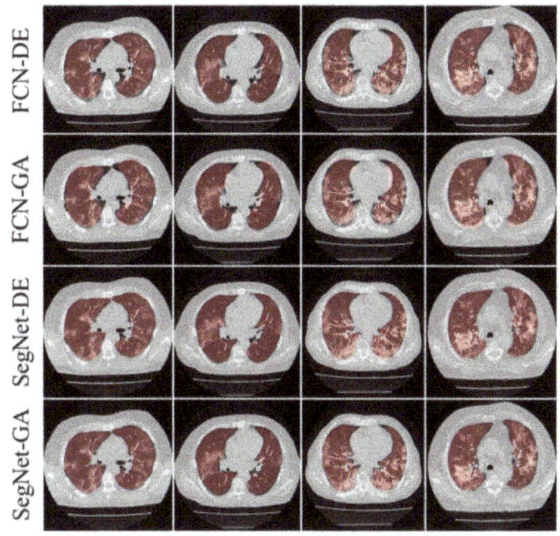

Figure 3.8. Overlays for optimized pruning networks over raw grayscale CT scans using CroMed data set. Top: FCN-DE and FCN-GA pruning combination and Bottom: SegNet-DE and SegNet-GA pruning combination.

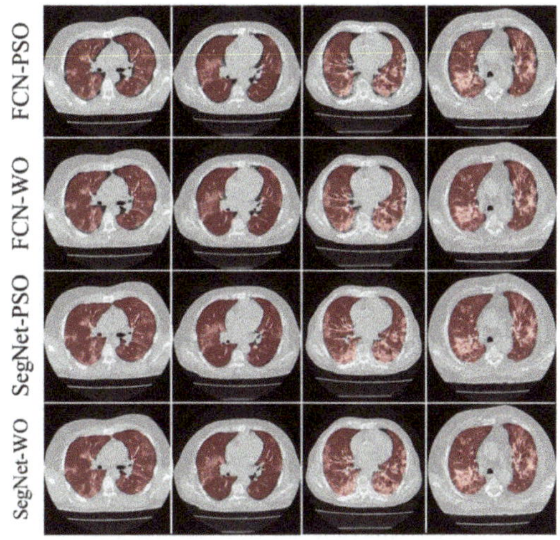

Figure 3.9. Overlays for optimized pruning networks over raw grayscale CT scans using CroMed data set. Top: FCN–PSO and FCN-WO pruning combination and Bottom: SegNet-PSO and SegNet-WO pruning combination.

WO, FCN + PSO, FCN + WO and SegNet + PSO. Figure 3.8 shows the combinations of SegNet + GA, FCN + DE, FCN + GA FCN + DE.

(B) Percentage storage reduction observed on the CroMed data set using pruning models

Using the percentage storage reduction (PSR) calculation as described in equations (3.4) and (3.5), table 3.1 displays the storage reduction for the eight pruning procedures.

Table 3.1. Percentage storage reduction (in MB) for the pruned AI models.

Models	Size (MB)	PSR	Models	Size (MB)	PSR
FCN	512	—	SegNet	20.9	—
FCN-DE	38.72	92.40%	SegNet-DE	0.6	97.10%
FCN-WO	23.85	95.30%	SegNet-GA	0.44	97.90%
FCN-GA	6.55	98.70%	SegNet-WO	0.25	98.80%
FCN–PSO	1.2	99.80%	SegNet-PSO	0.16	99.20%

$$PSR_{FCN}^{EA}\ (\%) = \left[\frac{S_{FCN} - S_{FCN-EA}}{S_{FCN}}\right] \times 100 \tag{3.4}$$

where the storage for the FCN model is represented by SFCN, and the storage for the EA algorithm is represented by SFCN-EA.

$$PSR_{SegNet}^{EA}\ (\%) = \left[\frac{S_{SegNet} - S_{SegNet-EA}}{S_{SegNet}}\right] \times 100 \tag{3.5}$$

where the storage for the SegNet model is represented by S_{SegNet}, and the storage for the EA algorithm is represented by $S_{SegNet\text{-}EA}$.

The PSR using equation (3.4) was 92%, 95%, 99%, and 100%, respectively, when compared against FCN for (i) FCN-DE, (ii) FCN-GA, (iii) FCN-PSO, and (iv) FCN-WO, and for (v) SegNet-DE, (vi) SegNet-GA, (vii) SegNet-PSO, and (viii) SegNet-WO, the PSR using Eq. This supports the idea that pruning significantly reduces the size of AI models, making the system quick and effective while, most crucially, maintaining clinically acceptable performance levels for the AI models. The PSR values are set to rise along the rows in table 3.1. We found that the SegNet-PSO and FCN-PSO pruning models had the greatest PSR. Models using SegNet-DE and FCN-DE had the lowest PSR. When FCN was utilised as a base, the intermediate PSR models were FCN-WO and FCN-GA. When SegNet was employed as a model, the intermediate models were SegNet-WO and SegNet-GA. Notably, compared to the basic model SegNet and FCN, all eight fused systems were five times quicker. PSR stands for percentage storage reduction, DE for differential evolution, GA for genetic algorithm, PSO for particle swarm optimization, and WO for whale optimization. Notably, Cr can have a maximum value equal to the entire number of hidden neurons in equation (3.2) and a minimum value of 1. One needs to make sure that the compressed hidden neurons are never 0 to prevent undesirable circumstances. After 10 iterations, the maximum Cr for FCN-PSO and SegNet-PSO were 962.47 and 133.12, respectively. After 4 iterations, the highest Cr for SegNet-GA and FCN-GA were 48.71 and 80.03, respectively. After 4 iterations, the Cr for FCN-DE was 13.54, but for SegNet-DE, it was 34.79. After 12 iterations, the Cr for SegNet-WO was 85.73 while the Cr for FCN-WO was 31.99.

Further compression would have resulted in performance reduction; hence it wasn't taken into account.

3.4.2 Performance evaluation

(A) Performance evaluation for eight EA models

The performance of the pruned AI models is compared in the proposed study using five primary types of performance assessment metrics: (i) ROC, (ii) AE, (iii) DS, (iv) BA, (v)JI, and (vi) CC. CroMed data having a cutoff of 80%, the cumulative frequency distribution (CFD) plot for AE is shown in figure 3.10. The estimated lung area is shown against the ground truth tracings and AI models in figure 3.11 using the mean and standard deviation (SD) lines. Similar to this, CC plots with an 80% cutoff are shown in figure 3.12. Figures 3.13 and 3.14 show CFD plots for JI and DS at an 80% CroMed threshold cutoff. ROC analysis and the variable threshold approach can be used to determine the COVIAS 2.0 diagnostic performance. SegNet-based four EA models with AUC values larger than ~0.94 (*P* 0.0001) and ~0.96 (*P* 0.0001), respectively, are shown using ROC curves in figure 3.15 for the CroMed data set.

(B) JI, CC, and DS for eight EA models for CroMed and NovMed data sets

Let DSFCN EA(m) and DSSegNet EA(m) be the DS for the fundamental SegNet and FCN models, respectively, corresponding to EA(m), where *m* might take the values 1, 2, 3, and 4, which, respectively, imply WO, GA, DE and PSO. Likewise, JIFCN EA(m) and JISegNet EA(m) be the JI for the base models SegNet and FCN, respectively, corresponding to *EA*(*m*), where *m* might take the values 1, 2, 3, and 4, which, respectively, imply DE, GA, PSO, and WO.

Let CCFCN *EA*(*m*) and CCSegNet *EA*(*m*) be the CC for the base models of SegNet and FCN, respectively, corresponding to *EA*(*m*), where *m* might take the values 1, 2, 3, and 4, which, respectively, imply DE, GA, PSO, and WO.

$$\Delta DS_{EA(m)}^{FCN}[d] = \frac{\left|DS_{EA(m)}^{FCN} - DS^{MedSeg}[d]\right|}{DS^{MedSeg}[d]},\ \Delta DS_{EA(m)}^{SegNet}[d] = \frac{\left|DS_{EA(m)}^{SegNet} - DS^{MedSeg}[d]\right|}{DS^{MedSeg}[d]} \tag{3.6}$$

$$\Delta JI_{EA(m)}^{FCN}[d] = \frac{\left|JI_{EA(m)}^{FCN} - JI^{MedSeg}[d]\right|}{JI^{MedSeg}[d]},\ \Delta JI_{EA(m)}^{SegNet}[d] = \frac{\left|JI_{EA(m)}^{SegNet} - JI^{MedSeg}[d]\right|}{JI^{MedSeg}[d]} \tag{3.7}$$

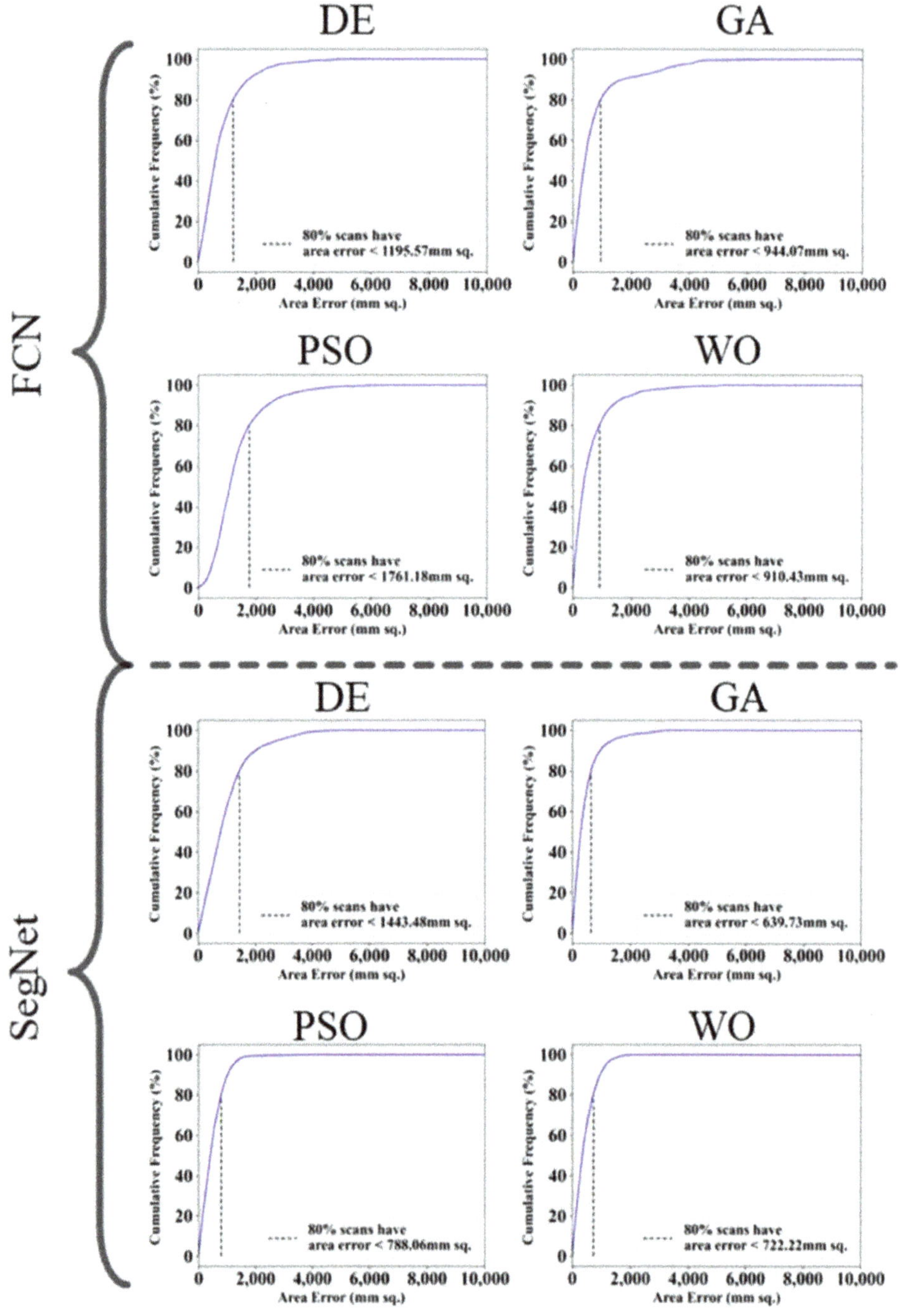

Figure 3.10. Cumulative frequency plot for lung Area Error for CroMed data. Top: FCN with four pruning (DE, GA, PSO, WO) and Bottom: SegNet with four pruning (DE, GA, PSO, WO).

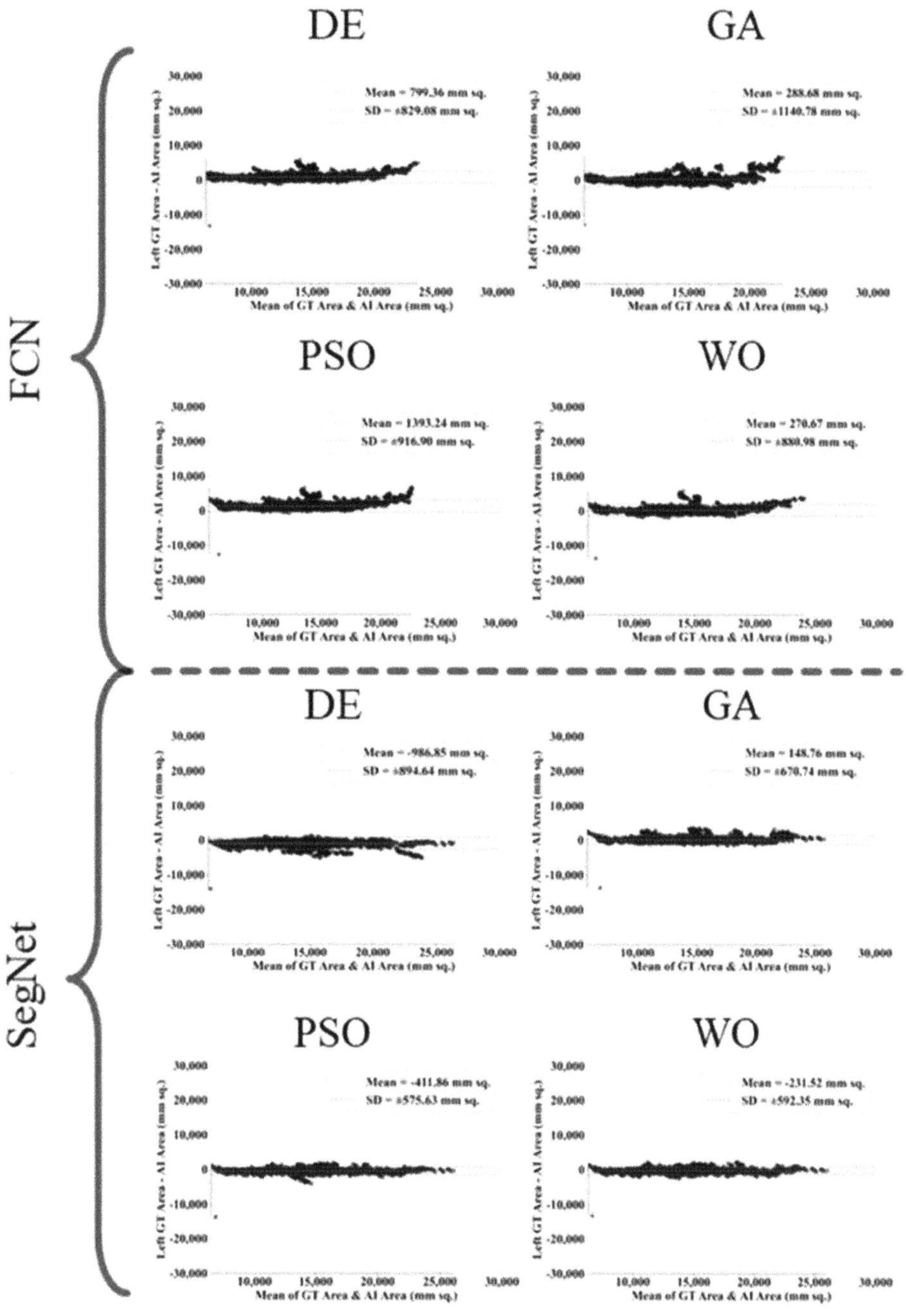

Figure 3.11. Bland-Altman plots for Area Error on Croatia data. Top: FCN with four pruning (DE, GA, PSO, WO) and Bottom: SegNet with four pruning (DE, GA, PSO, WO).

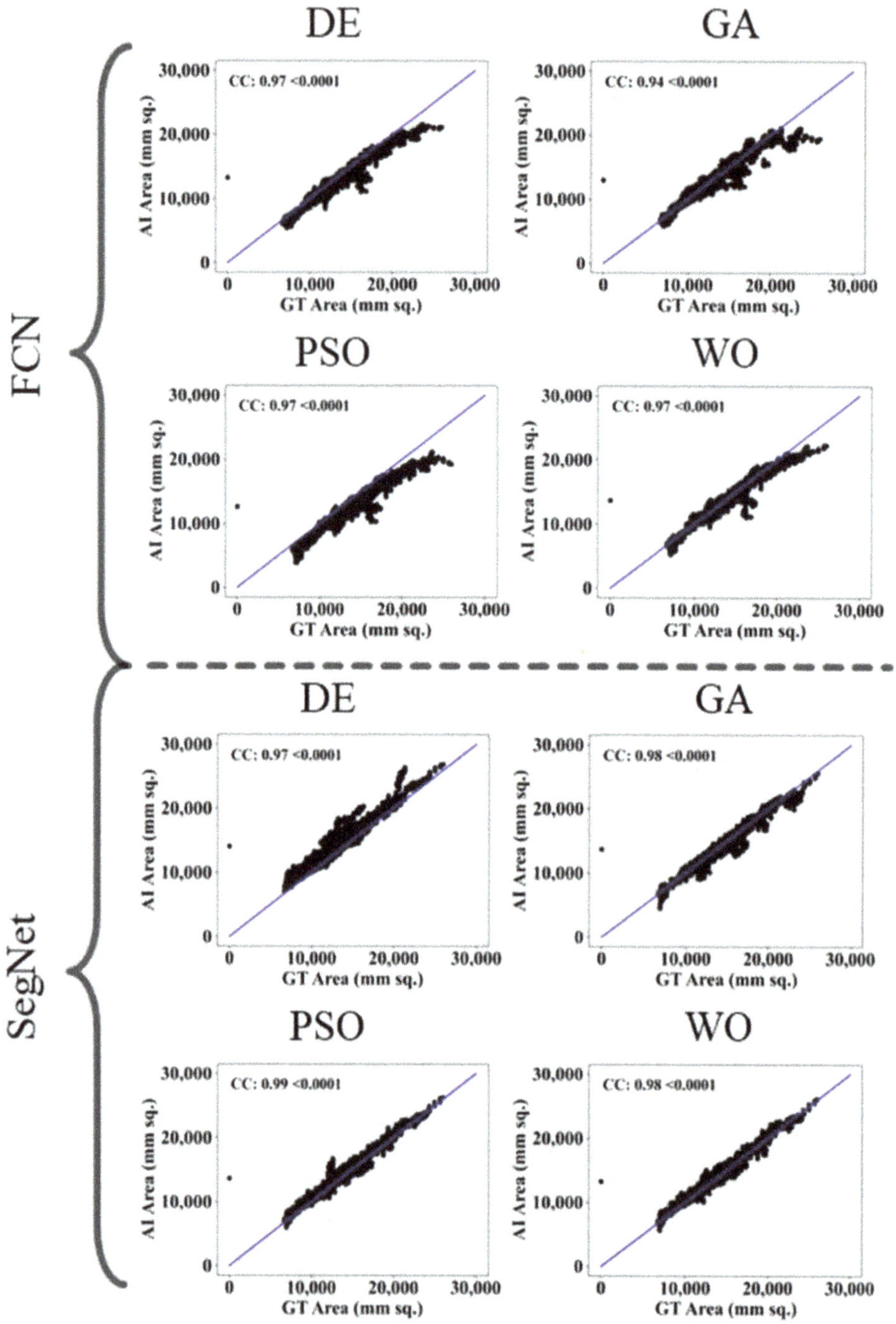

Figure 3.12. CC plots for Area Error on CroMed data. Top: FCN with four pruning (DE, GA, PSO, WO) and Bottom: SegNet with four pruning (DE, GA, PSO, WO).

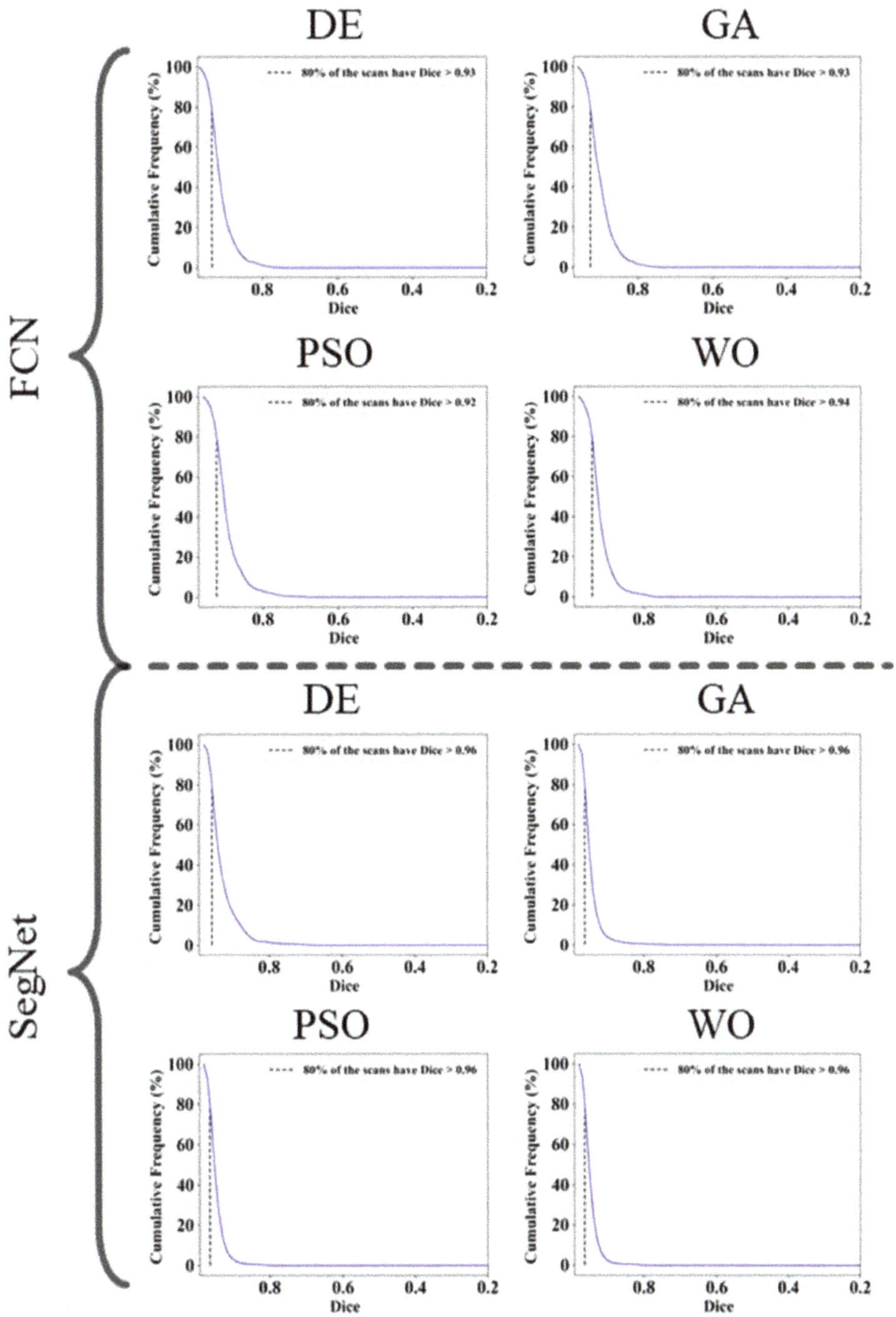

Figure 3.13. Dice Similarity on CroMed data. Top: FCN with four pruning (DE, GA, PSO, WO) and Bottom: SegNet with four pruning (DE, GA, PSO, WO).

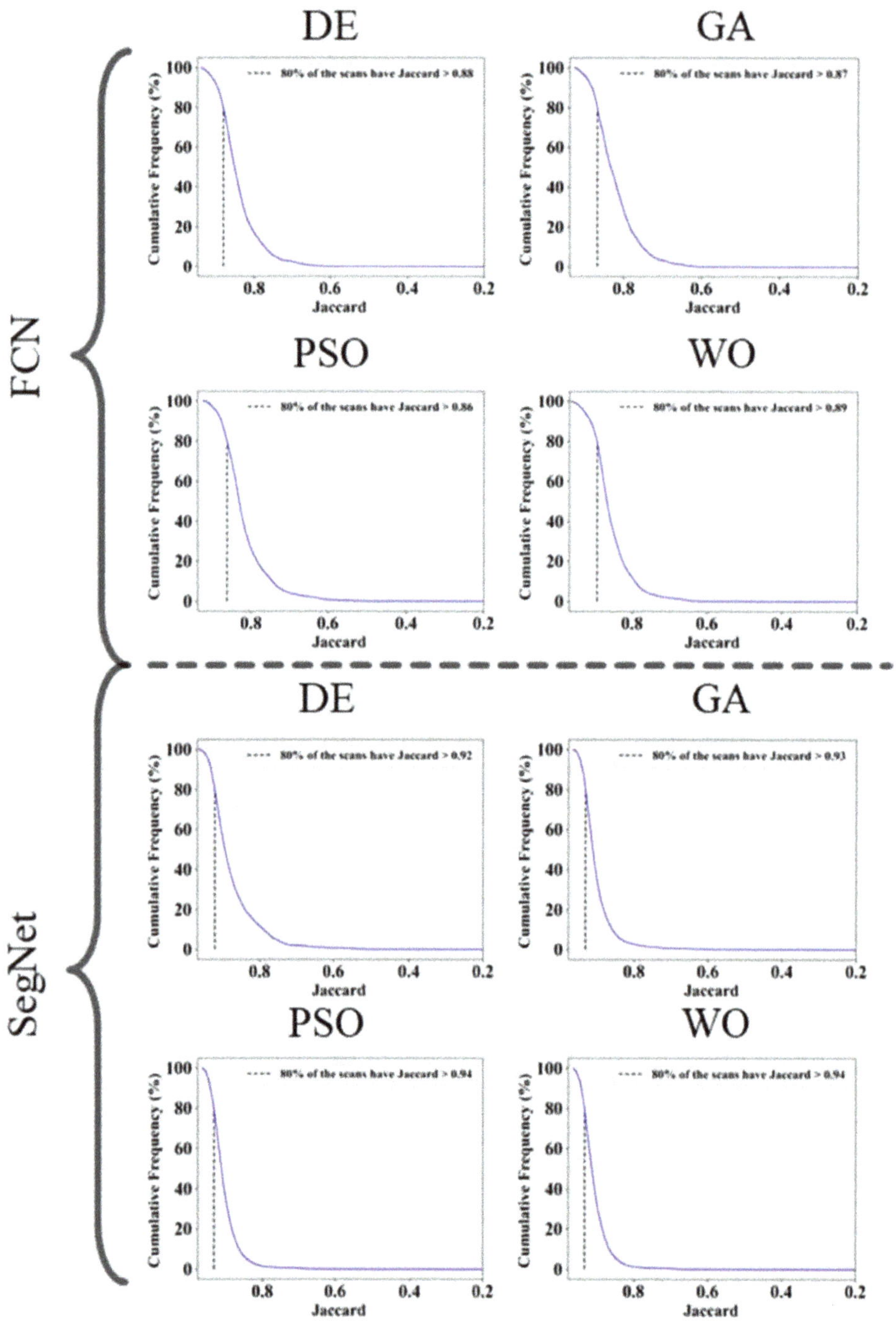

Figure 3.14. Jaccard Index plots for Area Error on CroMed data. Top: FCN with four pruning (DE, GA, PSO, WO) and Bottom: SegNet with four pruning (DE, GA, PSO, WO).

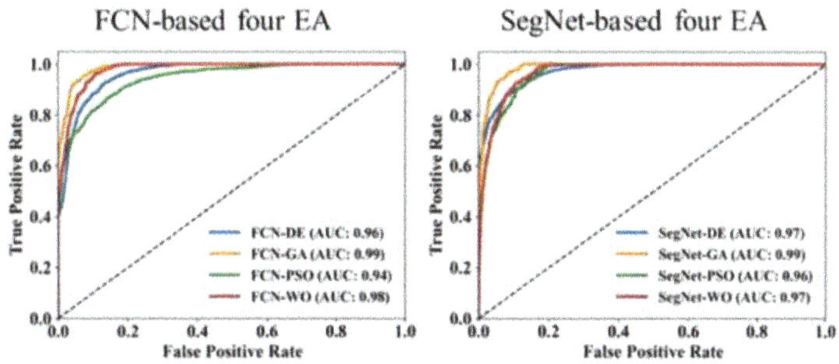

Figure 3.15. ROC using CroMed data set. Left: FCN-based four EA (DE, GA, PSO, WO) and Right: SegNet-based four EA (DE, GA, PSO, WO).

Table 3.2. DS, JI, and CC values for experimental CroMed and Unseen NovMed.

	CroMed			Unseen NovMed		
Models	DS	JI	CC	DS	JI	CC
FCN-DE	0.93	0.88	0.97	0.93	0.87	0.99
FCN-GA	0.93	0.87	0.94	0.94	0.88	0.98
FCN–PSO	0.92	0.86	0.97	0.91	0.84	0.98
FCN-WO	0.94	0.89	0.97	0.94	0.89	0.99
SegNet-DE	0.96	0.92	0.97	0.94	0.89	0.99
SegNet-GA	0.96	0.93	0.98	0.94	0.89	0.98
SegNet-PSO	0.96	0.94	0.99	0.95	0.91	0.99
SegNet-WO	0.96	0.94	0.98	0.95	0.91	0.99
MedSeg	0.96	0.92	0.99	0.95	0.9	0.99

$$\Delta CC_{EA(m)}^{FCN}[d] = \frac{\left|CC_{EA(m)}^{FCN} - CC^{MedSeg}[d]\right|}{CC^{MedSeg}[d]},\ \Delta CC_{EA(m)}^{SegNet}[d] = \frac{\left|CC_{EA(m)}^{SegNet} - CC^{MedSeg}[d]\right|}{CC^{MedSeg}[d]} \tag{3.8}$$

Using equations (3.6)–(3.8) one can compute the mean of all four EA models for FCN and SegNet base models as defined in equation (3.9).

$$\mu_{EA}^{FCN}[d] = \frac{\sum_{m=1}^{M=4} \Delta DS_{EA(m)}^{FCN}[d]}{M=4},\ \mu_{EA}^{SegNet}[d] = \frac{\sum_{m=1}^{M=4} \Delta DS_{EA(m)}^{SegNet}[d]}{M=4} \tag{3.9}$$

Table 3.2 presents the results obtained by MedSeg, including the CC, JI, and DS values for the eight trimmed AI models across the CroMed (experimental) and Unseen NovMed (validation) data sets. The left side

Table 3.3. Percentage difference in Dice similarity (against GT) for the eight EA models when compared against MedSeg using experimental CroMed and Unseen NovMed.

Models	CroMed	Unseen NovMed	Models	CroMed	Unseen NovMed
FCN-DE	3%	2%	SegNet-DE	0%	1%
FCN-GA	3%	1%	SegNet-GA	0%	1%
FCN–PSO	4%	4%	SegNet-PSO	0%	0%
FCN-WO	2%	1%	SegNet-WO	0%	0%
μ	**3%**	**2%**	**μ**	**0%**	**1%**
σ	**1%**	**1%**	**σ**	**0%**	**1%**

of table 3.3 corresponds to FCN-WO, FCN-PSO, FCN-GA, and FCN-DE, while the right side represents SegNet-WO, SegNet-PSO, SegNet-GA, and SegNet-DE. Using equation (3.9), table 3.3 displays the percentage difference in DS values for the eight AI pruned models.

For the CroMed (experimental) data set, the mean dice coefficients for the four trimmed FCN models were 3%, 8%, and 2%, respectively. On the other hand, the mean Dice coefficients for the four pruned SegNet models were 0%, 4%, and 1%, respectively. These findings highlight the performance variations between the FCN and SegNet models in terms of Dice similarity coefficients across the experimental data set. and using equation (3.9), unknown Novmed (validation) data sets. Additionally, table 3.4 shows the mean percentage difference using equation (3.7) for JI values, which for CroMed (experimental) is 5%, 15%, and 3% for trimmed FCN models and 1%, 8%, and 1% for pruned SegNet models and Unknown (validation) NovMed data sets. For CroMed (experimental), the CC value's mean percentage difference using equation (3.8) is shown in table 3.5 and is, respectively, 3%, 1%, and 1% for trimmed FCN models and 1%, 1%, and 0% for pruned SegNet models and Unseen NovMed (validation) data set.

(C) Comparison between eight EA models using figure-of-merit

We compared the eight pruned AI models based on FCN and SegNet. Using the standardised Figure-of-Merit (FoM) technique, we compared the SegNet and FCN-based eight pruned AI models [25, 115–117]. This may be modelled mathematically as follows. Let GT[d] represent the mean ground truth (GT) area for the datatype [d], and let AIEA[d] represent the mean area for EA models on the datatype [d], both obtained over 'n' pictures of the cohort datatype [d] comprising 'N' images, with 'd' denoting either NovMed or CroMed. These notations may be used to denote the FoM in % for the EA using datatype [d] as indicated in equation (3.10).

$$FoM_{EA}[d](\%) = \left[1 - \left(\frac{|\overline{AI_{EA}[d]} - \overline{GT[d]}|}{\overline{GT[d]}}\right)\right] \times 100 \tag{3.10}$$

Table 3.4. Percentage difference in Jaccard Index (against GT) for the eight EA models when compared against MedSeg using experimental CroMed and Unseen NovMed.

Models	CroMed	Unseen NovMed	Models	CroMed	Unseen NovMed
FCN-DE	4%	3%	SegNet-DE	0%	1%
FCN-GA	5%	2%	SegNet-GA	1%	1%
FCN–PSO	7%	7%	SegNet-PSO	2%	1%
FCN-WO	3%	1%	SegNet-WO	2%	1%
μ	**5%**	**3%**	**μ**	**1%**	**1%**
σ	**1%**	**2%**	**σ**	**1%**	**0%**

Table 3.5. Percentage difference in correlation coefficient (against GT) for the eight EA models when compared against MedSeg using experimental CroMed and Unseen NovMed.

Models	CroMed	Unseen NovMed	Models	CroMed	Unseen NovMed
FCN-DE	2%	0%	SegNet-DE	2%	0%
FCN-GA	5%	1%	SegNet-GA	1%	1%
FCN–PSO	2%	1%	SegNet-PSO	0%	0%
FCN-WO	2%	0%	SegNet-WO	1%	0%
μ	**3%**	**1%**	**μ**	**1%**	**0%**
σ	**2%**	**1%**	**σ**	**1%**	**1%**

Table 3.6. FoM comparison for CroMed using eight EA models.

Models	FCN	SegNet	Superior	2%	0%
DE	9433%	9300%	1.33	1%	1%
GA	9795%	9894%	0.99	0%	0%
PSO	9012%	9708%	6.96	1%	0%
WO	**9808%**	**9835%**	**0.28**	**1%**	**0%**
μ	95.12	96.84	2.39	**μ**	95.12
σ	3.76	2.68	3.08	**σ**	3.76

where $\overline{AI_{EA}[d]} = \frac{\sum_{n=1}^{N} A_{EA}(n)}{N}$ and $\overline{GT[d]} = \frac{\sum_{n=1}^{N} A_{GT}(n)}{N}$,

AI represents the area of the image 'n' using the evolutionary algorithm EA (DE, GA, PSO, and WO). / or (n) represents the GT area corresponding to image, '5' represents the summation of area of all the image in the cohort. Tables 3.6 and 3.7 show the FoM for the CroMed, and NovMed data sets. Table 3.6 shows the FoM for CroMed corresponding to FCN and SegNet-based EA models in column 2 and 3. 'The difference of the two columns is shown in the column 'Superior'. For the EA algorithm except

Table 3.7. FoM comparison for unseen NovMed using eight EA models.

Models	FCN	SegNet	Superior
DE	89.81	98.5	8.7
GA	93.21	89.25	3.96
PSO	85.62	94.32	8.7
WO	92.9	92.47	0.43
μ	90.38	93.64	5.45
σ	3.53	3.86	4.02

DE, SegNet-based EA is superior in the range 1%–10%. With a difference of 2.39%, the four EA's overall mean utilizing the two basic models is comparable. In columns 2 and 3, table 3.7 displays the FoM for unobserved NovMed corresponding to SegNet and FCN-based EA models. The 'Superior' column displays the difference between the two columns. SegNet-based EA outperforms all other EA algorithms, except GA and WO, by a margin of 1 to 10%. With a difference of 5.45%, the overall mean for EA utilizing the two basic models is comparable.

The mean FoM for FCN-based EA models across all three data sets (NovMed and CroMed) is around 94%, whereas for SegNet it was approximately 96%, as shown in tables 3.6 and 3.7. This proved unequivocally that the four EA models based on SegNet are better than the four EA models based on FCN. This is mostly due to the fact that the SegNet (figure 3.6) models have more layers than the FCN (figure 3.5) model does. Where FCN lacks skip connections, SegNet has them. SegNet outperforms FCN in the results, which are consistent with earlier applications [23, 25, 26, 28].

3.5 Validation and statistical tests

The proposed work includes a validation of lung segmentation utilizing (i) Unseen NovMed data set on the trained pruned AI models and (ii) a comparison of the outcomes with MedSeg, a web-based CT lung segmentation tool. The segmented lungs are shown in red in figures 3.16 and 3.17 with a grayscale backdrop utilizing Unseen NovMed data, respectively. Combining SegNet-GA, SegNet-DE, FCN-GA, and FCN-DE is displayed in figure 3.16. Combining SegNet-WO, SegNet-PSO, FCN-WO, and FCN-PSO using the Unseen NovMed data, respectively, is shown in figure 3.17. Additionally, keep in mind that 'Unseen NovMed' data sets were used to confirm the findings. The segmented lungs' CFD plots for JI, DS, CC, BA, and AE using COVLIAS 2.0 on the 'Unseen NovMed' data set are shown in figures 3.18–3.22. The Unseen NovMed ROC curves for the SegNet-based and FCN-based four EA models are shown in figure 3.23. The AUC values for each model are more than 0.86 (P 0.0001) and 0.86 (P 0.0001), respectively. The

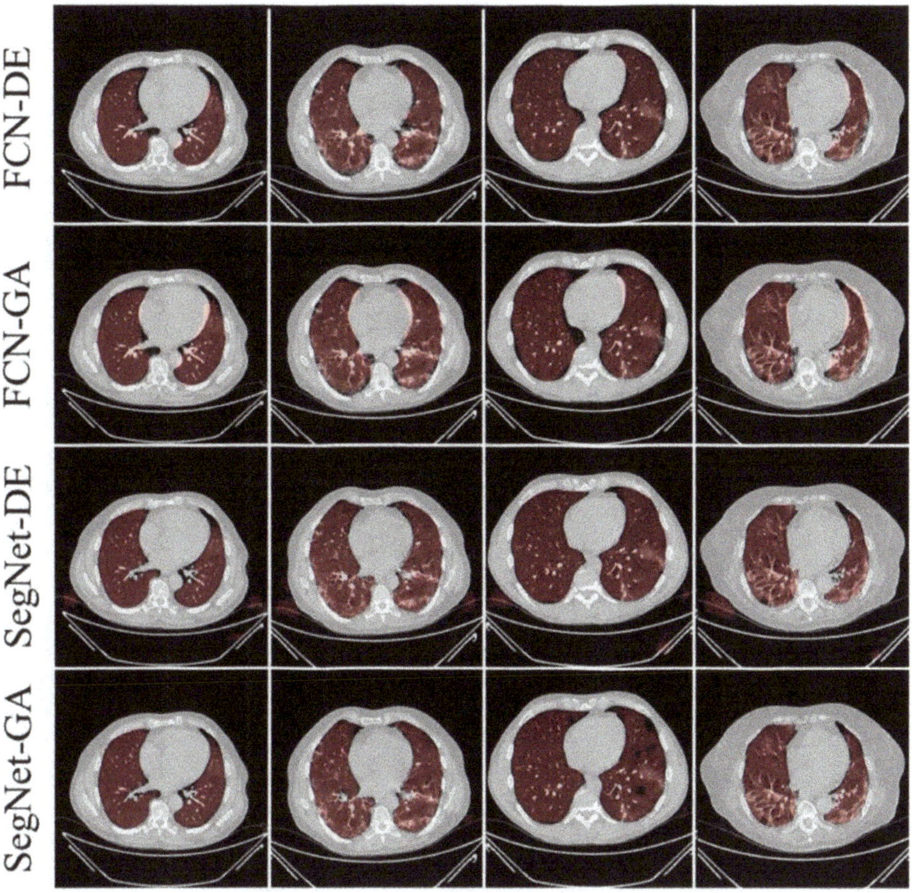

Figure 3.16. Overlays (red) for optimized pruning networks over raw grayscale CT scans for NovMed data set. Top: FCN-DE and FCN-GA pruning combination and Bottom: SegNet-DE and SegNet-GA pruning combination.

segmented lungs' CFD plots for JI, CC, AE, DS, and BA using COVLIAS 2.0 on the 'Unseen NovMed' data set are shown in figures 3.18–3.22. The Unseen NovMed ROC curves for the SegNet-based and FCN-based four EA models are shown in figure 3.23. The AUC values for each model are more than ~0.86 (*P* 0.0001) and ~0.86 (*P* 0.0001), respectively. Using COVLIAS 2.0 on the 'Unseen NovMed' data set, the segmented lungs' CFD plots for JI, DS, CC, BA, and AE are displayed in figures 3.18–3.22. Figure 3.23 displays the Unseen NovMed ROC curves for the SegNet- and FCN-based four EA models (as shown in figures 3.24–3.28). Each model's AUC values are more than 0.86 (*P* 0.0001) and less than 0.86 (*P* 0.0001), respectively. 0.99 (*P* 0.0001) and 0.89 (*P* 0.0001), respectively, are the values. The models used the following sizes, in decreasing order: FCN, FCN-DE, FCN-GA, figure 3.29 depicts FCN-WO and FCN-PSO, whereas figure 3.30 depicts SegNet-DE, SegNet-GA, SegNet-PSO, and SegNet-WO.

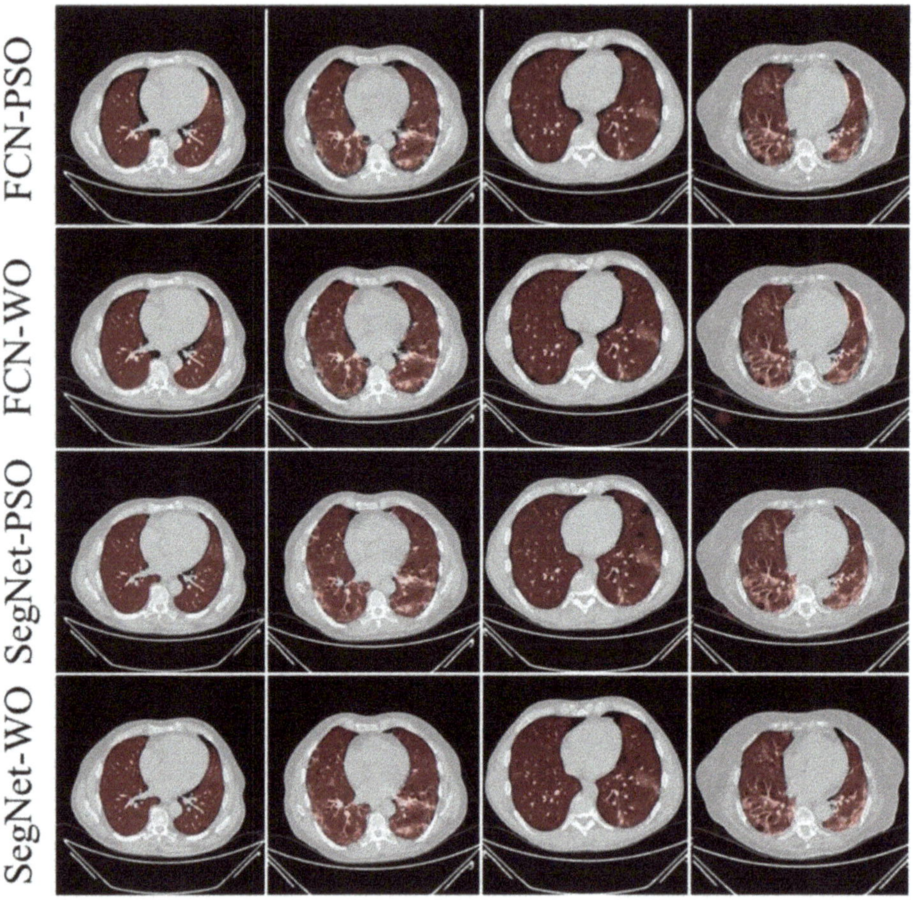

Figure 3.17. Overlays (red) for optimized pruning networks over raw grayscale CT scans for NovMed data set. Top: FCN-DE and FCN-GA pruning combination and Bottom: SegNet-DE and SegNet-GA pruning combination.

For the CC, JI and DS values utilizing the validation Unseen NovMed data sets and experimental CroMed, we show a summary and percentage improvement for all eight pruned AI models in tables 3.2–3.5, respectively. For validation and experimental data, the pruning model outperforms the basic model (FCN and SegNet) for all three performance assessment matrices when compared to the four pruning strategies. To assess the reliability of the AI-based segmentation system COVLIAS 2.0, statistical tests were performed using the experimental CroMed data set (table 3.8) and the validation Unseen NovMed data set (table 3.9). The statistical tests employed include the Wilcoxon test, Paired t-Test, and Mann-Whitney test. These tests were conducted to evaluate the performance and consistency of the COVLIAS 2.0 system based on the segmentation results obtained from the CroMed and Unseen NovMed data sets. MedCalc software (v18.2.1, Osteen, Belgium) was used for the whole analysis shown above.

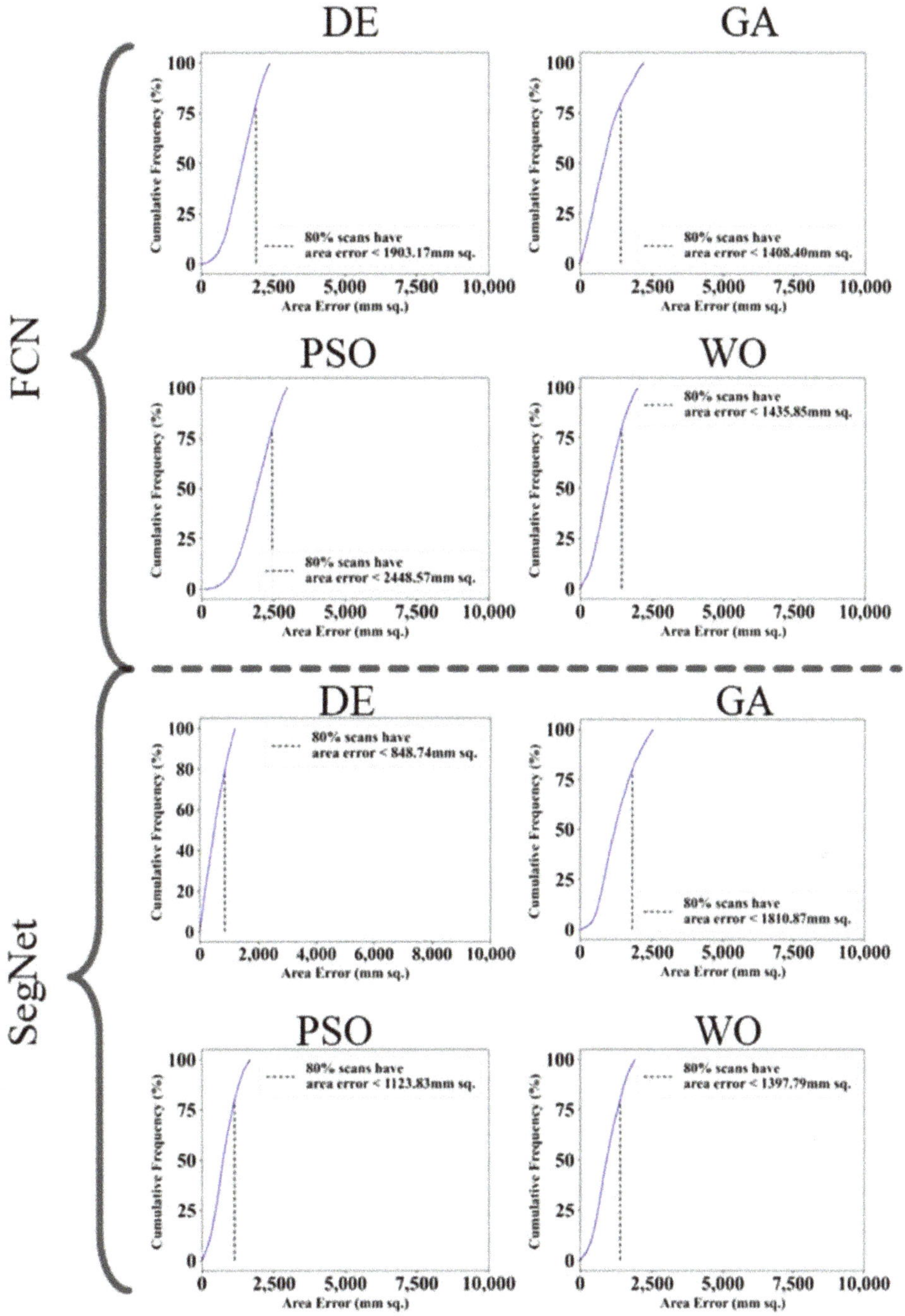

Figure 3.18. Cumulative frequency plot for Area Error using NovMed data.

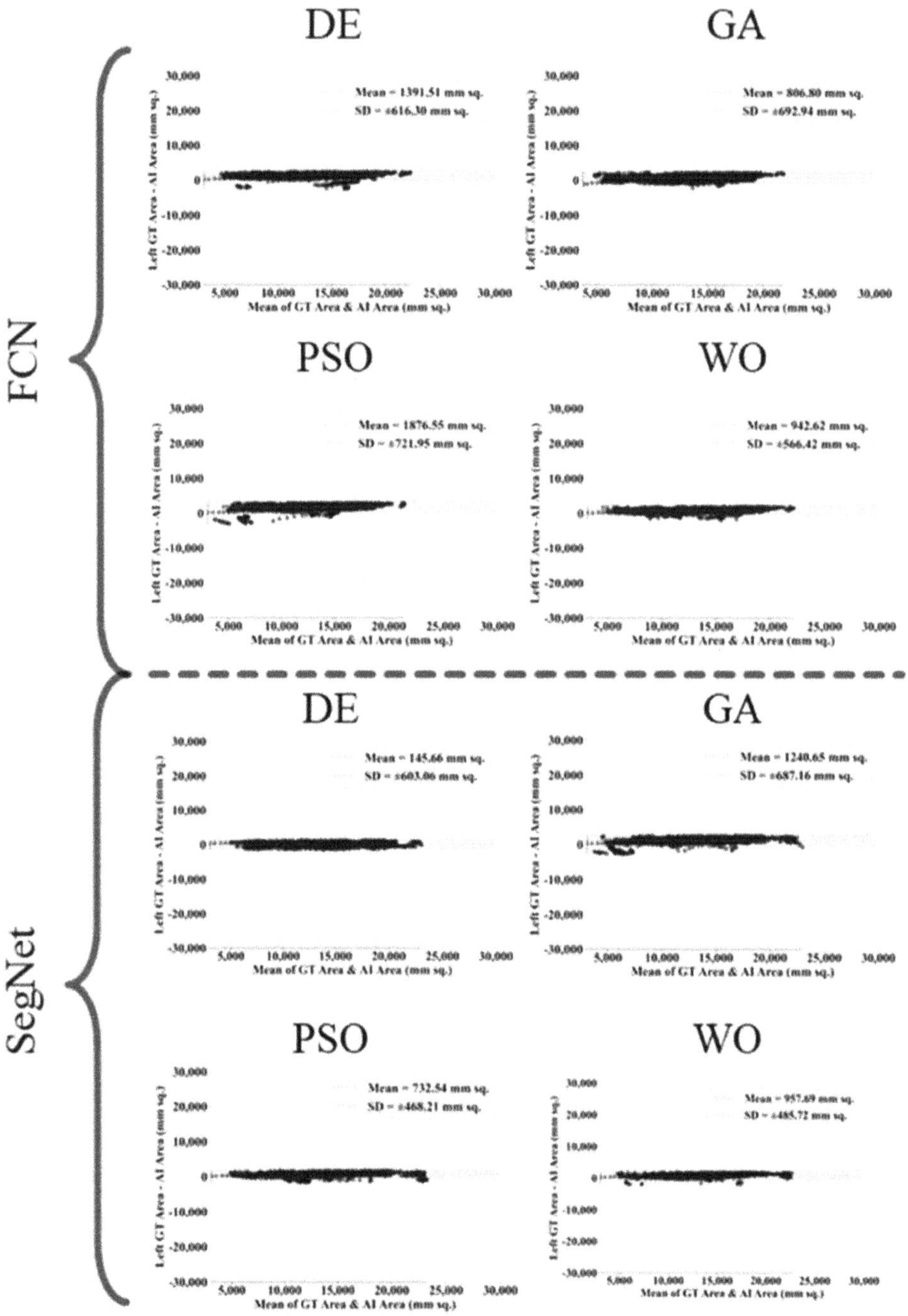

Figure 3.19. BA plot using NovMed data.

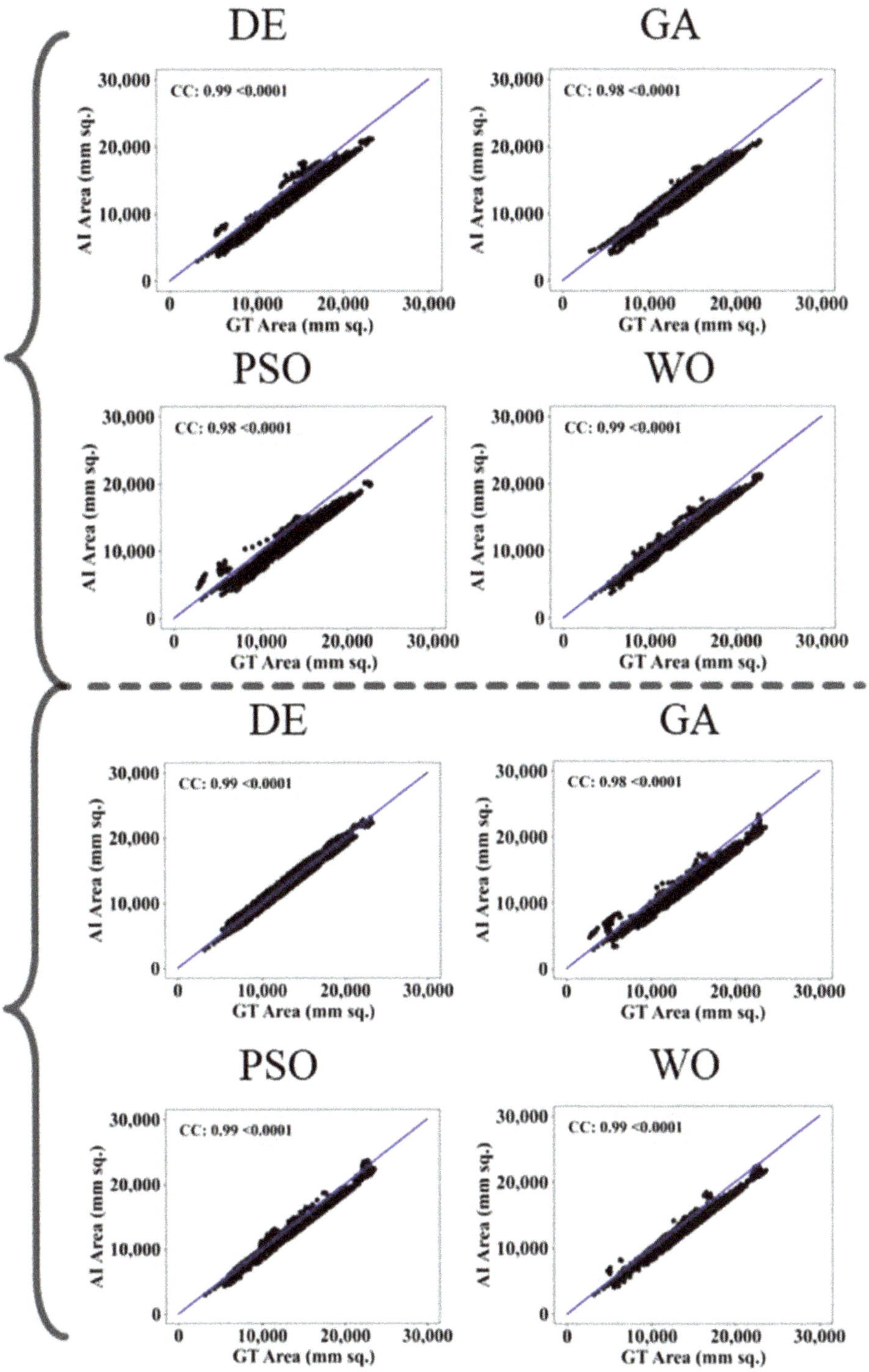

Figure 3.20. CC plot using NovMed data.

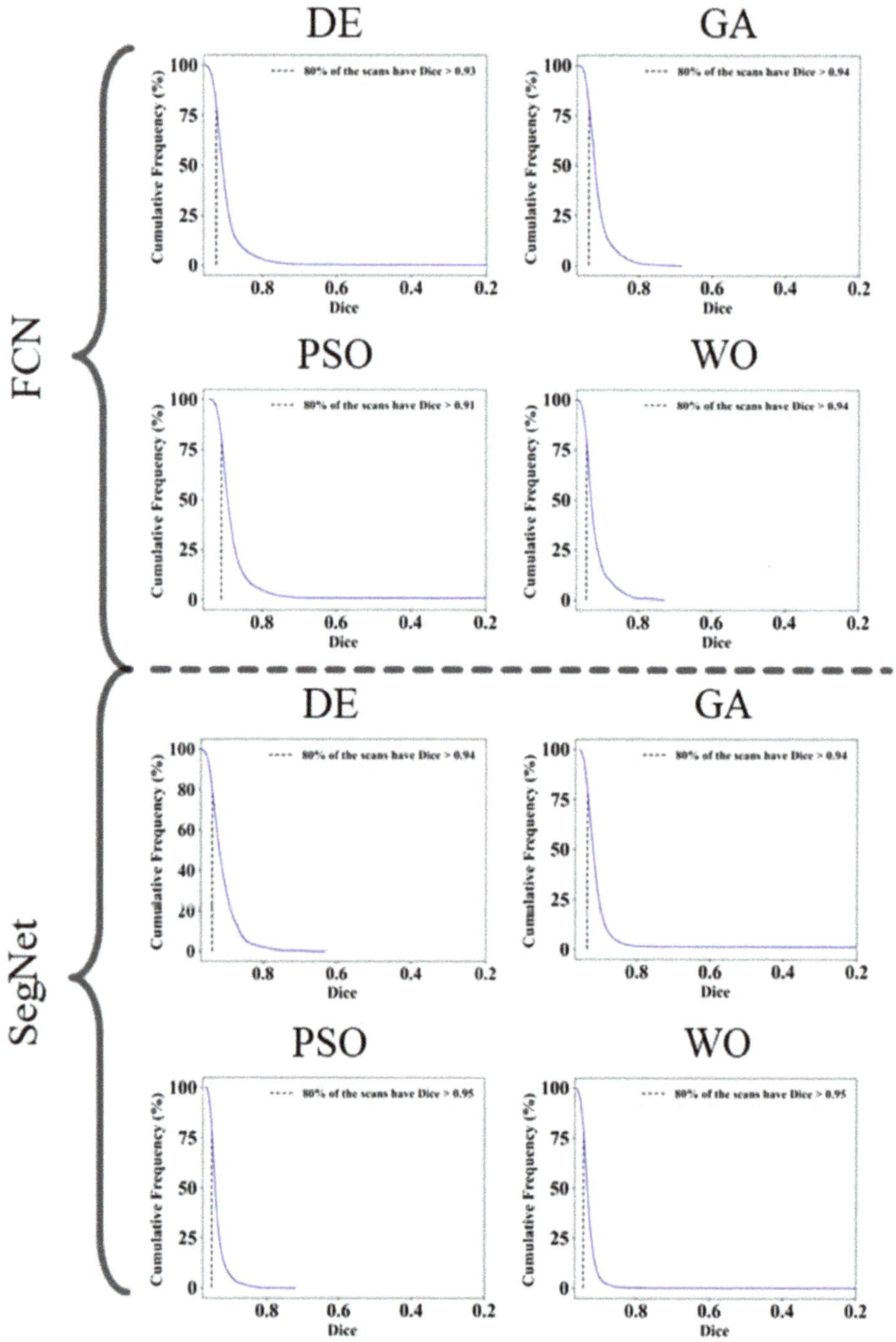

Figure 3.21. Cumulative frequency plot for Dice using NovMed data.

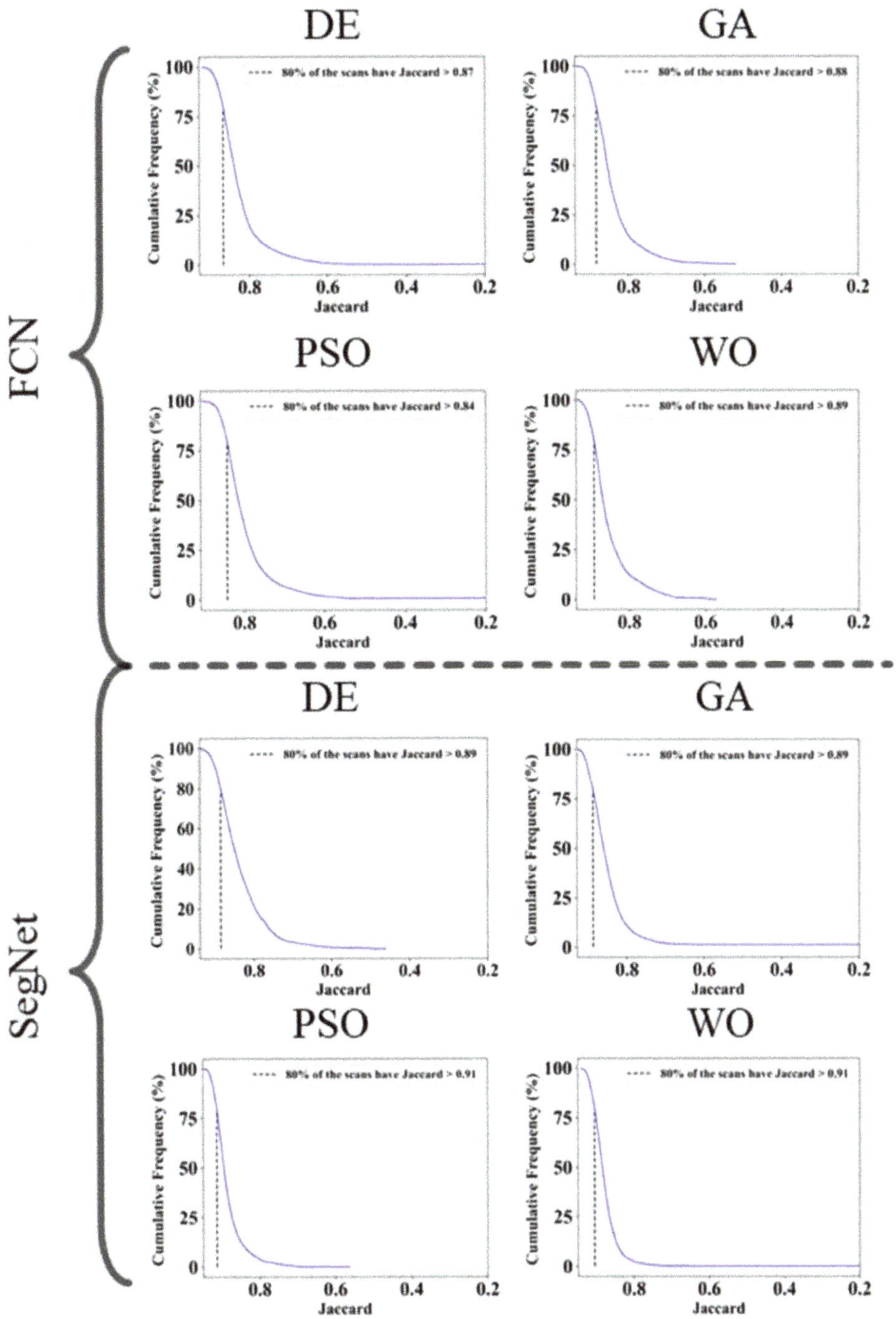

Figure 3.22. Cumulative frequency plot for Jaccard using NovMed data.

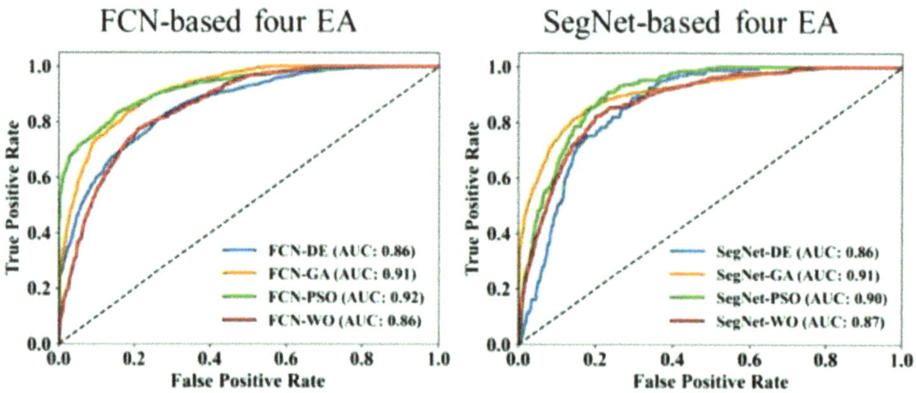

Figure 3.23. ROC using NovMed data set. Left: FCN-based four EA (DE, GA, PSO, WO) and Right: SegNet-based four EA (DE, GA, PSO, WO).

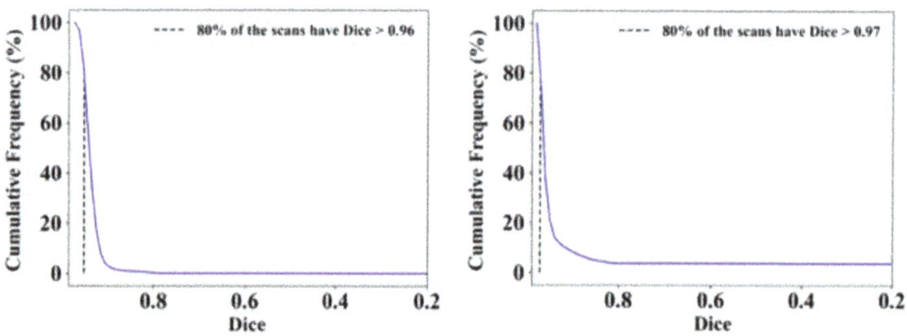

Figure 3.24. Cumulative frequency plot of DS for MedSeg using CroMed (left) and Unseen NovMed (right) data sets.

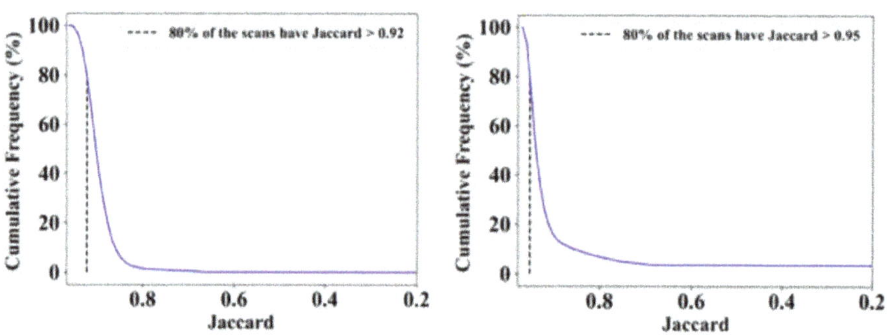

Figure 3.25. Cumulative frequency plot of JI for MedSeg using CroMed (left) and Unseen NovMed (right) data sets.

3.5.1 Lesion localization validation

Different properties of lesions include texture, contrast, intensity fluctuation, changes in density, and others [118]. The pipeline for lesion validation using

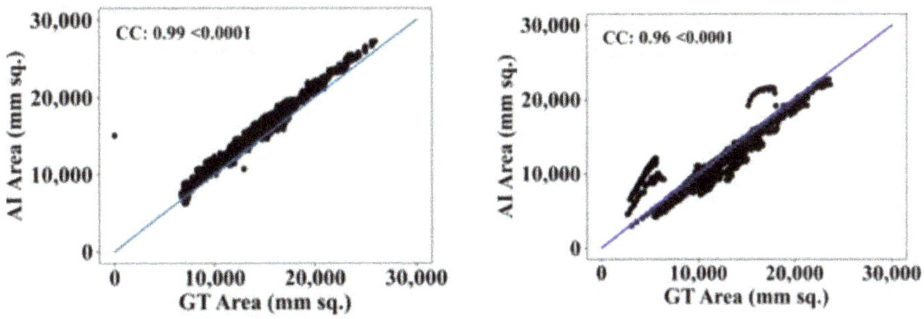

Figure 3.26. CC plot for MedSeg vs. GT using CroMed (left) and Unseen NovMed (right) data sets

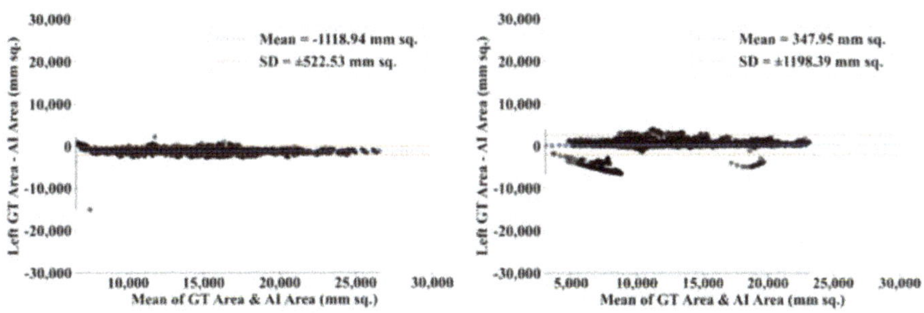

Figure 3.27. BA plot for MedSeg vs. GT using CroMed (left) and Unseen NovMed (right) data sets.

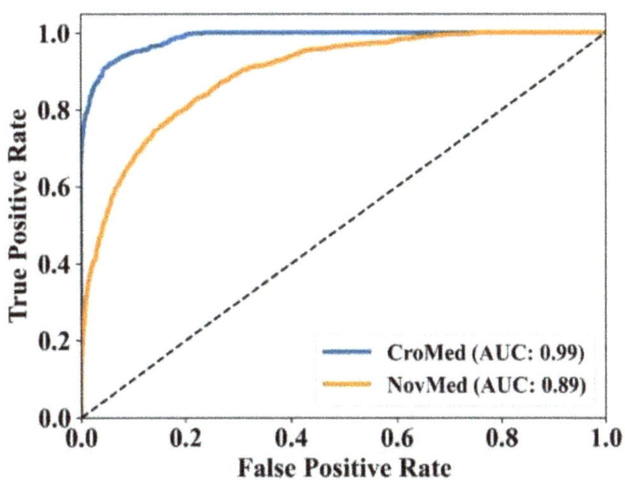

Figure 3.28. ROC plot for MedSeg vs. GT using CroMed and Unseen NovMed data sets.

heatmaps is shown in figure 3.31 [118]. The CT image is the input to the eight-trimmed segmentation model, which creates the segmented lungs. This segmented lung is sent to DenseNet-121 for categorization into COVID-19 and Controls, two classes. I apologize if it seemed that way. Here's the revised version: The Grad-CAM

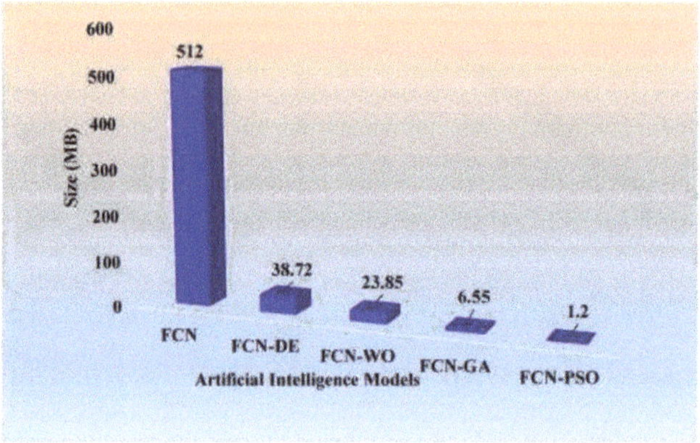

Figure 3.29. Size (in Megabyte) of the FCN-based models in descending order.

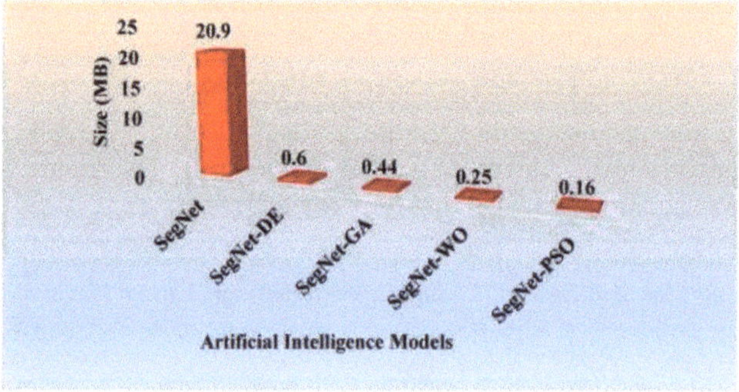

Figure 3.30. Size (in Megabyte) of the SegNet-based models in descending order.

Table 3.8. Statistical test for CroMed data set.

Models	Paired *t*-Test	Mann-Whitney	Wilcoxon
FCN-DE	$P < 0.0001$	$P < 0.0001$	$P < 0.0001$
FCN-GA	$P < 0.0001$	$P < 0.0001$	$P < 0.0001$
FCN–PSO	$P < 0.0001$	$P < 0.0001$	$P < 0.0001$
FCN-WO	$P < 0.0001$	$P < 0.0001$	$P < 0.0001$
SegNet-DE	$P < 0.0001$	$P < 0.0001$	$P < 0.0001$
SegNet-GA	$P < 0.0001$	$P < 0.0001$	$P < 0.0001$
SegNet-PSO	$P < 0.0001$	$P < 0.0001$	$P < 0.0001$
SegNet-WO	$P < 0.0001$	$P < 0.0001$	$P < 0.0001$
MedSeg	$P < 0.0001$	$P < 0.0001$	$P < 0.0001$

technique, which stands for Gradient-weighted Class Activation Mapping, was utilized to generate the lesion heatmap (figure 3.32). Grad-CAM operates by leveraging the gradients of a target concept, such as 'COVID-19' in this classification network, to create a coarse feature map that highlights the important regions in the

Table 3.9. Statistical test for NovMed data set

Models	Paired *t*-Test	Mann-Whitney	Wilcoxon
FCN-DE	$P < 0.0001$	$P < 0.0001$	$P < 0.0001$
FCN-GA	$P < 0.0001$	$P < 0.0001$	$P < 0.0001$
FCN–PSO	$P < 0.0001$	$P < 0.0001$	$P < 0.0001$
FCN-WO	$P < 0.0001$	$P < 0.0001$	$P < 0.0001$
SegNet-DE	$P < 0.0001$	$P < 0.0001$	$P < 0.0001$
SegNet-GA	$P < 0.0001$	$P < 0.0001$	$P < 0.0001$
SegNet-PSO	$P < 0.0001$	$P < 0.0001$	$P < 0.0001$
SegNet-WO	$P < 0.0001$	$P < 0.0001$	$P < 0.0001$
MedSeg	$P < 0.0001$	$P < 0.0001$	$P < 0.0001$

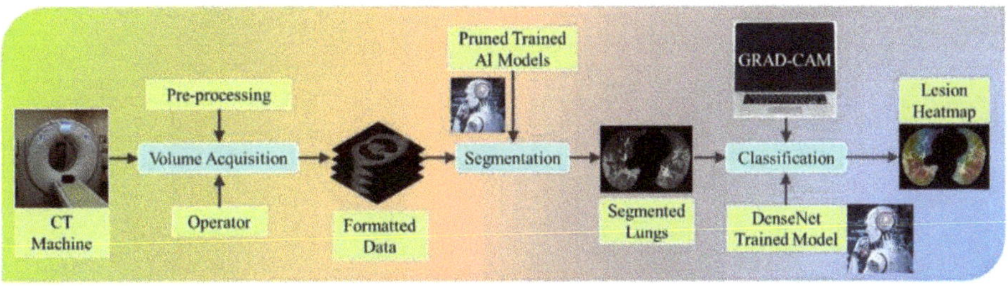

Figure 3.31. COVLIAS 2.0 lesion heatmap pipeline using DenseNet model using the infrastructure of pruned AI models during lung segmentation.

image for concept prediction. By analyzing the gradients of each target concept flowing into the final convolutional layer, Grad-CAM generates a coarse localization map that identifies the key areas in the image relevant to the concept being predicted. To obtain the Grad-CAM heatmap for a specific model, an input image is first provided. Then, the gradient of the output of the chosen model layer is computed with respect to the model's loss. The regions contributing to the gradient are subsequently extracted, resized, and scaled to generate the heatmap, which can be overlaid onto the original image. The ReLU activation function is commonly used in this process as it emphasizes features that have a positive impact on the target class. DenseNet-121 (figure 3.33) is composed of 120 convolutions and four Avg-Pool layers. One of the key characteristics of DenseNet-121 is the presence of transition layers in each layer, including those within a single thick block. These transition layers distribute their weights among multiple inputs, enabling deeper layers to leverage features extracted earlier in the process. This architecture of DenseNet allows for more compact models, achieves state-of-the-art performance, and demonstrates superior results when applied to various datasets compared to traditional ResNet or CNN equivalents. Furthermore, DenseNets require fewer parameters while facilitating feature reuse [119–121]. In figures 3.34–3.36, the significant areas that distinguish COVID-19 CT scans from Control CT scans are highlighted.

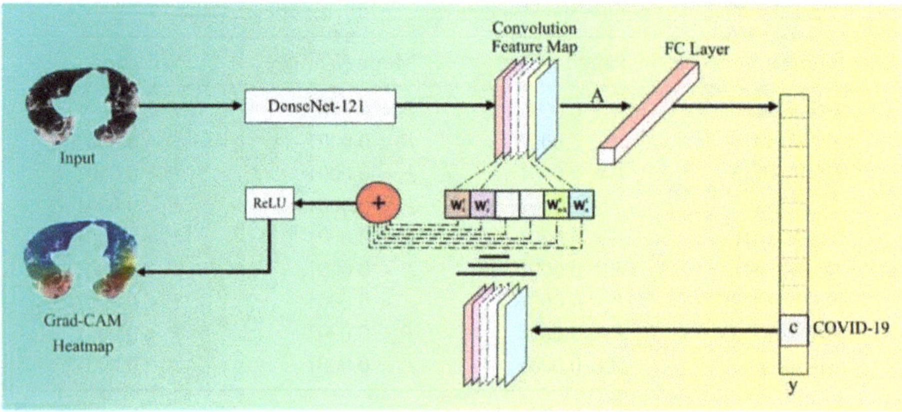

Figure 3.32. Grad-CAM process using DenseNet-121 model utilized for lesion localization.

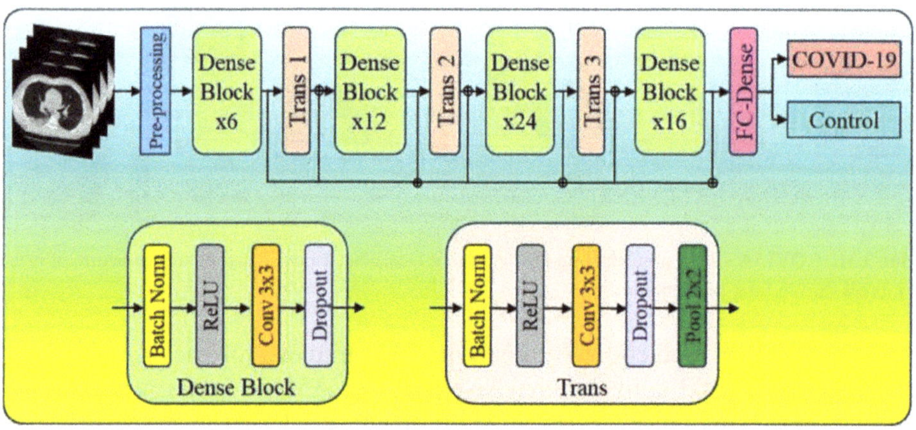

Figure 3.33. DenseNet-121 model utilized for lesion localization.

3.6 Discussion

3.6.1 Major contributions and study findings

The proposed study introduced eight pruned AI algorithms for CT lung segmentation, namely FCN-GA, FCN-WO, SegNet-GA, FCN-DE, SegNet-DE, FCN-PSO, SegNet-WO, and SegNet-PSO. These algorithms were trained using a dataset of 5000 COVID-19 CT images from 80 patients in the CroMed dataset. The preprocessing of the CroMed dataset involved modifying the Hounsfield unit (HU) range to focus on the lung area (1600, 400), which enhanced the training of the models. The second major contribution of the study was the validation of the pruned AI models using the Unseen NovMed dataset, which consisted of 4000 CT scans from 72 patients. The Unseen NovMed dataset was annotated by senior radiologists, similar to the CroMed dataset. This validation process aimed to assess the performance of the models on previously unseen data. The third novelty of the study was the design of lesion localization, which involved superimposing heatmaps over the segmented lungs generated by the pruned AI models. The heatmaps highlighted the presence of lesions by displaying them in red. This approach utilized the Grad-CAM heatmap technique, which was originally developed for generating the classification model using DenseNet-121 (figures 3.33–3.35). By applying Grad-CAM to the grayscale lung images, the regions of interest corresponding to the lesions were visually emphasized.

Overall, the study contributed novel pruned AI algorithms for CT lung segmentation, validated their performance on unseen data, and introduced a lesion localization method using Grad-CAM heatmaps on the segmented lungs. In addition to the previously mentioned contributions, the study also made a fourth significant contribution by comparing the evolutionary-based COVLIAS 2.0 to MedSeg, a web-based lung segmentation program [41]. This comparison was conducted using a low-storage and high-speed infrastructure, which was a novel approach in this context. The suggested AI models in COVLIAS 2.0 were evaluated for their clinical relevance by comparing their performance to MedSeg. The evaluation considered metrics such as Dice Similarity (DS), Jaccard Index (JI), and Correlation Coefficient (CC). The results showed a 5% difference in these metrics between the COVLIAS 2.0 AI models and MedSeg, indicating the clinical relevance of the suggested models. Tables 3.3–3.5 in the study provided a summary of the performance metrics and the percentage improvement achieved by each of the eight pruned AI models in terms of CC, JI, and DS values. These tables offer a comprehensive overview of the performance comparison between the different models and highlight the improvements achieved by the COVLIAS 2.0 AI models. Overall, the study's fourth contribution involved comparing COVLIAS 2.0 to MedSeg, assessing the clinical relevance of the suggested AI models, and presenting a summary of their performance and improvement in DS, JI, and CC values in tables 3.3–3.5.

Our fifth significant contribution in the study focused on the storage reductions achieved by the pruned AI models compared to their base models. Specifically, the pruned AI models, including FCN-WO, FCN-PSO, FCN-GA, and FCN-DE, showed storage reductions of 100%, 99%, 95%, and 92%, respectively, when compared to the base FCN model. Similarly, SegNet-WO, SegNet-PSO, SegNet-DE, and SegNet-GA exhibited storage reductions of 99%, 99%, 98%, and 97%, respectively, compared to the standard SegNet model. These significant storage reductions validate our theory and demonstrate the exceptional clinical abilities of COVLIAS 2.0. To further support the dependability and stability of COVLIAS 2.0, we conducted rigorous clinical and statistical studies. Our data analysis encompassed a comprehensive range of assessment metrics, including ROC, AE, CC, BA, JI, and DS plots, ensuring a thorough 360-degree evaluation of the system. Through these studies and analyses, we were able to establish the reliability and stability of COVLIAS 2.0, further validating our theory and highlighting its potential clinical applications.

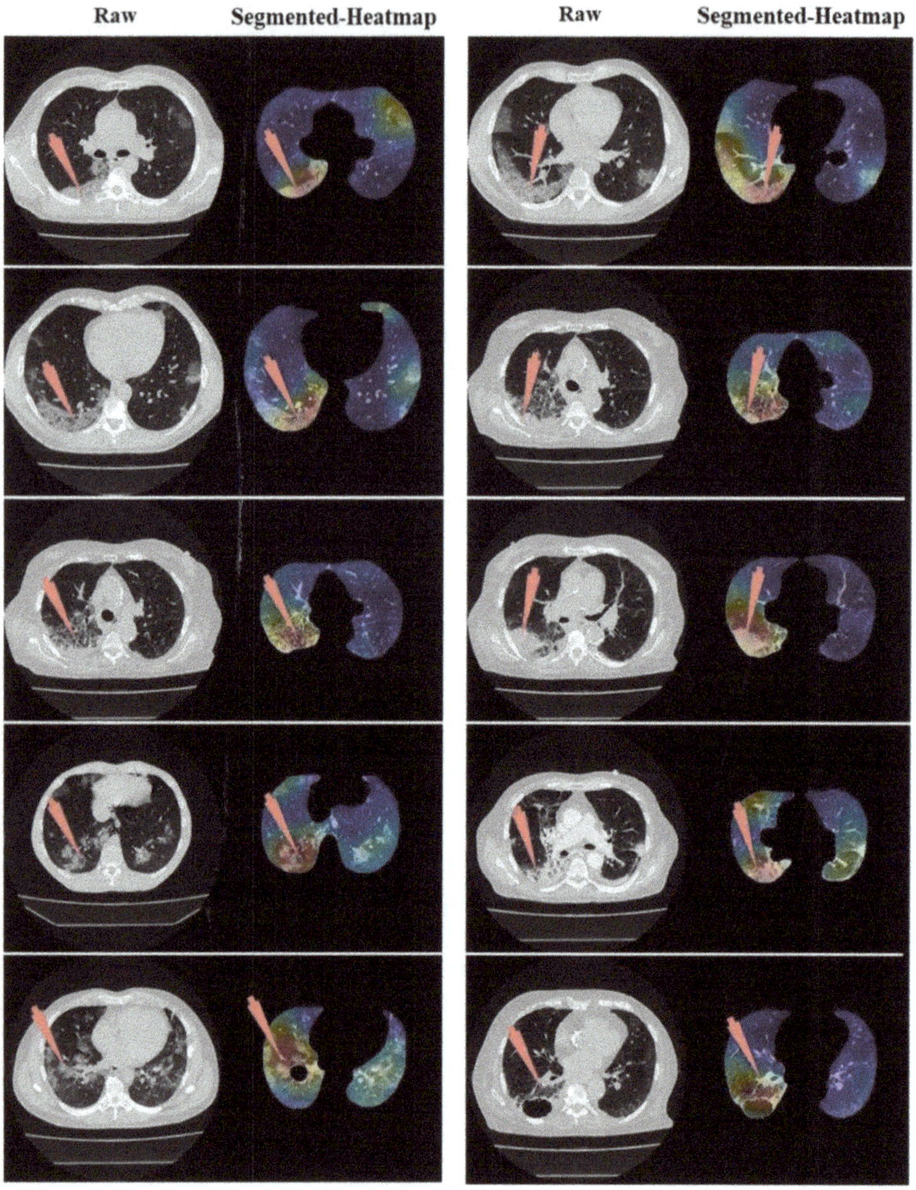

Figure 3.34. Left: Raw grayscale CT slice. Right: Segmented lungs with colored heatmap where red arrow indicates the right lung damage only.

3.6.2 Benchmarking

In our research, we encountered a limited number of publications specifically focused on the segmentation of CT lungs using pruning techniques. However, we found that pruning methods have been widely employed in various papers for the categorization of COVID-19 pneumonia on x-ray images. Despite this, we decided to incorporate CT lung segmentation in our benchmarking approach, drawing

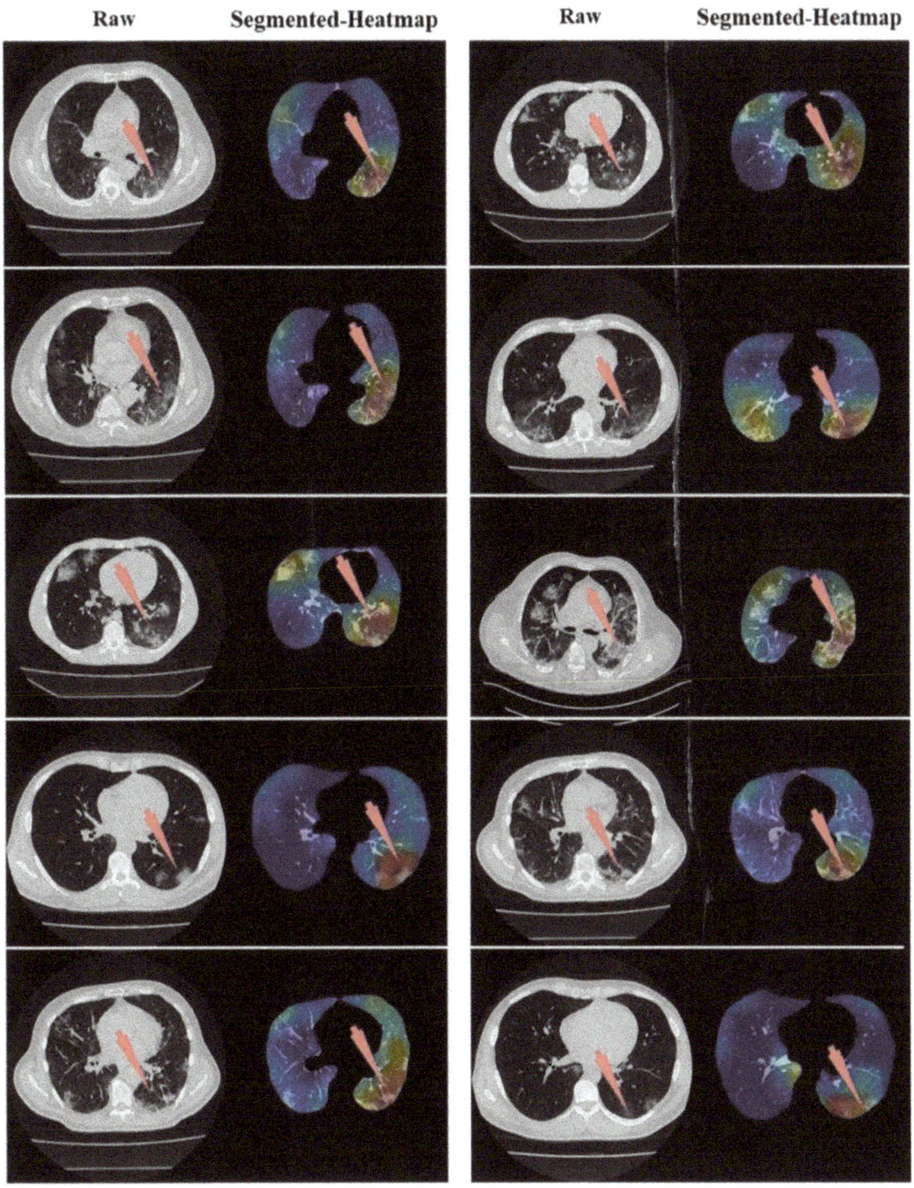

Figure 3.35. Left: Raw grayscale CT slice. Right: Segmented lungs with heatmap where red arrow indicated the left lung damage.

inspiration from studies such as Jiang *et al* [82] and Kogilavani *et al* [84], which utilized lung classification techniques with models like VGG16 [85] and DenseNet [86]. Furthermore, we referred to the dataset presented in Ref [92], where Paluru *et al* [91] developed a CT-based COVID-19 detection model called AnamNet. We also considered the works of Saood *et al* [97] and Cai *et al* [96] among others. It is worth mentioning that these studies did not provide detailed analysis such as Jaccard Index (JI)

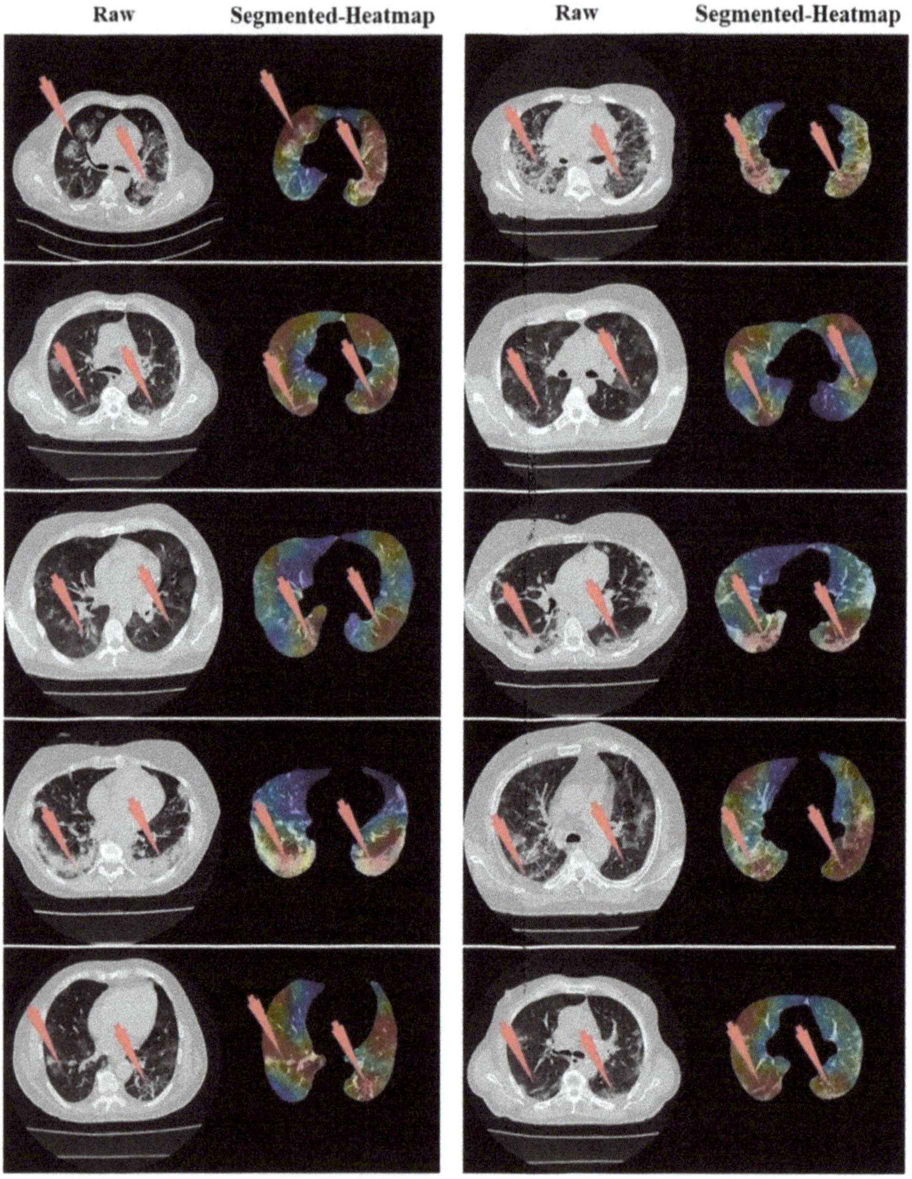

Figure 3.36. Left: Raw grayscale CT slice. Right: Segmented lungs with heatmap where red arrow indicated the damage in both left and right lungs.

or Boundary Accuracy (BA) plots, nor did they compare lung area errors. Interestingly, the authors in these studies reduced the size of the training models without employing any specific model pruning techniques, which resulted in shorter running times. While the focus of our study was on CT lung segmentation, we acknowledge the importance of these previous works in the field of COVID-19 detection and classification using various techniques and models (table 3.10).

Table 3.10. Benchmarking table.

	C1	C2	C3	C4	C5	C6	C7	C8	C9	C10	C11	C12	C13	C14
R#	Author	# Patients	# Images	Image Dim	Model Types	Classification vs Segmentation	Pruning	Model type	Dim	AE	DS	JI	BA	ACC
R1	Jiang *et al*	—	1168	5122	VGG16	Classification	×	SDL	2D	×	×	×	×	0.94
					ResNet-50									0.95
					Inception v3									0.96
					Inception ResNet v2									0.96
					DenseNet-169									0.99
R2	Kogilavani *et al*	—	~3873	2242	VGG16	Classification	×	SDL	2D	×	×	×	×	0.97
					MobileNet									0.96
					Densenet-121									0.97
					Xception									0.92
					Efficientnet									0.8
					NASNet									0.89
R3	Paluru *et al*	69	~4339	5122	AnamNet	Segmentation	×	SDL	2D	×	0.75	×	×	0.99
R4	Saood *et al*	—	~100	2562	UNet	Segmentation	×	SDL	2D	×	0.73	×	×	0.91
					SegNet						0.74			0.95
R5	Cai *et al*	99	~250	—	UNet	Segmentation	×	SDL	2D	×	0.98	0.96	×	×
R6	**Suri *et al* (Proposed)**	**~152**	**~9000**	**5122**	**FCN**	**Segmentation**	✓	**SDL**	**2D**	✓	**0.78**	**0.65**	✓	**0.96**
					FCN-DE						**0.93**	**0.88**		**0.97**
					FCN-GA						**0.93**	**0.87**		**0.97**
					FCN–PSO						**0.92**	**0.86**		**0.97**
					FCN-WO						**0.94**	**0.89**		**0.98**
					SegNet						**0.96**	**0.93**		**0.98**
					SegNet-DE						**0.96**	**0.92**		**0.98**
					SegNet-GA						**0.96**	**0.93**		**0.98**
					SegNet-PSO SegNet-WO						**0.96**	**0.94**		**0.98**
											0.96	**0.94**		**0.99**

3.6.3 Strengths, weakness, and extensions

According to our study, the application of the eight pruning strategies, namely SegNet-WO, SegNet-PSO, SegNet-GA, SegNet-DE, FCN-WO, FCN-PSO, FCN-GA, and FCN-DE, resulted in a remarkable reduction of storage by approximately 97%. This finding supports our hypothesis that pruning techniques effectively decrease the size of AI models while maintaining their performance at clinically acceptable levels, making the system fast and efficient. The overall performance of the AI models remained unaffected by the trimming methods and exhibited significant advantages. Firstly, it demonstrated a five-fold increase in speed compared to the base models. Secondly, the system was validated using the 'Unseen NovMed' dataset, ensuring its reliability and generalization capability. Thirdly, the lesion localization technique produced impressive results, and the assessment by board-certified radiologists further confirmed its effectiveness. Lastly, the system exhibited impressive clinical statistical outcomes. The results of our pilot study are encouraging and hold promise. Further investigations are warranted, particularly regarding the effectiveness of hybrid deep learning networks in classification and segmentation tasks. We plan to explore non-statistical or statistical pruning and optimization methods, as well as the potential benefits of UNet's recent advancements and the utilization of big data frameworks for incorporating data from multiple sources. Additionally, we intend to evaluate the capabilities of UNet-EA, FCN-EA, and SegNet-EA by comparing them within the expanded scope of our study.

It is important to note that while we applied Hounsfield Unit (HU) modifications to magnify the lung region in CT images, alternative techniques such as noise reduction or a combination of window-level and noise smoothing could also be employed [122–126]. Obtaining the lesion region using traditional methods like stochastic segmentation or level sets, followed by appropriate adjustments to the loss function, could be explored as well. Furthermore, considering the lack of clinical validations, the use of AI systems tends to result in better accuracy, and thus, it may be beneficial to incorporate AI bias calculations, such as AP (ai) Bias, to assess the potential biases associated with AI algorithms [127–129].

3.7 Conclusions

Our project represents a pioneering pilot study that combines two distinct deep learning paradigms with eight different evolutionary approaches. The primary objective of this study was to develop eight specialized techniques for COVID-19 CT lung segmentation that offer minimal storage requirements and high-speed performance. Remarkably, despite reducing storage by 97%, COVLIAS 2.0 successfully maintained performance levels. Compared to the basic models, the eight pruning techniques exhibited an average speed improvement of approximately five times. In order to validate our hypothesis, we conducted a comparison between the open-source web-based lung segmentation tool, MedSeg, and our eight trimmed AI models. The results confirmed the clinical applicability of our hypothesis, with the overall system error rate being less than 6%. The online COVLIAS 2.0 system demonstrated an impressive processing time of approximately 0.25 seconds. This reliability and speed make COVLIAS 2.0 suitable for real-world clinical settings, where accuracy and stability are of utmost importance. The findings of this study highlight the effectiveness and dependability of COVLIAS 2.0 as a robust solution for COVID-19 CT lung segmentation [130].

References

[1] Cucinotta D and Vanelli M 2020 WHO declares COVID-19 a pandemic *Acta Biomed. Atenei Parm.* **91** 157–60

[2] WHO coronavirus (COVID-19) dashboard, (https://covid19.who.int/) (24 Janurary 2022

[3] Saba L *et al* 2020 Molecular pathways triggered by COVID-19 in different organs: ACE2 receptor-expressing cells under attack? A review *Eur. Rev. Med. Pharmacol. Sci.* **24** 12609–22

[4] Suri J S *et al* 2020 COVID-19 pathways for brain and heart injury in comorbidity patients: a role of medical imaging and artificial intelligence-based COVID severity classification: a review *Comput. Biol. Med.* **124** 103960

[5] Cau R, Bassareo P P, Mannelli L, Suri J S and Saba L 2021 Imaging in COVID-19-related myocardial injury *Int. J. Cardiovasc. Imaging.* **37** 1349–60

[6] Onnis C *et al* 2022 Non-invasive coronary imaging in patients with COVID-19: a narrative review *Eur. J. Radiol.* **149** 110188

[7] Viswanathan V, Puvvula A, Jamthikar A D, Saba L, Johri A M, Kotsis V, Khanna N N, Dhanjil S K, Majhail M and Misra D P 2021 Bidirectional link between diabetes mellitus and coronavirus disease 2019 leading to cardiovascular disease: a narrative review *World J. Diabetes* **12** 215

[8] Fanni D *et al* 2021 Vaccine-induced severe thrombotic thrombocytopenia following COVID-19 vaccination: a report of an autoptic case and review of the literature *Eur. Rev. Med. Pharmacol. Sci.* **25** 5063–9

[9] Gerosa C, Faa G, Fanni D, Manchia M, Suri J, Ravarino A, Barcellona D, Pichiri G, Coni P and Congiu T 2021 Fetal programming of COVID-19: may the Barker hypothesis explain the susceptibility of a subset of young adults to develop severe disease? *Eur. Rev. Med. Pharmacol Sci.* **25** 5876–84

[10] Congiu T, Demontis R, Cau F, Piras M, Fanni D, Gerosa C, Botta C, Scano A, Chighine A and Faedda E 2021 Scanning electron microscopy of lung disease due to COVID-19-a case report and a review of the literature *Eur. Rev. Med Pharmacol. Sci.* **25** 7997–8003

[11] Suri J S and Laxminarayan S 2003 *Angiography and Plaque Imaging: Advanced Segmentation Techniques* (Boca Raton, FL: CRC Press)

[12] Faa G, Gerosa C, Fanni D, Barcellona D, Cerrone G, Orrù G, Scano A, Marongiu F, Suri J and Demontis R 2021 Aortic vulnerability to COVID-19: is the microvasculature of vasa vasorum a key factor? A case report and a review of the literature *Eur. Rev. Med. Pharmacol. Sci.* **25** 6439–42

[13] Munjral S, Ahluwalia P, Jamthikar A D, Puvvula A, Saba L, Faa G, Singh I M, Chadha P S, Turk M and Johri A M 2021 Nutrition, atherosclerosis, arterial imaging, cardiovascular risk stratification, and manifestations in COVID-19 framework: a narrative review *Front. Biosci. (Landmark Edition)* **26** 1312–39

[14] Congiu T, Fanni D, Piras M, Gerosa C, Cau F, Barcellona D, D'Aloja E, Demontis R, Chighine F and Nioi M 2022 Ultrastructural findings of lung injury due to vaccine-induced

immune thrombotic thrombo-cytopenia (VITT) following COVID-19 vaccination: a scanning electron microscopic study *Eur. Rev. Med. Pharmacol Sci.* **26** 270–7

[15] Fang Y, Zhang H, Xie J, Lin M, Ying L, Pang P and Ji W 2020 Sensitivity of chest CT for COVID-19: comparison to RT-PCR *Radiology* **296** E115–7

[16] Dram´e M, Teguo M T, Proye E, Hequet F, Hentzien M, Kanagaratnam L and Godaert L 2020 Should RT-PCR be considered a gold standard in the diagnosis of COVID-19? *J. Med. Virol.* **92** 2312–3

[17] Xiao A T, Tong Y X and Zhang S 2020 False negative of RT-PCR and prolonged nucleic acid conversion in COVID-19: rather than recurrence *J. Medi. Virol.* **92** 1755–6

[18] Suri J S, Agarwal S, Gupta S K, Puvvula A, Biswas M, Saba L, Bit A, Tandel G S, Agarwal M and Patrick A 2021 A narrative review on characterization of acute respiratory distress syndrome in COVID-19-infected lungs using artificial intelligence *Comput. Biol. Med.* **130** 104210

[19] Biswas M *et al* 2019 State-of-the-art review on deep learning in medical imaging *Front. Biosci. (Landmark Ed)* **24** 392–426

[20] Saba L *et al* 2019 The present and future of deep learning in radiology *Eur. J. Radiol.* **114** 14–24

[21] Kuppili V, Biswas M, Sreekumar A, Suri H S, Saba L, Edla D R, Marinhoe R T, Sanches J M and Suri J S 2017 Extreme learning machine framework for risk stratification of fatty liver disease using ultrasound tissue characterization *J. Med. Syst.* **41** 1–20

[22] Chen X, Tang Y, Mo Y, Li S, Lin D, Yang Z, Yang Z, Sun H, Qiu J and Liao Y 2020 A diagnostic model for coronavirus disease 2019 (COVID-19) based on radiological semantic and clinical features: a multi-center study *Eur. Radiol.* **30** 4893–902

[23] Jain P K, Sharma N, Giannopoulos A A, Saba L, Nicolaides A and Suri J S 2021 Hybrid deep learning segmentation models for atherosclerotic plaque in internal carotid artery B-mode ultrasound *Comput. Biol. Med.* **136** 104721

[24] Jena B, Saxena S, Nayak G K, Saba L, Sharma N and Suri J S 2021 Artificial intelligence-based hybrid deep learning models for image classification: the first narrative review *Comput. Biol. Med.* **137** 104803

[25] Suri J S, Agarwal S, Pathak R, Ketireddy V, Columbu M, Saba L, Gupta S K, Faa G, Singh I M and Turk M 2021 Covlias 1.0: lung segmentation in COVID-19 computed tomography scans using hybrid deep learning artificial intelligence models *Diagnostics* **11** 1405

[26] Jain P K, Sharma N, Saba L, Paraskevas K I, Kalra M K, Johri A, Nicolaides A N and Suri J S 2022 Automated deep learning-based paradigm for high-risk plaque detection in B-mode common carotid ultrasound scans: an asymptomatic Japanese cohort study *Int. Angiol.* **41** 9–23

[27] Suri J S, Agarwal S, Carriero A, Pasch'e A, Danna P S, Columbu M, Saba L, Viskovic K, Mehmedovi´c A and Agarwal S 2021 COVLIAS 1.0 vs. MedSeg: artificial intelligence-based comparative study for automated COVID-19 computed tomography lung segmentation in Italian and Croatian cohorts *Diagnostics* **11** 2367

[28] Suri J S *et al* 2021 Inter-variability study of COVLIAS 1.0: hybrid deep learning models for COVID-19 lung segmentation in computed tomography *Diagnostics (Basel)* **11** 2025

[29] Skandha S S, Nicolaides A, Gupta S K, Koppula V K, Saba L, Johri A M, Kalra M S and Suri J S 2022 A hybrid deep learning paradigm for carotid plaque tissue characterization

and its validation in multicenter cohorts using a supercomputer framework *Comput. Biol. Med.* **141** 105131
[30] Gupta N, Gupta S K, Pathak R K, Jain V, Rashidi P and Suri J S 2022 Human activity recognition in artificial intelligence framework: a narrative review *Artif. Intell. Rev.* **55** 4755–808
[31] Saba L, Biswas M, Suri H S, Viskovic K, Laird J R, Cuadrado-Godia E, Nicolaides A, Khanna N N, Viswanathan V and Suri J S 2019 Ultrasound-based carotid stenosis measurement and risk stratification in diabetic cohort: a deep learning paradigm *Cardiovasc. Diagn. Ther.* **9** 439–61
[32] Agarwal M, Saba L, Gupta S K, Johri A M, Khanna N N, Mavrogeni S, Laird J R, Pareek G, Miner M and Sfikakis P P 2021 Wilson disease tissue classification and characterization using seven artificial intelligence models embedded with 3D optimization paradigm on a weak training brain magnetic resonance imaging datasets: a supercomputer application *Med. Biol. Eng. Comput.* **59** 511–33
[33] Sanagala S S, Nicolaides A, Gupta S K, Koppula V K, Saba L, Agarwal S, Johri A M, Kalra M S and Suri J S 2021 Ten fast transfer learning models for carotid ultrasound plaque tissue characterization in augmentation framework embedded with heatmaps for stroke risk stratification *Diagnostics* **11** 2109
[34] LeCun Y, Denker J and Solla S 1989 Optimal brain damage *Adv. Neural Inf. Process. Systems* 2
[35] Zhu M and Gupta S 2017 To prune, or not to prune: exploring the efficacy of pruning for model compression Arxiv Preprint arXiv:1710.01878
[36] Band S S, Janizadeh S, Chandra Pal S, Saha A, Chakrabortty R, Shokri M and Mosavi A 2020 Novel ensemble approach of deep learning neural network (DLNN) model and particle swarm optimization (PSO) algorithm for prediction of gully erosion susceptibility *Sensors* **20** 5609
[37] Brodzicki A, Piekarski M and Jaworek-Korjakowska J 2021 The whale optimization algorithm approach for deep neural networks *Sensors* **21** 8003
[38] Ashraf N M, Mostafa R R, Sakr R H and Rashad M 2021 Optimizing hyperparameters of deep reinforcement learning for autonomous driving based on whale optimization algorithm *PLoS One* **16** e0252754
[39] Acharya U R, Mookiah M R, Vinitha Sree S, Yanti R, Martis R J, Saba L, Molinari F, Guerriero S and Suri J S 2014 Evolutionary algorithm-based classifier parameter tuning for automatic ovarian cancer tissue characterization and classification *Ultraschall Med.* **35** 237–45
[40] Horry M, Chakraborty S, Pradhan B, Paul M, Zhu J, Loh H W, Barua P D and Arharya U R 2022 Debiasing pipeline improves deep learning model generalization for X-ray based lung nodule detection Arxiv Preprint arXiv:2201.09563
[41] MedSeg 2022 https://htmlsegmentation.s3.eu-north-1.amazonaws.com/index.html
[42] Eelbode T, Bertels J, Berman M, Vandermeulen D, Maes F, Bisschops R and Blaschko M B 2020 Optimization for medical image segmentation: theory and practice when evaluating with dice score or jaccard index *IEEE Trans. Med. Imaging* **39** 3679–90
[43] Giavarina D 2015 Understanding bland Altman analysis *Biochem. Med.* **25** 141–51
[44] Dewitte K, Fierens C, Stockl D and Thienpont L M 2002 Application of the bland–altman plot for interpretation of method-comparison studies: a critical investigation of its practice *Clin. Chem.* **48** 799–801

[45] Asuero A G, Sayago A and Gonzalez A 2006 The correlation coefficient: an overview *Crit. Rev. Anal. Chem.* **36** 41–59

[46] Taylor R 1990 Interpretation of the correlation coefficient: a basic review *J. Diagn. Med. Sonogr.* **6** 35–9

[47] El-Baz A, Gimel'farb G and Suri J S 2015 *Stochastic Modeling for Medical Image Analysis* 1st edn (Boca Raton, FL: CRC Press)

[48] El-Baz A, Jiang X and Suri J S 2016 *Biomedical Image Segmentation: Advances and Trends* (Boca Raton, FL: CRC Press)

[49] El-Baz A S, Acharya R, Mirmehdi M and Suri J S 2011 *Multi Modality State-Of-The-Art Medical Image Segmentation and Registration Methodologies* vol 2 (Berlin: Springer Science & Business Media)

[50] Maniruzzaman M, Rahman M J, Al-MehediHasan M, Suri H S, Abedin M M, El-Baz A and Suri J S 2018 Accurate diabetes risk stratification using machine learning: role of missing value and outliers *J. Med. Syst.* **42** 1–17

[51] Maniruzzaman M, Kumar N, Abedin M M, Islam M S, Suri H S, El-Baz A S and Suri J S 2017 Comparative approaches for classification of diabetes mellitus data: machine learning paradigm *Comput. Methods Programs Biomed.* **152** 23–34

[52] Maniruzzaman M, Suri H S, Kumar N, Abedin M M, Rahman M J, El-Baz A, Bhoot M, Teji J S and Suri J S 2018 Risk factors of neonatal mortality and child mortality in Bangladesh *J. Glob. Health* **8**

[53] Maniruzzaman M, Jahanur Rahman M, Ahammed B, Abedin M M, Suri H S, Biswas M, El-Baz A, Bangeas P, Tsoulfas G and Suri J S 2019 Statistical characterization and classification of colon microarray gene expression data using multiple machine learning paradigms *Comput. Methods Programs Biomed.* **176** 173–93

[54] Noor N M, Than J C, Rijal O M, Kassim R M, Yunus A, Zeki A A, Anzidei M, Saba L and Suri J S 2015 Automatic lung segmentation using control feedback system: morphology and texture paradigm *J. Med. Syst.* **39** 1–18

[55] Acharya R U, Faust O, Alvin A P, Sree S V, Molinari F, Saba L, Nicolaides A and Suri J S 2012 Symptomatic vs. asymptomatic plaque classification in carotid ultrasound *J. Med. Syst.* **36** 1861–71

[56] Acharya U R, Faust O, Alvin V S S A P, Krishnamurthi G, Seabra J C, Sanches J and Suri J S 2013 Understanding symptomatology of atherosclerotic plaque by image-based tissue characterization *Comput. Methods Programs Biomed.* **110** 66–75

[57] Acharya U R, Faust O, Sree S V, Alvin A P C, Krishnamurthi G, Sanches J and Suri J S 2011 Atheromatic™: symptomatic vs. asymptomatic classification of carotid ultrasound plaque using a combination of HOS, DWT and texture *2011 Annual Int. Conf. of the IEEE Engineering in Medicine and Biology Society* (Piscataway, NJ: IEEE) pp 4489–92

[58] Acharya U R, Mookiah M R, Vinitha Sree S, Afonso D, Sanches J, Shafique S, Nicolaides A, Pedro L M, Fernandes E F J and Suri J S 2013 Atherosclerotic plaque tissue characterization in 2D ultrasound longitudinal carotid scans for automated classification: a paradigm for stroke risk assessment *Med. Biol. Eng. Comput.* **51** 513–23

[59] Molinari F, Liboni W, Pavanelli E, Giustetto P, Badalamenti S and Suri J S 2007 Accurate and automatic carotid plaque characterization in contrast enhanced 2-D ultrasound images *2007 29th Annual Int. Conf. of the IEEE Engineering in Medicine and Biology Society* (Piscataway, NJ: IEEE) pp 335–8

[60] Acharya U, Vinitha Sree S, Mookiah M, Yantri R, Molinari F, Ziele´znik W, Małyszek-Tumidajewicz J, Stępie´n B, Bardales R and Witkowska A 2013 Diagnosis of Hashimoto's thyroiditis in ultrasound using tissue characterization and pixel classification *Proc. Inst. Mech. Eng. Part H J. Eng. Med.* **227** 788–98

[61] Biswas M, Kuppili V, Edla D R, Suri H S, Saba L, Marinhoe R T, Sanches J M and Suri J S 2018 Symtosis: a liver ultrasound tissue characterization and risk stratification in optimized deep learning paradigm *Comput. Methods Programs Biomed.* **155** 165–77

[62] Saba L *et al* 2021 Multimodality carotid plaque tissue characterization and classification in the artificial intelligence paradigm: a narrative review for stroke application *Ann. Transl. Med.* **9** 1206

[63] Banchhor S K, Londhe N D, Araki T, Saba L, Radeva P, Laird J R and Suri J S 2017 Wall-based measurement features provides an improved IVUS coronary artery risk assessment when fused with plaque texture-based features during machine learning paradigm *Comput. Biol. Med.* **91** 198–212

[64] Acharya U R, Saba L, Molinari F, Guerriero S and Suri J S 2012 Ovarian tumor characterization and classification: a class of GyneScan™ systems *2012 Annual Int. Conf. of the IEEE Engineering in Medicine and Biology Society* (Piscataway, NJ: IEEE) pp 4446–9

[65] Pareek G, Acharya U R, Sree S V, Swapna G, Yantri R, Martis R J, Saba L, Krishnamurthi G, Mallarini G and El-Baz A 2013 Prostate tissue characterization/classification in 144 patient population using wavelet and higher order spectra features from transrectal ultrasound images *Technol. Cancer Res. Treat.* **12** 545–57

[66] Shrivastava V K, Londhe N D, Sonawane R S and Suri J S 2015 Exploring the color feature power for psoriasis risk stratification and classification: a data mining paradigm *Comput. Biol. Med.* **65** 54–68

[67] Shrivastava V K, Londhe N D, Sonawane R S and Suri J S 2017 A novel and robust Bayesian approach for segmentation of psoriasis lesions and its risk stratification *Comput. Methods Prog. Biomed.* **150** 9–22

[68] Shrivastava V K, Londhe N D, Sonawane R S and Suri J S 2015 Reliable and accurate psoriasis disease classification in dermatology images using comprehensive feature space in machine learning paradigm *Expert Syst. Appl.* **42** 6184–95

[69] Acharya U R, Kannathal N, Ng E, Min L C and Suri J S 2006 Computer-based classification of eye diseases *2006 Int. Conf. of the IEEE Engineering in Medicine and Biology Society* (Piscataway, NJ: IEEE) pp 6121–4

[70] Saba L and Suri J S 2013 *Multi-Detector CT Imaging: Principles, Head, Neck, and Vascular Systems* (Boca Raton, FL: CRC Press)

[71] Murgia A, Erta M, Suri J S, Gupta A, Wintermark M and Saba L 2020 CT imaging features of carotid artery plaque vulnerability *Ann. Transl. Med.* **8**

[72] Saba L, di Martino M, Siotto P, Anzidei M, Argiolas G M, Porcu M, Suri J S and Wintermark M 2018 Radiation dose and image quality of computed tomography of the supra-aortic arteries: a comparison between single-source and dual-source CT scanners *J. Neuroradiol.* **45** 136–41

[73] Wu J, Pan J, Teng D, Xu X, Feng J and Chen Y -C 2020 Interpretation of CT signs of 2019 novel coronavirus (COVID-19) pneumonia *Eur. Radiol.* **30** 5455–62

[74] De Wever W, Meersschaert J, Coolen J, Verbeken E and Verschakelen J A 2011 The crazy-paving pattern: a radiological-pathological correlation *Insights into Imaging* **2** 117–32

[75] Niu R, Ye S, Li Y, Ma H, Xie X, Hu S, Huang X, Ou Y and Chen J 2021 Chest CT features associated with the clinical characteristics of patients with COVID-19 pneumonia *Ann. Med.* **53** 169–80
[76] Salehi S, Abedi A, Balakrishnan S and Gholamrezanezhad A 2020 Coronavirus disease 2019 (COVID-19): a systematic review of imaging findings in 919 patients *AJR Am. J. Roentgenol* **215** 87–93
[77] Xie X, Zhong Z, Zhao W, Zheng C, Wang F and Liu J 2020 Chest CT for typical coronavirus disease 2019 (COVID-19) pneumonia: relationship to negative RT-PCR testing *Radiology* **296** E41–5
[78] Gozes O, Frid-Adar M, Greenspan H, Browning P D, Zhang H, Ji W, Bernheim A and Siegel E 2020 Rapid AI development cycle for the coronavirus (Covid-19) pandemic: initial results for automated detection and patient monitoring using deep learning CT image analysis Arxiv Preprint arXiv:2003.05037
[79] Shalbaf A and Vafaeezadeh M 2021 Automated detection of COVID-19 using ensemble of transfer learning with deep convolutional neural network based on CT scans *Int. J. Comput. Assist. Radiol. Surg.* **16** 115–23
[80] Yang X, He X, Zhao J, Zhang Y, Zhang S and Xie P 2020 COVID-CT-dataset: a CT Scan Dataset about COVID-19 arXiv Preprint arXiv:2003.13865
[81] Cau R *et al* 2021 Computed tomography findings of COVID-19 pneumonia in intensive care unit-patients *J. Public Health Res.* **10**
[82] Yang X, He X, Zhao J, Zhang Y, Zhang S and Xie P 2020 COVID-CT-dataset: a CT Scan Dataset about COVID-19 arXiv Preprint arXiv:2003.13865
[83] Setio A A A, Traverso A, De Bel T, Berens M S, Van Den Bogaard C, Cerello P, Chen H, Dou Q, Fantacci M E and Geurts B 2017 Validation, comparison, and combination of algorithms for automatic detection of pulmonary nodules in computed tomography images: the LUNA16 challenge *Med. Image Anal.* **42** 1–13
[84] Kogilavani S, Prabhu J, Sandhiya R, Kumar M S, Subramaniam U, Karthick A, Muhibbullah M and Imam S B S 2022 COVID-19 detection based on lung ct scan using deep learning techniques *Comput. Math. Methods Med.* **2022** 7672196
[85] Simonyan K and Zisserman A 2014 very deep convolutional networks for large-scale image recognition arXiv preprint arXiv:
[86] Iandola F, Moskewicz M, Karayev S, Girshick R, Darrell T and Keutzer K 2014 Densenet: implementing efficient convnet descriptor pyramids arXiv Preprint arXiv:1404.1869
[87] Howard A G, Zhu M, Chen B, Kalenichenko D, Wang W, Weyand T, Andreetto M and Adam H 2017 Mobilenets, efficient convolutional neural networks for mobile vision applications arXiv Preprint arXiv:1704.04861
[88] Chollet F 2017 Xception: deep learning with depthwise separable convolutions *Proc. of the IEEE Conf. on Computer Vision and Pattern Recognition* pp 1251–8
[89] Zoph B, Vasudevan V, Shlens J and Le Q V 2018 Learning transferable architectures for scalable image recognition *Proc. of the IEEE Conf. on Computer Vision and Pattern Recognition* pp 8697–710
[90] Tan M and Le Q 2019 Efficientnet: rethinking model scaling for convolutional neural networks *Int. Conf. on Machine Learning, PMLR* pp 6105–14
[91] Paluru N, Dayal A, Jenssen H B, Sakinis T, Cenkeramaddi L R, Prakash J, Yalavarthy P K and Systems L 2021 Anam-Net: anamorphic depth embedding-based lightweight CNN for

segmentation of anomalies in COVID-19 chest CT images *IEEE Transact. Neural Networks Learn. Syst.* **32** 932–46 3rd ed
[92] COVID-19 database, 10 February 2022, 2022 (https://radiopaedia.org/articles/covid-19-4?lang=gb)
[93] Paszke A, Chaurasia A, Kim S and Culurciello E 2016 Enet: a deep neural network architecture for real-time semantic segmentation arXiv preprint arXiv: 1606.02147
[94] Zhou Z, Siddiquee M M R, Tajbakhsh N and Liang J 2020 UNet++: redesigning skip connections to exploit multiscale features in image segmentation *IEEE Trans. Med. Imaging* **39** 1856–67
[95] Wang Y, Zhou Q, Liu J, Xiong J, Gao G, Wu X and Latecki L J 2019 Lednet: a lightweight encoder-decoder network for real-time semantic segmentation *2019 IEEE Int. Conf. on Image Processing (ICIP)* (Piscataway, NJ: IEEE) pp 1860–4
[96] Cai W, Liu T, Xue X, Luo G, Wang X, Shen Y, Fang Q, Sheng J, Chen F and Liang T 2020 CT quantification and machine-learning models for assessment of disease severity and prognosis of COVID-19 patients *Acad. Radiol.* **27** 1665–78
[97] Saood A and Hatem I 2021 COVID-19 lung CT image segmentation using deep learning methods: U-Net versus SegNet *BMC Med. Imaging* **21** 1–10
[98] Biswas M, Kuppili V, Saba L, Edla D R, Suri H S, Sharma A, Cuadrado- Godia E, Laird J R, Nicolaides A and Suri J S 2019 Deep learning fully convolution network for lumen characterization in diabetic patients using carotid ultrasound: a tool for stroke risk *Med. Biol. Eng. Comput.* **57** 543–64
[99] Badrinarayanan V, Kendall A and Cipolla R 2017 SegNet: a deep convolutional encoder-decoder architecture for image segmentation *IEEE Trans. Pattern Anal. Mach. Intell.* **39** 2481–95
[100] Fleetwood K 2004 An introduction to differential evolution *Proc. of Mathematics and Statistics of Complex Systems (MASCOS) One Day Symp.* pp 785–91 *(Brisbane, Australia)*
[101] Price K V 2013 *Differential Evolution, Handbook of Optimization* (Berlin: Springer) pp 187–214
[102] Singh D, Kumar V and Kaur M 2020 Classification of COVID-19 patients from chest CT images using multi-objective differential evolution–based convolutional neural networks *Eur. J. Clin. Microbiol. Infect Dis.* **39** 1379–89
[103] Bas¸türk A and Günay E 2009 Efficient edge detection in digital images using a cellular neural network optimized by differential evolution algorithm *Expert Syst. Appl.* **36** 2645–50
[104] Ruse M 1975 Charles Darwin's theory of evolution: an analysis *J. Hist. Biol.* **8** 219–41
[105] Kozek T, Roska T and Chua L O 1993 Genetic algorithm for CNN template learning *IEEE Trans. Circuits Syst.* I **40** 392–402
[106] Sun Y, Xue B, Zhang M, Yen G G and Lv J 2020 Automatically designing CNN architectures using the genetic algorithm for image classification *IEEE Trans. Cybern.* **50** 3840–54
[107] Kennedy J and Eberhart R 1995 Particle swarm optimization *Proc. of ICNN'95-Int. Conf. on Neural Networks* (Piscataway, NJ: IEEE) pp 1942–8
[108] Navaneeth B and Suchetha M 2019 PSO optimized 1-D CNN-SVM architecture for real-time detection and classification applications *Comput. Biol. Med.* **108** 85–92

[109] Wang Y, Zhang H and Zhang G 2019 cPSO-CNN, An efficient PSO-based algorithm for fine-tuning hyper-parameters of convolutional neural networks *Swarm. Evol. Comput.* **49** 114–23
[110] Mirjalili S and Lewis A 2016 The whale optimization algorithm *Adv. Eng. Software* **95** 51–67
[111] Dixit U, Mishra A, Shukla A and Tiwari R 2019 Texture classification using convolutional neural network optimized with whale optimization algorithm *SN Appl. Sci.* **1** 1–11
[112] Rana N, Latiff M S A, Abdulhamid S I M and Chiroma H 2020 Whale optimization algorithm: a systematic review of contemporary applications, modifications and developments *Neural Comput. Appl.* **32** 16245–77
[113] Yuan T, Liu W, Han J and Lombardi F 2020 High performance CNN accelerators based on hardware and algorithm co-optimization *IEEE Trans. Circuits Syst.* I **68** 250–63
[114] Zhang J, Raj P, Zarar S, Ambardekar A and Garg S 2019 CompAct: on-chip com pression of act ivations for low power systolic array based CNN acceleration *ACM Trans. Embed. Comput. Syst. (TECS)* **18** 1–24
[115] Saba L, Banchhor S K, Londhe N D, Araki T, Laird J R, Gupta A, Nicolaides A and Suri J S 2017 Web-based accurate measurements of carotid lumen diameter and stenosis severity: an ultrasound-based clinical tool for stroke risk assessment during multicenter clinical trials *Comput. Biol. Med.* **91** 306–17
[116] Saba L, Than J C, Noor N M, Rijal O M, Kassim R M, Yunus A, Ng C R and Suri J S 2016 Inter-observer variability analysis of automatic lung delineation in normal and disease patients *J. Med. Syst.* **40** 142
[117] Molinari F, Meiburger K M, Saba L, Acharya U R, Famiglietti L, Georgiou N, Nicolaides A, Mamidi R S, Kuper H and Suri J S 2014 Automated carotid IMT measurement and its validation in low contrast ultrasound database of 885 patient indian population epidemiological study: results of atheroedge® software *Multi-Modality Atherosclerosis Imaging and Diagnosis* (Berlin: Springer) pp 209–19
[118] Mirmehdi M 2008 *Handbook of Texture Analysis* (Imperial College Press)
[119] He G, Ping A, Wang X and Zhu Y 2019 Alzheimer's disease diagnosis model based on three-dimensional full convolutional DenseNet *2019 10th Int. Conf. on Information Technology in Medicine and Education (ITME)* (Piscataway, NJ: IEEE) pp 13–7
[120] Ouhami M, Es-Saady Y, Hajji M E, Hafiane A, Canals R and Yassa M E 2020 Deep Transfer learning models for tomato disease detection *Int. Conf. on Image and Signal Processing (Berlin)* (Springer) pp 65–73
[121] Ruiz J, Mahmud M, Modasshir M, Shamim Kaiser M and Alzheimer's Disease Neuroimaging Initiative 2020 3D densenet ensemble in 4-way classification of Alzheimer's disease *Int. Conf. on Brain Informatics* (Berlin: Springer) 85–96
[122] Saba L, Sanagala S S, Gupta S K, Koppula V K, Laird J R, Viswanathan V, Sanches M J, Kitas G D, Johri A M and Sharma N 2021 A multicenter study on carotid ultrasound plaque tissue characterization and classification using six deep artificial intelligence models: a stroke application *IEEE Trans. Instrum. Meas.* **70** 1–12
[123] Jain P K, Sharma N, Saba L, Paraskevas K I, Kalra M K, Johri A, Laird J R, Nicolaides A N and Suri J S 2021 Unseen artificial intelligence—deep learning paradigm for segmentation of low atherosclerotic plaque in carotid ultrasound: a multicenter cardiovascular study *Diagnostics* **11** 2257

[124] El-Baz A and Suri J S 2019 *Big Data in Multimodal Medical Imaging* (Boca Raton, FL: CRC Press)

[125] Sudeep P, Palanisamy P, Rajan J, Baradaran H, Saba L, Gupta A and Suri J S 2016 Speckle reduction in medical ultrasound images using an unbiased non-local means method *Biomed. Signal Process. Control* **28** 1–8

[126] Suri J S, Liu K, Singh S, Laxminarayan S N, Zeng X and Reden L 2002 Shape recovery algorithms using level sets in 2-D/3-D medical imagery: a state-of-the-art review *IEEE Trans. Inf. Technol. Biomed.* **6** 8–28

[127] Suri J S *et al* 2021 Systematic review of artificial intelligence in acute respiratory distress syndrome for COVID-19 lung patients: a biomedical imaging perspective *IEEE J. Biomed. Health Inform.* **25** 4128–39

[128] Paul S, Maindarkar M, Saxena S, Saba L, Turk M, Kalra M, Krishnan P R and Suri J S 2022 Bias investigation in artificial intelligence systems for early detection of Parkinson's disease: a narrative review *Diagnostics* **12** 166

[129] Suri J S, Bhagawati M, Paul S, Protogeron A, Sfikakis P P, Kitas G D, Khanna N N, Ruzsa Z, Sharma A M and Saxena S 2022 Understanding the bias in machine learning systems for cardiovascular disease risk assessment: the first of its kind review *Comput. Biol. Med.* **142** 105204

[130] Agarwal M, Agarwal S, Saba L, Chabert G L, Gupta S, Carriero A and Suri J S 2022 Eight pruning deep learning models for low storage and high-speed COVID-19 computed tomography lung segmentation and heatmap-based lesion localization: a multicenter study using COVLIAS 2.0 *Comput. Biol. Med.* **146** 105571

IOP Publishing

Multimodality Imaging, Volume 2
Heart, lungs and peripheral organs
Mainak Biswas and Jasjit S Suri

Chapter 4

An investigation of the inter-variability in COVLIAS 1.0: hybrid deep learning approaches for segmenting COVID-19 lungs in CT scans

Venkateshh Moningi, Sushant Agarwal, Mainak Biswas and Jasjit S Suri

The initial and crucial step for assessing the severity of COVID-19 in the lungs involves segmenting them on computed tomography (CT) scans. However, current deep learning-based artificial intelligence (AI) models used for segmentation may exhibit bias due to the evaluation of a single set of ground truth (GT) annotations. To address this, we propose an inter-variability analysis that utilizes two GT tracers for lung segmentation on CT scans of COVID-19 patients. We trained three AI models (PSP Net, VGG-SegNet, and ResNet-SegNet) using these GT annotations. Our hypothesis is that if AI models trained on GT tracings from multiple experience levels exhibit performance within a 5% range, they can be considered robust and unbiased. We employed the K5 protocol (80% training, 20% testing) and evaluated the performance using ten different metrics. Our database consisted of 5000 CT chest images from 72 COVID-19-infected patients. By computing the coefficient of correlations (CC) between the output of the two AI models trained with different GT tracers and analyzing the differences in CC, we found that the differences were 0%, 0.51%, and 2.04% (all <5%). This validates our hypothesis. Although the performance was comparable, the order of performance was as follows: ResNet-SegNet > PSP Net > VGG-SegNet. During the inter-variability analysis of CT lung segmentation in COVID-19 patients, the AI models demonstrated clinical robustness and stability.

4.1 Introduction

On January 30, 2020, the World Health Organization's International Health Regulations and Emergency Committee (IHREC) declared COVID-19 as a 'public health emergency of international significance' or 'pandemic' [1]. Globally, COVID-19 has resulted in over 231 million infections and more than 4.7 million deaths [1].

doi:10.1088/978-0-7503-2352-9ch4

While the 'severe acute respiratory syndrome coronavirus 2' (SARS-CoV-2) primarily affects the pulmonary and vascular systems, it can also lead to complications such as pulmonary embolism [2], myocardial infarction, stroke, or mesenteric ischemia [3–5]. Comorbidities like obesity, hypertension, and diabetes mellitus significantly increase the mortality and severity of COVID-19 [6, 7]. The recommended diagnostic method is real-time reverse transcription-polymerase chain reaction (RT-PCR) [8]. In patients with moderate to severe illness or underlying comorbidities, the presence and extent of pulmonary opacities, including ground-glass opacities (GGO), mixed opacities observed in CT scans and consolidation [7, 12–14], are utilized to evaluate the severity of the disease using chest radiographs and computed tomography (CT) [9–11]. The assessment of COVID-19 pneumonia severity is predominantly based on a subjective semantic description of the type and extent of pulmonary opacities by radiologists. This process is time-consuming, subjective, and labor-intensive for semi-quantitative evaluation of opacities [15–18]. Consequently, there is a demand for real-time prognostic solutions and efficient techniques for early detection of COVID-19. Machine learning (ML) offers a viable solution to address this challenge by providing a diverse range of techniques [19]. ML has previously been utilized for the detection of various malignancies, including breast [20], liver [21, 22], thyroid [23–25], skin [26, 27], prostate [28, 29], ovary [30], and lung [31]. Segmentation plays a crucial role in the disease detection process and consists of two primary components: classification and segmentation [32–35]. Deep learning (DL), an extension of ML, leverages multiple layers of data to automatically extract and classify relevant image features [38–43]. Hybrid DL (HDL), a technique that combines two DL systems, addresses some challenges associated with individual DL models, such as overfitting and hyperparameter optimization, thus mitigating bias [44, 45]. The process of segmenting ground truth (GT) annotations for organs is a crucial step in training AI models. However, it is a time-consuming and costly procedure as it requires the expertise of radiologists and trained workers who are scarce and expensive to hire. Relying on a single tracer for these annotations can introduce bias into the AI system. To address this issue, the GT annotated dataset is generated using a diverse set of tracers, enhancing the system's robustness and reducing AI bias [46–49]. This approach allows the AI model to understand and adapt to the variations in tracing techniques employed by different tracers. Therefore, it is essential to have an AI-based system that is automated and incorporates multiple tracers to prevent AI bias. To validate the effectiveness of such automated AI systems, it is necessary to conduct inter-variability studies involving two or more observers. These studies help demonstrate the validity and reliability of the automated AI system. To validate the effectiveness of an AI system for segmenting lungs in COVID-19 CT scans, two requirements are proposed. The first requirement is that the performance of two human observers should be within a 5% margin of agreement. This ensures consistency and reliability among the observers in annotating the lung regions. The second requirement states that the AI system, utilizing the ground truth tracings from these observers, should also fall within the 5% threshold. This requirement evaluates how well the AI system can replicate the annotations made by the human observers. The COVID-19 CT lung

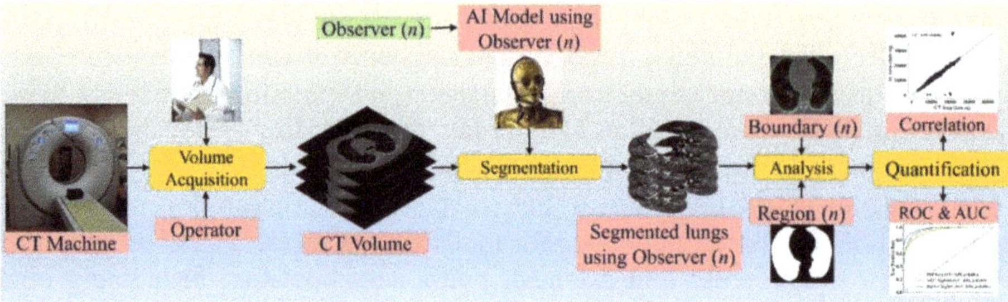

Figure 4.1. COVLIAS 1.0: Inter-variability analysis of CT-based lung segmentation and quantification system for COVID-19 patients. ROC: Receiver operating characteristic; AUC: Area under the curve.

segmentation system illustrated in figure 4.1 involves obtaining CT volumes from a CT scanner. Multiple observers annotate the lung regions, and AI models are created based on these annotations for lung segmentation. The system generates binary masks of the lung, its border, and boundary overlays to evaluate its performance. The objective of the study is to develop a reliable AI system using an inter-observer paradigm, ensuring consistent and accurate lung segmentations in COVID-19 CT scans. The findings of this study can have significant implications for clinical diagnosis and treatment planning. The inter-variability research is structured as follows. Section 4.2 is Methodology; this section provides a description of the methodology used in the study. It includes details about the AI architectures employed, demographic data utilized, the loss functions employed, and the COVLIAS 1.0 pipeline. Section 4.3 is Experimental Procedure; in this section the experimental procedure is explained. It outlines the steps taken to conduct the study, including data collection, preprocessing, and implementation of the AI models. Section 4.4 is for Results and Performance Assessment. The results of the study and the assessment of performance are presented in this section. It includes an analysis of the outcomes obtained from the experiments and an evaluation of the AI system's performance based on the predefined criteria. Section 4.5 is Discussions; in this section the findings of the study are discussed in detail. It provides an in-depth analysis and interpretation of the results, highlighting their significance and implications. Section 4.6 is Conclusions; the conclusions drawn from the study are summarized in this section. It provides a summary of the key findings, their implications, and potential future research directions. Each section contributes to the overall understanding of the inter-variability research, from the methodology and experimental procedure to the results, discussions, and conclusions.

4.2 Methodology

4.2.1 Patient demographics, image acquisions, and data preparation

4.2.1.1 Demographics

The dataset consists of 72 adult Italian patients, with 46 of them being male and the remaining patients being female. On average, the individuals in the dataset had a height of 173 cm and a weight of 79 kg. Among the patients, 12 individuals had their

RT-PCR results confirmed through broncho-alveolar lavage, resulting in a total of 60 patients with positive COVID-19 test results [50]. The average ground-glass opacities (GGO) score in the cohort was 4.1, indicating a generally poor overall condition.

4.2.1.2 Image acquisition

For all chest CT scans, a 128-slice multidetector-row CT scanner called ‘Philips Ingenuity Core’ from Philips Healthcare was utilized. The scans were performed with patients in a supine position, holding a single full inspiratory breath. No intravenous or oral contrast material was administered during the scans. The Philips Z-DOM automated tube current modulation was employed, using a 120 kV voltage, 226 mAs/slice, 0.5-second gantry rotation time, 1.08 spiral pitch factor, and a 64 0.625 detector configuration [47, 48, 51]. The reconstructed images were created with a thickness of 1 mm using soft tissue kernels with a 512 × 512 matrix for the mediastinal window and lung kernels with a 768 × 768 matrix for the lung window. The Picture Archiving and Communication System (PACS) workstation employed two Eizo 35 × 43 cm monitors with a 2048 × 1536 matrix to evaluate the CT images. Figure 4.2 showcases representative CT scans of COVID-19 patients, demonstrating the challenges posed by variations in lung diameters and irregular intensity patterns among the patients.

4.2.2 Architecture

The COVLIAS 1.0 system incorporates three models: one solo DL (SDL) model and two HDL models. The solo DL model operates independently, while the hybrid DL models combine the capabilities of two DL systems. In the proposed study, three AI models are included: VGG-SegNet, ResNet-SegNet [55], and PSP Net. These models are utilized for the task of lung segmentation in the context of COVID-19. Each model has its own architecture and characteristics, which contribute to their respective performance and effectiveness in the segmentation process (figure 4.3).

4.2.2.1 Three AI models: PSP Net, VGG-SegNet, and ResNet-SegNet

The Pyramid Scene Parsing Network (PSP Net) is a semantic segmentation network that incorporates the overall context of an image [52]. The architecture of PSP Net consists of four components: input, feature map, pyramid pooling module, and output (figure 4.4). The input to the network is the image that needs to be segmented. The feature map is obtained by applying a set of dilated convolution and pooling blocks, which extract features from the input image. To preserve more prominent features, the dilated convolution layer is introduced in the last two blocks of the network. The pyramid pooling module plays a critical role in the network as it captures the overall context of the image or feature map from the previous stage. This module consists of four sections, each with a different scaling capacity (1, 2, 3, and 6). The scaling factors allow the module to capture spatial features and enhance the resolution of recorded features (1 × 1 scaling) as well as capture higher-resolution details (6 × 6 scaling). By combining the outputs from all four sections using global

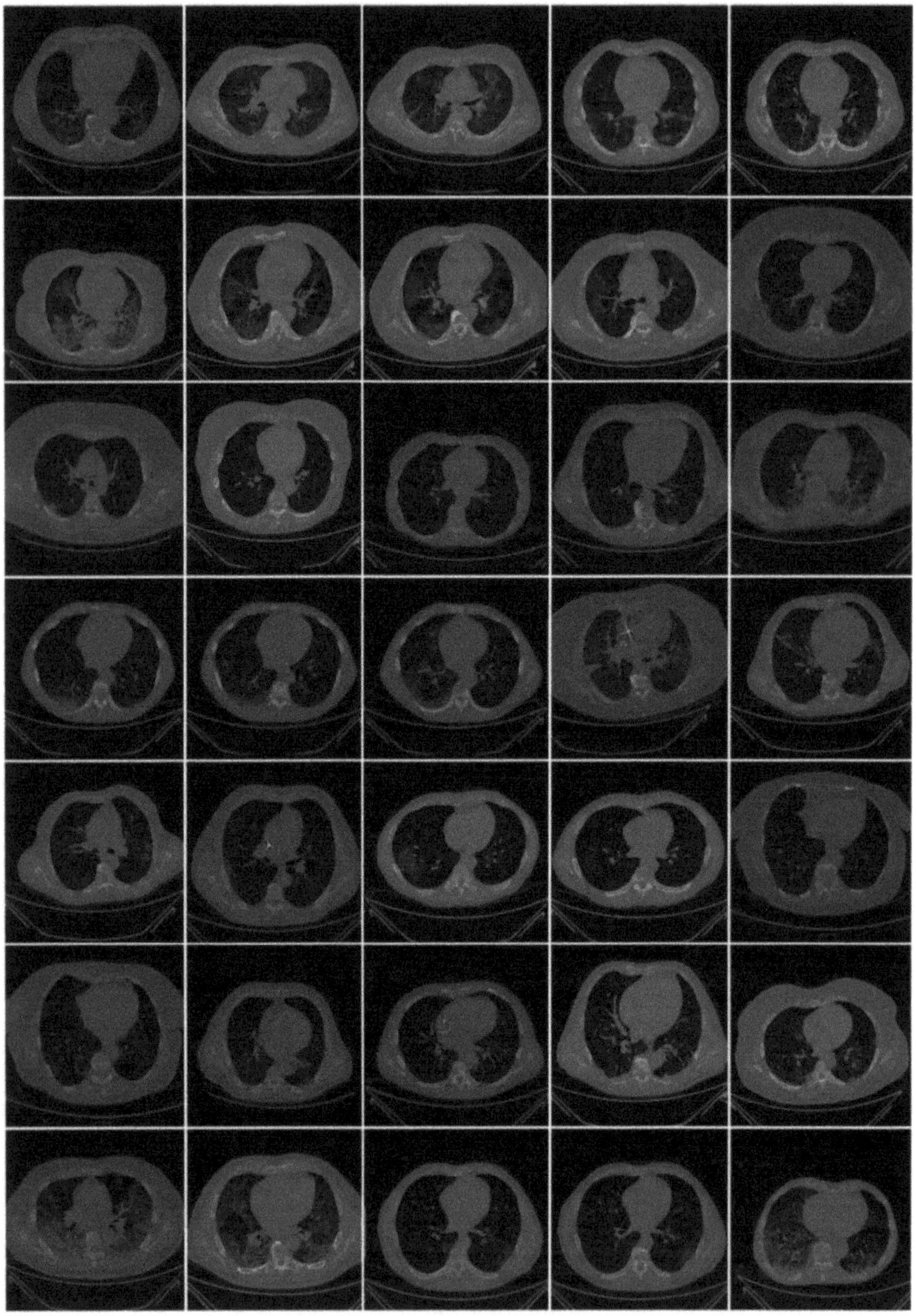

Figure 4.2. Raw lung COVID-19 CT scans taken from different patients in the database.

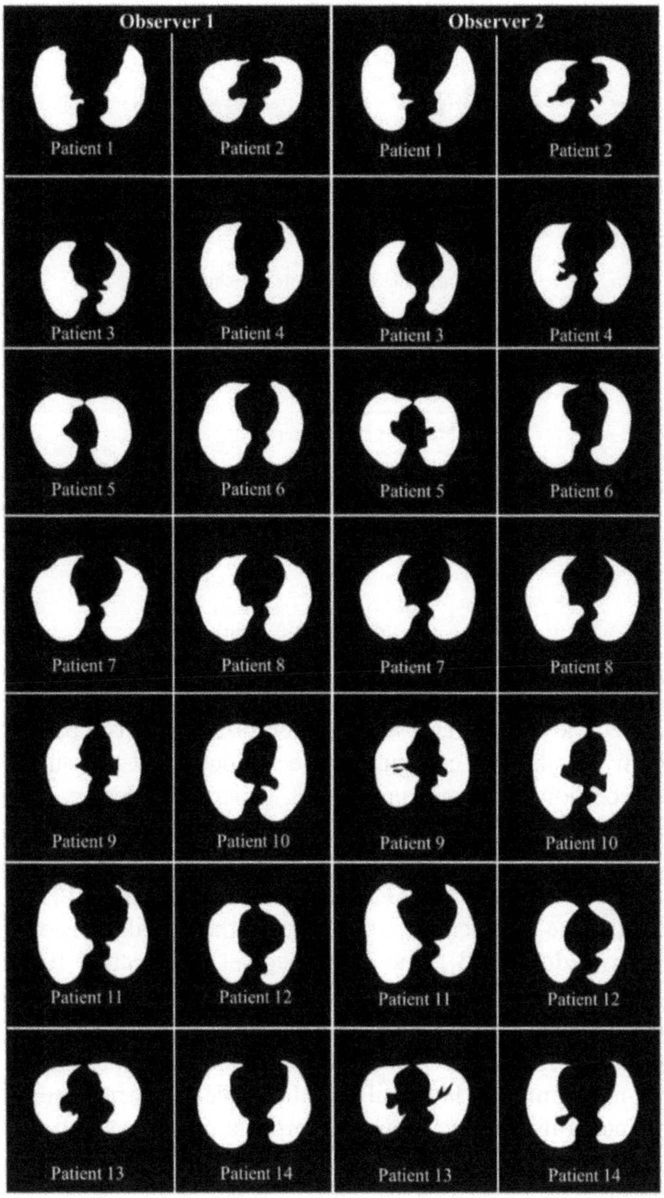

Figure 4.3. GT white binary mask for AI model training for Observer 1 vs. Observer 2.

average pooling, the module effectively incorporates contextual information. The output of the global average pooling is then fed into several convolutional layers in the final part of the network. These layers further process the pooled features to generate a set of predicted classes. Ultimately, the network produces an output binary mask that represents the predicted classes. In summary, the PSP Net architecture utilizes the pyramid pooling module to capture holistic contextual

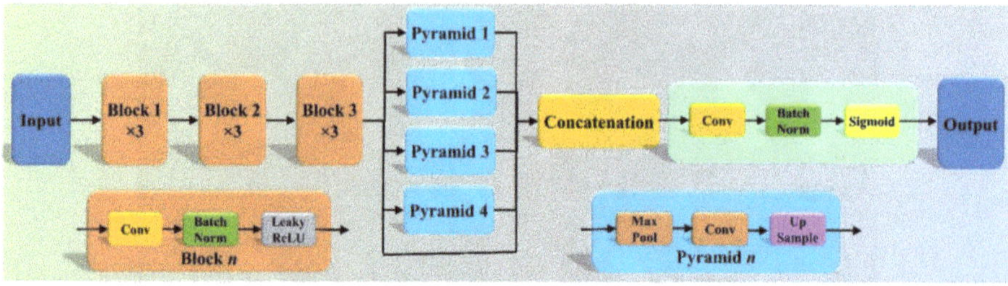

Figure 4.4. PSP Net architecture.

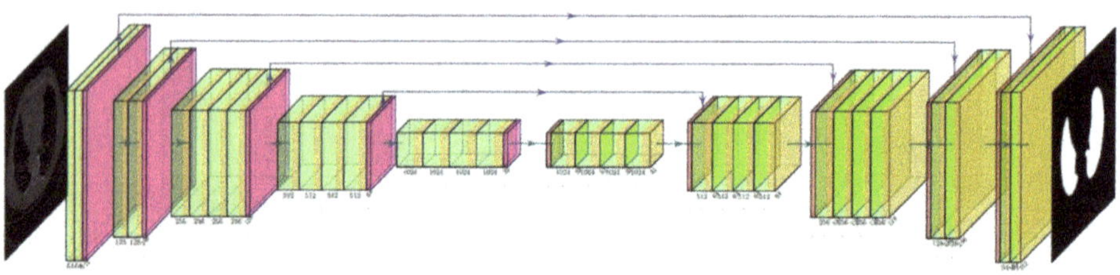

Figure 4.5. VGG-SegNet architecture.

information. It employs different scaling factors to capture spatial features and higher-resolution details. By combining the outputs and passing them through subsequent convolutional layers, the network generates a binary mask that represents the predicted classes for effective image segmentation. To reduce training times, the VGGNet architecture (figure 4.5) was developed by replacing the kernel filter in the first layer with an 11 × 11 filter and in the subsequent layer with a 5 × 5 filter. Additionally, the number of parameters in the two-dimensional convolution (Conv) layers was reduced [53]. The VGG-SegNet architecture used in this study consists of three main components: an encoder, a decoder, and a pixel-wise SoftMax classifier. In comparison to the SegNet design, which utilizes 13 Conv layers in the encoder component [54], the VGG-SegNet architecture incorporates 16 Conv layers. This increase in the number of layers allows the model to extract more features from the image, enhancing its ability to capture important characteristics.

4.2.2.2 Loss functions for AI models

The suggested system trains the AI models using cross-entropy (CE)-loss. For each of the three AI models, the CE-loss is shown in equation (4.1) below as lCE:

$$l_{CE} = -[(x_i \times \log p_i) + (1 - x_i) \times \log(1 - p_i)] \quad (4.1)$$

where, GT (table 4.1) is represented by x_i, label 0 is for representing $(1-x_i)$, the classifier's (SoftMax) probability we represent as p_i which is used at the AI model's last layer, and the two-term product is represented by x. The three AI architectures that have been trained using the CE-loss function are shown in figures 4.4–4.6.

Table 4.1. Comparison of the CC values obtained between AI model area and the GT area corresponding to Observer 1 and Observer 2.

	Left	PSP Net Right	Mean	Left	VGG-SegNet Right	Mean	Left	ResNet-SegNet Right	Mean
Observer 1	0.98	0.98	0.98	0.98	0.99	0.99	0.98	0.98	0.98
Observer 2	0.98	0.98	0.98	0.98	0.98	0.98	1.00	1.00	1.00
% Difference	0.00	0.00	0.00	0.00	1.01	0.51	2.04	2.04	2.04

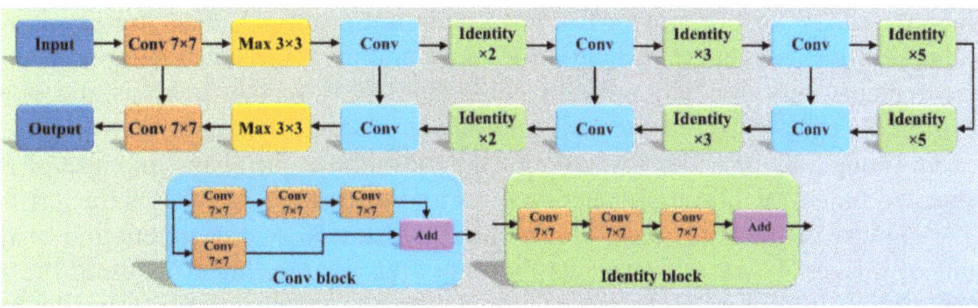

Figure 4.6. ResNet-SegNet architecture.

4.3 Experimental protocol

4.3.1 Accuracy estimation of AI models using cross-validation

In order to evaluate the accuracy of the AI models, a standardized cross-validation (CV) protocol was adopted. Our group has previously published various CV-based protocols using an AI framework [27, 30, 37, 56, 57]. For this study, the K5 protocol was utilized, considering the moderate size of the data. The K5 protocol involves splitting the data into five folds, where each fold consists of 20% testing data (1000 CT images) and 80% training data (4000 CT images). It is important to note that each of the five folds may have a different test set due to the way the folds were created. Within the K5 protocol, there was an internal validation mechanism where 10% of the data was set aside for validation purposes. By employing this CV protocol, the AI models could be evaluated and validated using multiple test sets while ensuring a sufficient amount of training and validation data for robust model training and assessment.

4.3.2 Lung quantification

The quantification of segmented lungs was performed using AI models employing two different techniques inspired by form analysis. In the first technique, which focuses on the balloon-shaped region of the lungs, the lung area (LA) was computed using the area parameter [58, 59]. This approach is suitable due to the shape of the region. In the second technique, considering that the lung's form is more longitudinal than circular, the lung's long axis (LLA) was calculated. A similar approach was used in cardiac calculations for the long-axis perspective, as described in reference [60]. In this case, the objective was to determine the lung long axis (LLA) by measuring the farthest distance segment connecting the anterior and posterior parts of the lungs. To compute the lung area (LA), the number of white pixels in the binary mask of the segmented lungs was counted. To convert these pixel measurements into more meaningful units, a resolution factor of 0.52 was applied. Specifically, the pixel count for LA was converted to mm^2, while the LLA measurement was converted to mm. By using this resolution factor, the pixel-based measurements obtained from the binary mask were transformed into meaningful values in terms of area (mm^2) for LA and length (mm) for LLA. This strategy allows for quantitative assessment of lung dimensions and area by translating pixel-based measurements into clinically relevant units, facilitating further analysis and interpretation. By employing these techniques and conversions, the quantification of lung area and long axis provides valuable information about the shape and size of the segmented lungs, facilitating further analysis and assessment of lung characteristics in the context of the study. If N in the database stands for the summation of the count of image, $A_{ai}(m, n)$ represents the lung area for the image 'n' using the AI model 'm', $A_{ai}(m)$ refers to the mean lung area corresponding to the AI model 'm,' and A_{gt} stands for the mean area of the GT binary mask, then mathematically $A_{ai}(m)$ and A_{gt} can be computed as shown in equation (4.2).

$$\left.\begin{aligned} \overline{A}_{ai}(m) &= \frac{\sum_{n=1}^{N} A_{ai}(m, n)}{N} \\ \overline{A}_{gt} &= \frac{\sum_{n=1}^{N} A_{gt}(n)}{N} \end{aligned}\right\} \tag{4.2}$$

Similar to this, $LA_{ai}(m, n)$ represents LLA in the image 'n' using the AI model 'm', LLA corresponds to the AI model 'm', which is represented by $LA_{ai}(m)$ and LA_{gt} refers the corresponding mean LLA of the GT binary lung mask. $LA_{ai}(m)$ and LA_{gt} may then be determined analytically as stated in equation (4.3).

$$\left.\begin{aligned} \overline{LA}_{ai}(m) &= \frac{\sum_{n=1}^{N} LA_{ai}(m, n)}{N} \\ \overline{LA}_{gt} &= \frac{\sum_{n=1}^{N} LA_{gt}(n)}{N} \end{aligned}\right\} \tag{4.3}$$

4.3.3 AI model accuracy computation

To evaluate the accuracy of the AI system, a comparison was made between the predicted output and the ground truth pixel values. Since the output lung mask only produced black or white pixels, these data were converted into binary integers (0 or 1). The final step involved summing all these binary integers and dividing them by the total number of pixels in the image.

In the context of evaluating accuracy, the terms TP (true positive), TN (true negative), FN (false negative), and FP (false positive) are used. These terms represent different outcomes in the comparison between the predicted output and the ground truth. To calculate the accuracy of the AI system, equation (4.4) [61] can be utilized:

$$\mathrm{ACC\ (ai)\ (\%)} = \left(\frac{\mathrm{TP} + \mathrm{TN}}{\mathrm{TP} + \mathrm{FN} + \mathrm{TN} + \mathrm{FP}}\right) \times 100 \tag{4.4}$$

By applying this equation, the accuracy of the AI system can be determined, providing a quantitative measure of its performance in terms of correctly predicting true positive and true negative instances while minimizing false positives and false negatives.

4.4 Results and performance

4.4.1 Results

In the previous version of COVLIAS (COVLIAS 1.0) [54], the intention was to work with a dataset consisting of 5000 images, with a testing ratio of 2:3. However, the focus of this current study is on inter-observer variability research using the K5 protocol within a cross-validation (CV) framework. For this study, two sets of annotations, referred to as Observer 1 and Observer 2, were utilized for training the AI models. The output results of the AI models are binary masks that segment the lungs, which can be compared to the results obtained in earlier published research. Figures 4.7–4.9 visually present the three AI models, showcasing their overlay with the AI-generated binary mask, segmented lung, and colored segmented lung. These visual representations provide insights into the performance and output of the AI models, highlighting the accuracy and effectiveness of the lung segmentation process carried out by the AI system.

4.4.2 Performance evaluation

In this section, the performance evaluations (PE) of the three AI models are discussed separately for Observer 1 and Observer 2. The findings are visually compared in section 4.4.2.1, where the lung long axis and boundary overlays are compared with the ground truth boundary and axis, respectively. This comparison provides insights into the accuracy and similarity of the AI models' output in relation to the annotations provided by the two observers. Furthermore, section 4.4.2.2 presents the PE for lung area error. This evaluation includes various analyses such as cumulative frequency (CF) plots, Bland–Altman plots, Jaccard Index (JI)

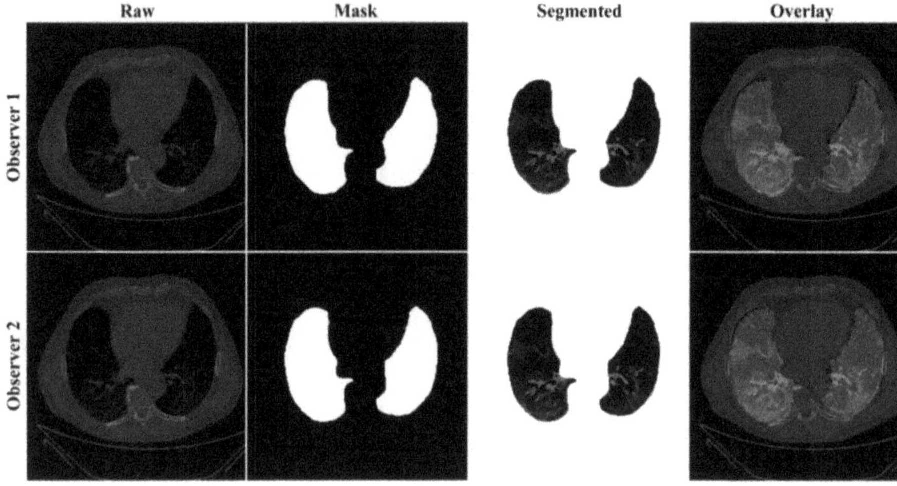

Figure 4.7. Results from PSP Net while using Observers 1 and 2. Columns are the raw, binary mask output, segmented lung region, and overlay of the estimated lung region vs. ground truth region.

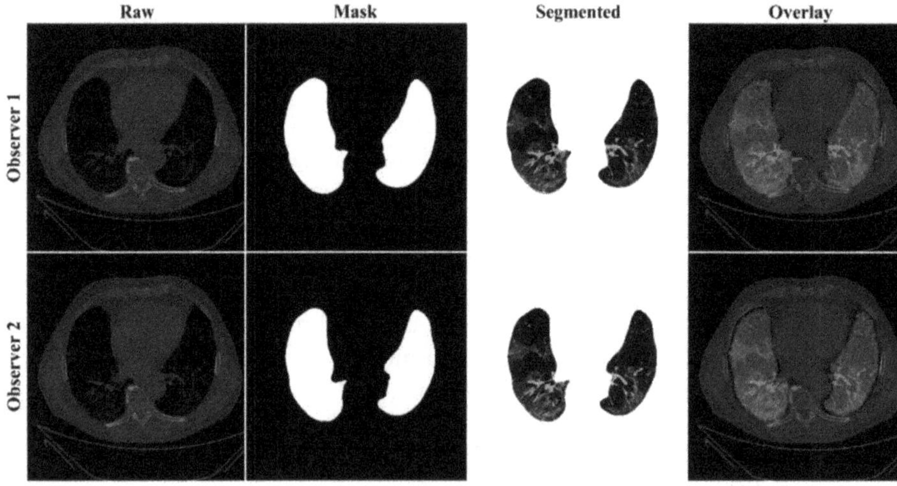

Figure 4.8. Results from VGG-SegNet while using Observers 1 and 2. Columns are the raw, binary mask output, segmented lung region, and overlay of the estimated lung region vs. ground truth region.

and Dice Similarity (DS) plots, as well as ROC and AUC curves. These analyses assess the performance of the three AI-based models specifically for Observer 1 vs. Observer 2, providing a comprehensive understanding of the accuracy and agreement between the AI models and the annotations made by the observers. Similarly, the lung long axis error (LLAE) is also evaluated using cumulative plots, correlation coefficient (CC), and Bland–Altman plots. These assessments shed light on the

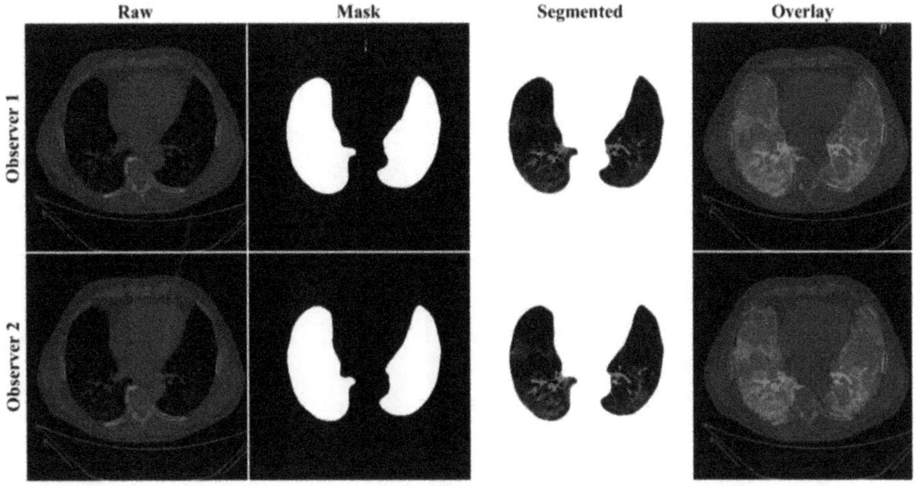

Figure 4.9. Results from ResNet-SegNet while using Observers 1 and 2. Columns are the raw, binary mask output, segmented lung region, and overlay of the estimated lung region vs. ground truth region.

performance of the AI models in accurately determining the lung's long axis measurement, again comparing it to the ground truth values provided by the observers. Overall, these performance evaluations provide a comprehensive analysis of the AI models' accuracy and agreement with the annotations made by the observers, allowing for a thorough understanding of their effectiveness in lung segmentation and measurement.

4.4.2.1 Lung boundary and long axis visualization

Figure 4.10 illustrates the overlay of the boundaries generated by the three AI models (green) and the ground truth boundaries (red) for Observer 1 (left) and Observer 2 (right) on a COVID-19 CT slice shown in grayscale. This visual comparison allows for a qualitative assessment of how well the AI models align with the annotations provided by the observers. The three AI models used in this study are PSP Net, VGG-SegNet, and ResNet-SegNet, respectively. Figure 4.11, on the other hand, displays the AI-long axis (green) and the GT-long axis (red) between Observer 1 and Observer 2 for the three AI models. This comparison focuses specifically on the measurement of the lung's long axis. By visualizing the alignment of the AI-generated long axis with the ground truth long axis, it provides insights into the accuracy and agreement of the AI models in capturing this aspect of lung morphology. The GT boundary (white) corresponding to Observer 1 (left) and Observer 2 (right) is also depicted in figure 4.11. This boundary, generated using the ImgTracerTM tool, represents the extent of the left and right lungs from anterior to posterior. It provides additional context for understanding the coverage and accuracy of the AI models' segmentations in relation to the ground truth annotations. These visualizations aid in the assessment and comparison of the AI models' performance in capturing the boundaries and long axis of the lungs, demonstrating how well they align with the annotations provided by the observers.

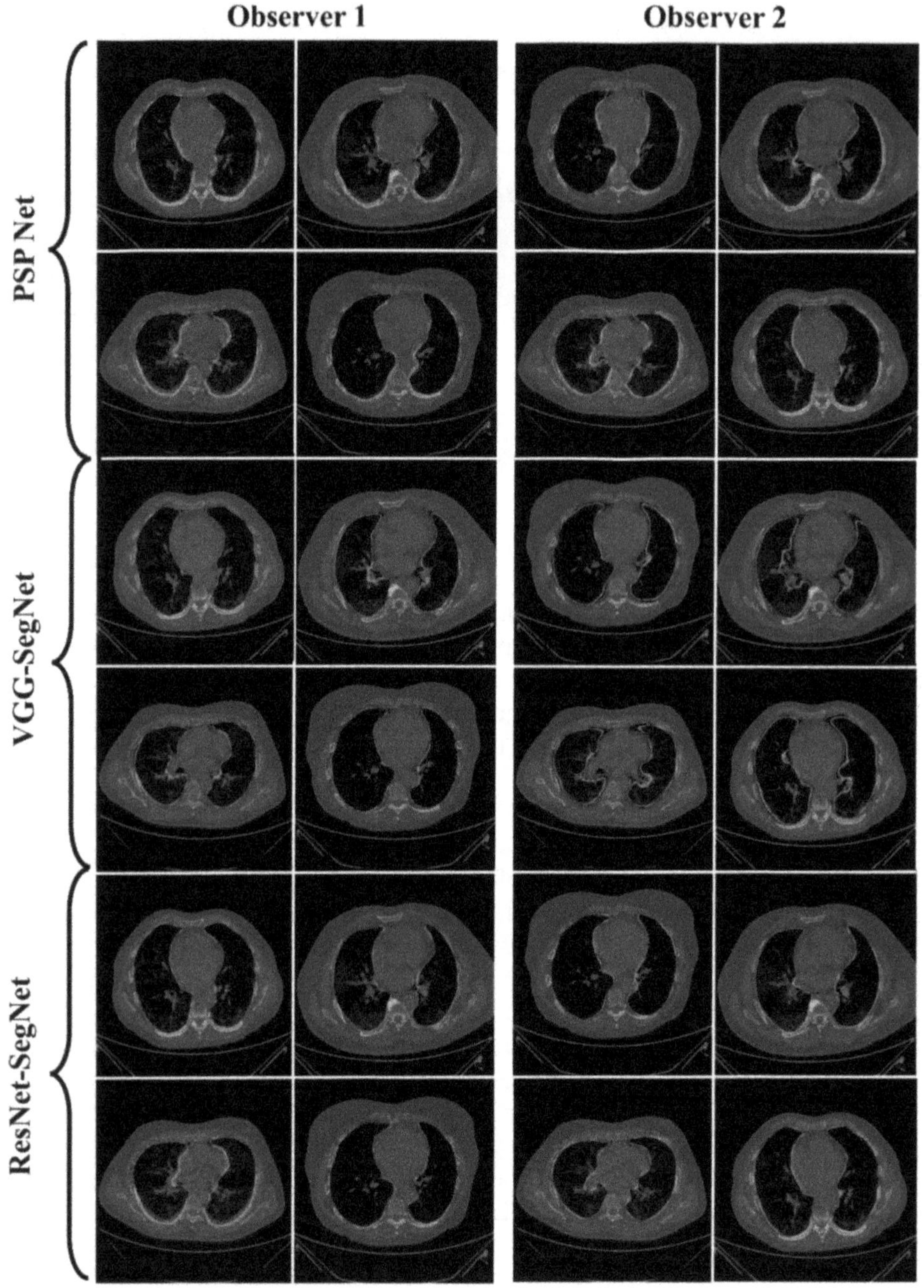

Figure 4.10. AI-model segmented boundary (green) vs. GT boundary (red) for Observer 1 and Observer 2.

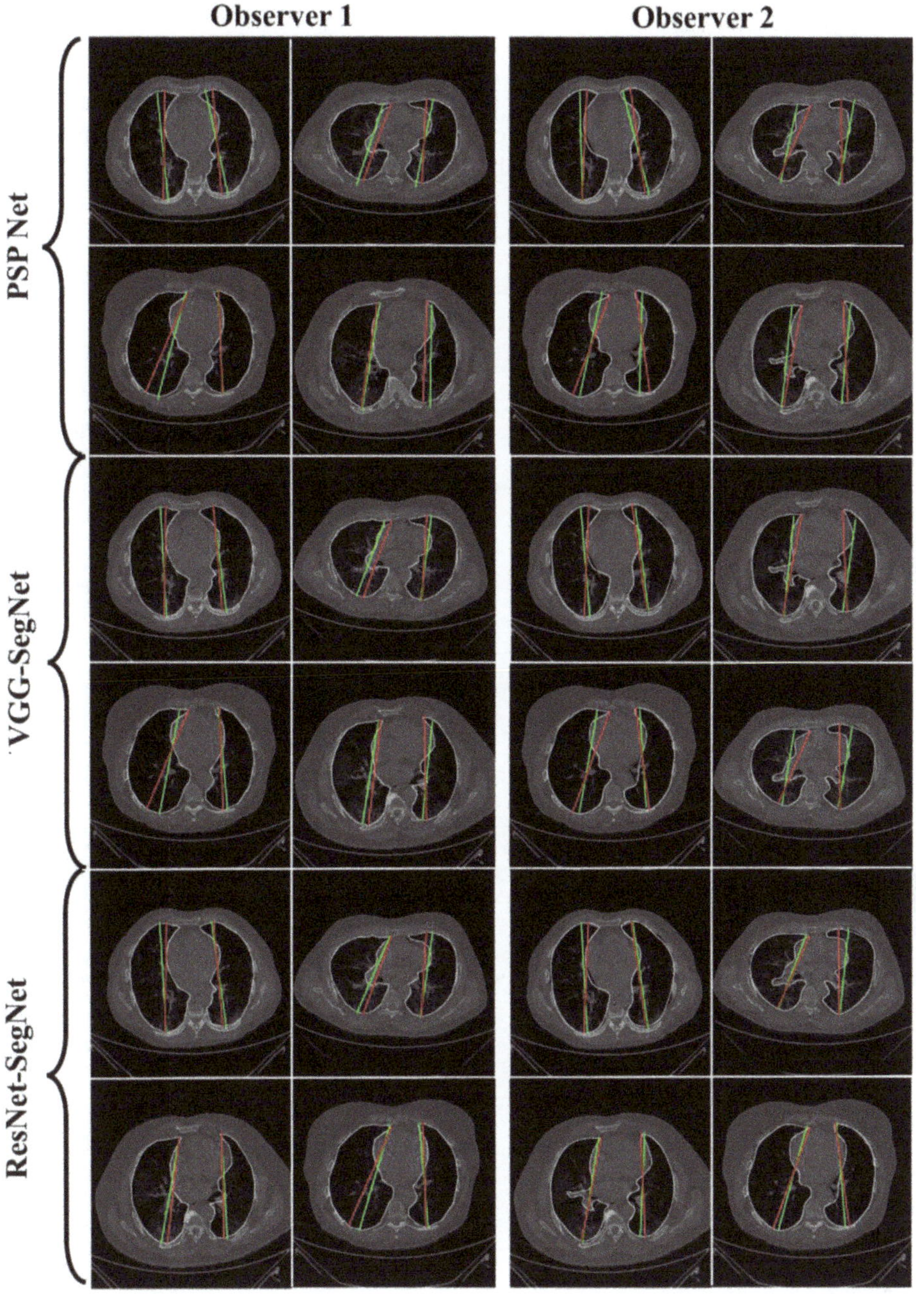

Figure 4.11. AI-model long axis (green) vs. GT-long axis (red) for Observer 1 and Observer 2.

4.4.2.2 Performance metrics for the lung area error

Lung area error cumulative frequency plot

Figures 4.12 (for the left lung) and 4.13 (for the right lung) present the cumulative frequency analysis of the lung area error (LAE) for the three AI models compared to a reference value. The purpose of this analysis is to compare the frequency of

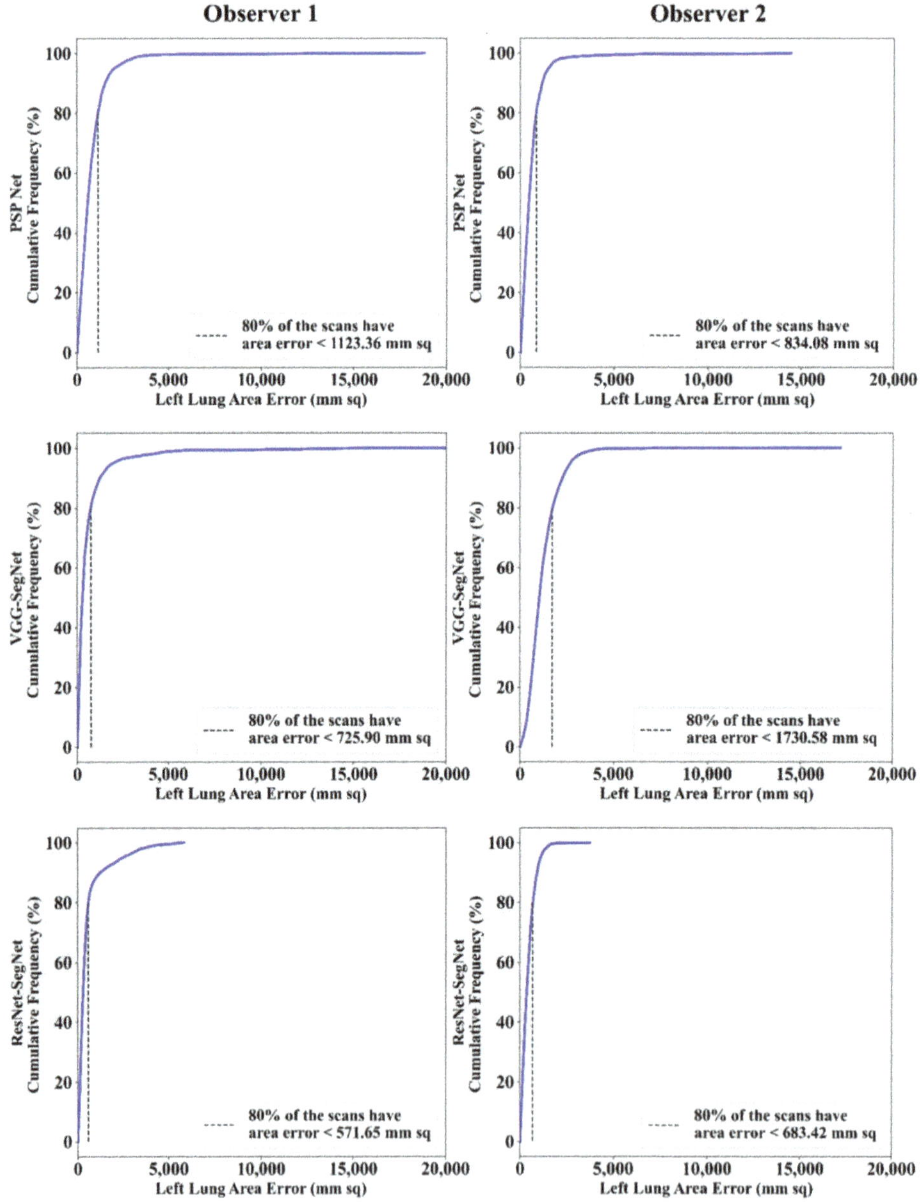

Figure 4.12. Cumulative frequency plot of left LAE using three AI models: Observer 1 vs. Observer 2.

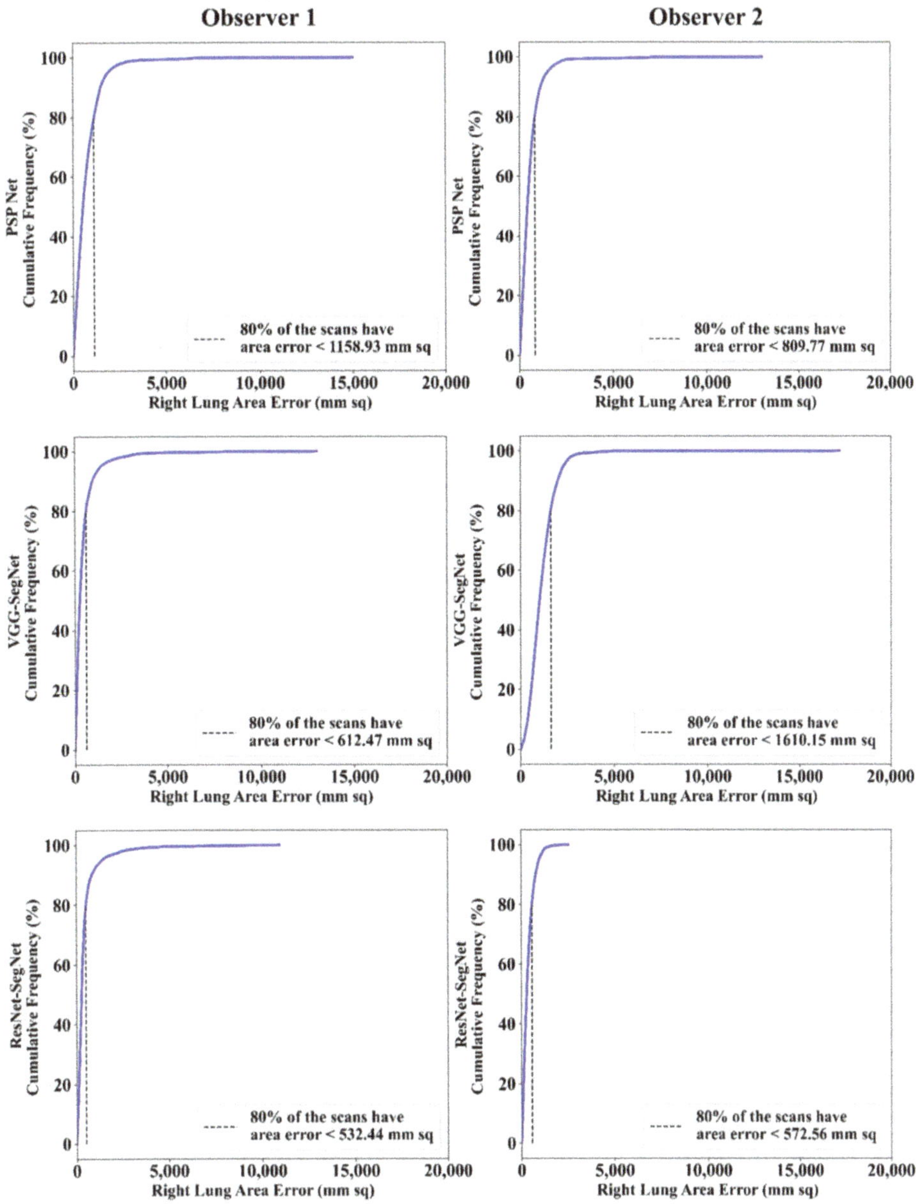

Figure 4.13. Cumulative frequency plot of right LAE using three AI models: Observer 1 vs. Observer 2.

occurrence of LAE values between Observer 1 and Observer 2 and assess how the three AI models perform in relation to the observers' annotations. To highlight the differences among the AI models, a threshold score of 80% was chosen. The LAE values at this cutoff point are as follows: For the three AI models using Observer 1: AI Model 1: LAE = 1123.36 mm2—AI; Model 2: LAE = 725.90 mm2|—AI; Model 3: LAE = 571.65 mm2. For the three AI models using Observer 2: AI Model 1:

LAE = 834.08 mm2—AI; Model 2: LAE = 1730.58 mm2—AI; Model 3: LAE = 683.42 mm2. These values represent the differences between the AI models' estimated lung area and the ground truth annotations provided by the observers. The cumulative frequency analysis helps visualize the distribution of these differences and provides insights into the performance of the AI models in estimating the lung area compared to the observers' measurements.

Correlation plot for lung area error

Figures 4.14 and 4.15 present the CC plots for the LA estimated by the three AI models compared to the GT annotations, corresponding to the left and right lung regions between Observers 1 and 2. Table 4.1 provides an overview of the CC values and the percentage difference between Observers 1 and 2 for each AI model. The CC value measures the strength and direction of the linear relationship between the LA

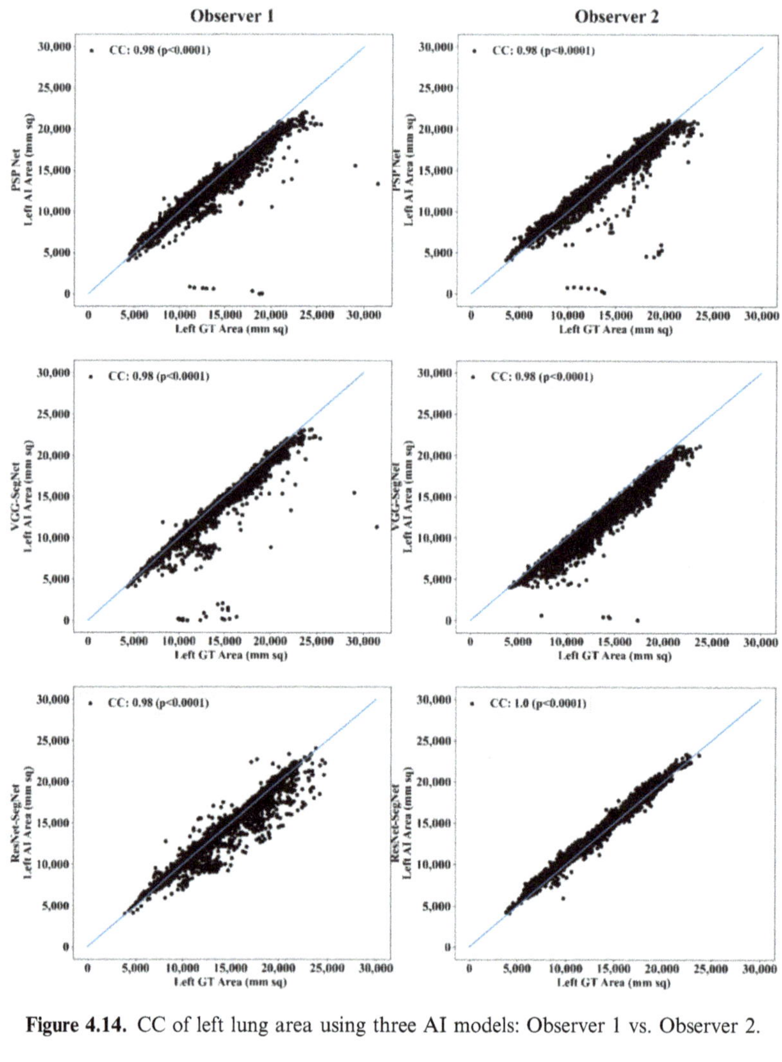

Figure 4.14. CC of left lung area using three AI models: Observer 1 vs. Observer 2.

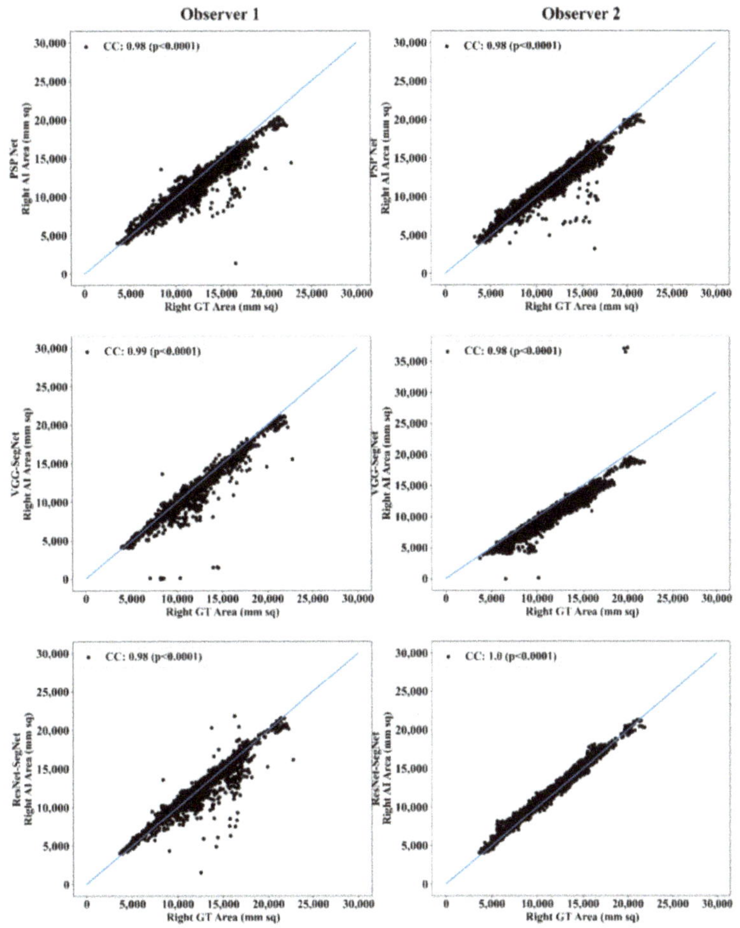

Figure 4.15. CC of right lung area using three AI models: Observer 1 vs. Observer 2.

estimated by the AI models and the GT annotations. The percentage difference represents the variation in the CC values between the two observers. Based on the hypothesis's specified error level ($p < 0.001$), the range of the percentage difference for the CC value is 0% to 2.04%, or 5%. This indicates that the AI models are well-suited for the suggested scenario of the inter-observer variability study and demonstrate clinical validity. The small percentage difference suggests a high level of agreement between the AI models and the observers in estimating the lung area. These findings support the conclusion that the AI models can effectively capture and reproduce the variability observed between the two observers, validating their utility in assisting with inter-observer variability studies in clinical settings.

Jaccard index and dice similarity

Figure 4.16 displays a cumulative frequency histogram for the Dice Similarity (DS) index between Observers 1 and 2, using the three AI models (PSP Net,

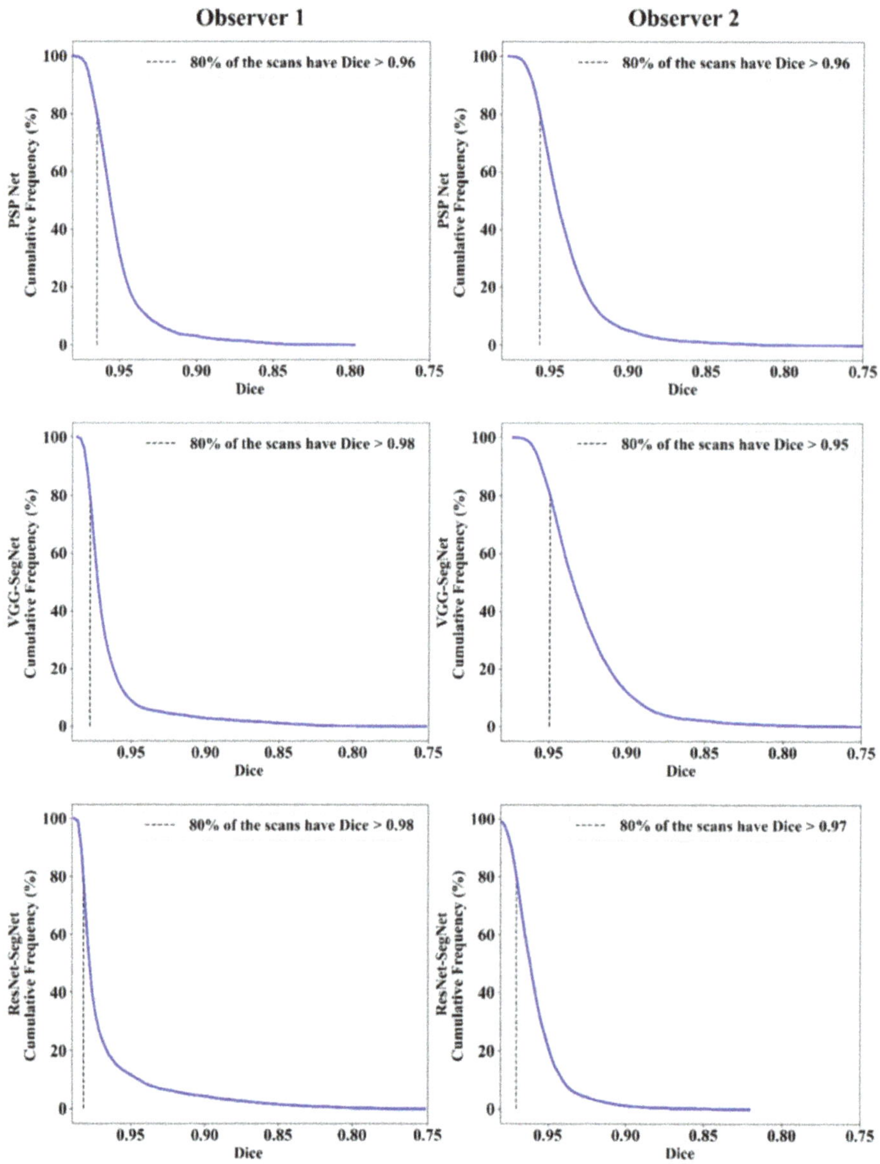

Figure 4.16. DS for combined lung using three AI models: Observer 1 vs Observer 2.

VGG-SegNet, and ResNet-SegNet). The DS index measures the overlap between the segmented lung regions produced by the AI models and the ground truth annotations. The histogram reveals that in 80% of the CT scans, the DS index exceeded 0.95. This indicates a high level of similarity and agreement between the AI models and the two observers in delineating the lung boundaries. Similarly, figure 4.17 presents a cumulative frequency plot for the Jaccard Index (JI) between

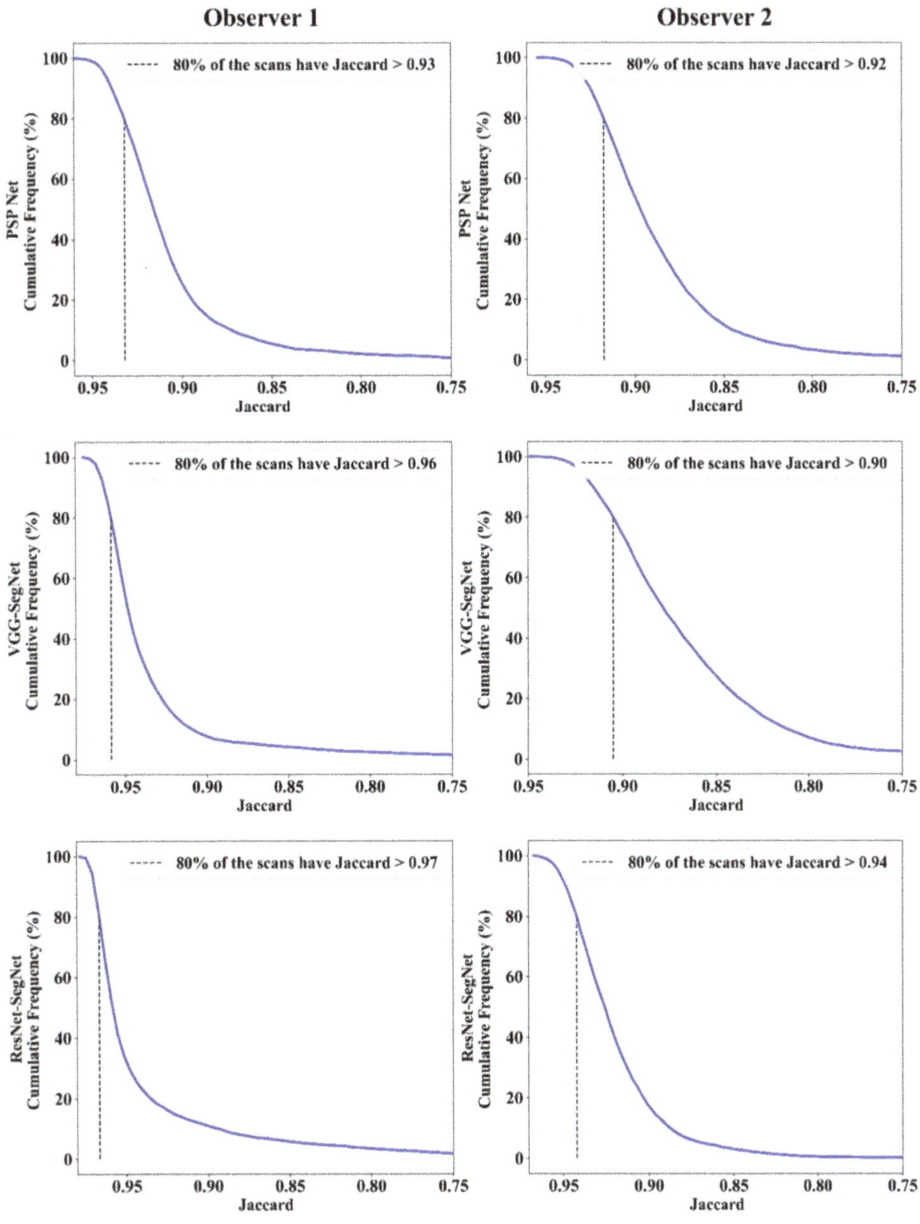

Figure 4.17. JI for combined lung using three AI models: Observer 1 vs. Observer 2.

Observers 1 and 2, using the three AI models. The JI is another metric that evaluates the overlap and similarity between the segmented lung regions. The plot demonstrates that in 80% of the CT scans, the JI exceeded 0.90. This further confirms the strong agreement between the AI models and the observers in terms of lung segmentation accuracy. The consistent high values of DS and JI indicate that the AI models are capable of accurately segmenting the lungs and producing results that closely align with the annotations of the two observers. These results reinforce the reliability and effectiveness of the AI models in capturing inter-observer variability.

Bland–Altman plot for lung area

Figures 4.18 and 4.19 illustrate Bland–Altman plots, following the computation approach described in our previous works [48, 62], to assess the consistency of lung area differences between the AI models and the GT for Observer 1 and Observer 2, respectively. The Bland–Altman plot is a statistical method, known as the Bland–Altman analysis, used to examine the agreement between two measurement techniques or observers.

In these plots, we plot the differences in lung area (AI model—GT) on the y-axis against the average lung area (AI model + GT divided by 2) on the x-axis. The mean difference is represented by the center line, and the upper and lower dashed lines

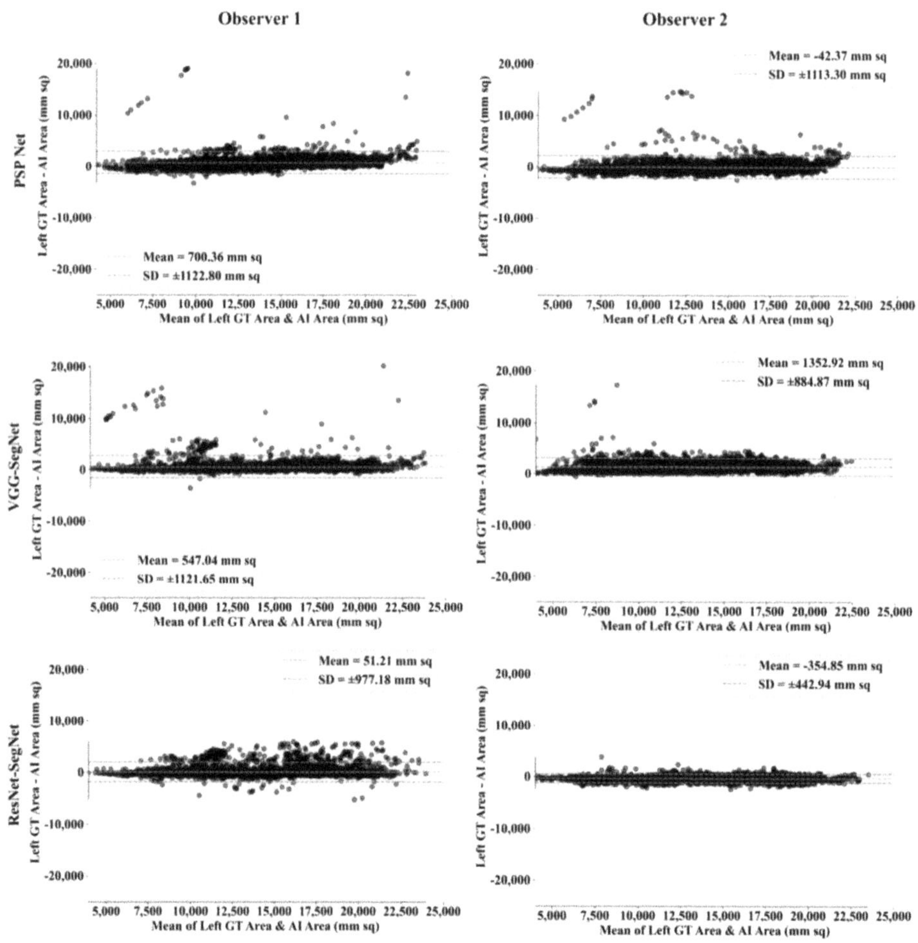

Figure 4.18. BA for left LA for three AI models: Observer 1 vs. Observer 2.

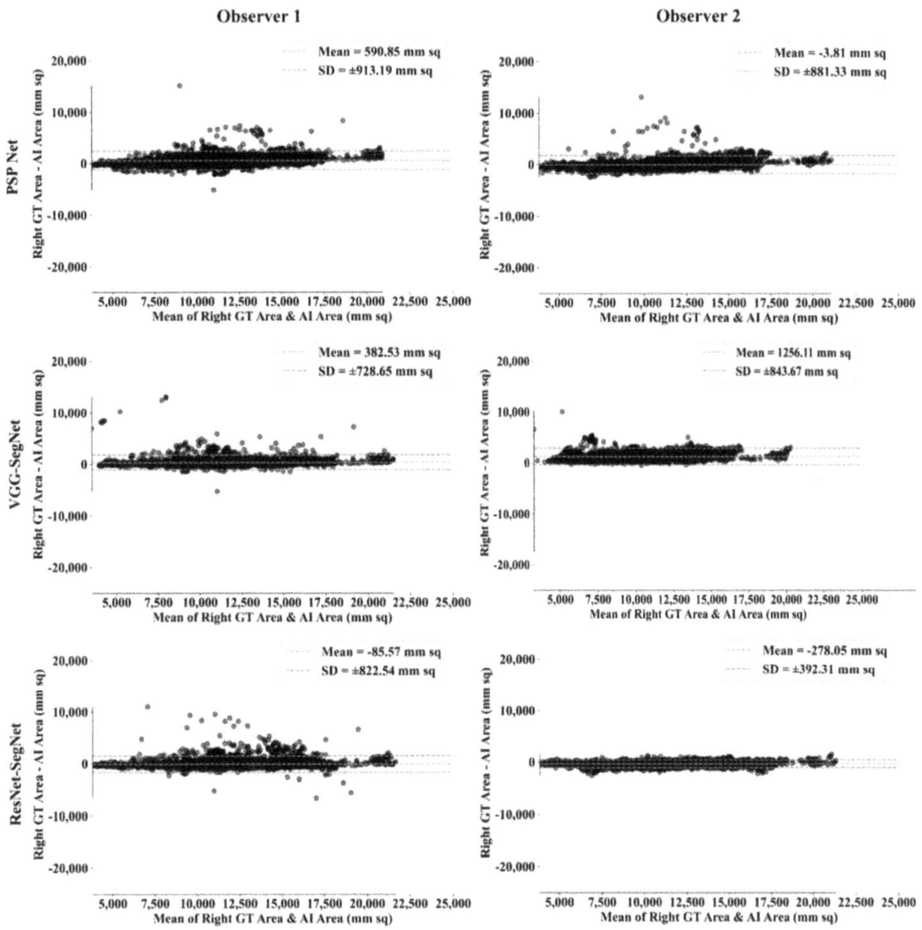

Figure 4.19. BA for right LA for three AI models: Observer 1 vs. Observer 2.

correspond to the limits of agreement, calculated as the mean difference ± 1.96 times the standard deviation. The Bland–Altman plots provide valuable insights into the agreement and consistency between the AI models and the GT regarding lung area measurements. By referring to figures 4.18 and 4.19, we can observe the mean lung area difference and its standard deviation for both Observer 1 and Observer 2. The inclusion of these reference numbers aids in linking the results to the relevant studies [48, 62]. These measures allow us to evaluate the level of agreement and reliability of the AI models in accurately estimating the lung area when compared to the GT. By visually examining the Bland–Altman plots, we can assess the systematic bias, if any, between the AI models and the GT, as well as the range within which most of the differences fall. This analysis contributes to the overall understanding of the agreement between the two approaches and provides valuable insights into the performance of the AI models in quantifying lung area.

ROC plots for lung area

Figure 4.20 displays the receiver operating characteristic (ROC) curve and corresponding area under the curve (AUC) values for the three AI models (PSP Net, VGG-SegNet, and ResNet-SegNet) when comparing Observer 1 and Observer 2. The ROC curve is a graphical representation that illustrates the performance of a diagnostic system, such as an AI model, at different discrimination thresholds. In the ROC curve, the true positive rate (sensitivity) is plotted on the y-axis, and the false positive rate (1—specificity) is plotted on the x-axis. Each point on the curve represents the performance of the AI model at a particular discrimination threshold. The AUC value, which ranges from 0 to 1, quantifies the overall performance of the AI model. A higher AUC value indicates better discrimination and diagnostic accuracy. By examining figure 4.20 and referring to the mentioned AI models, we can assess how well each model performs in distinguishing between Observer 1 and Observer 2. The ROC curve allows us to visualize the trade-off between sensitivity

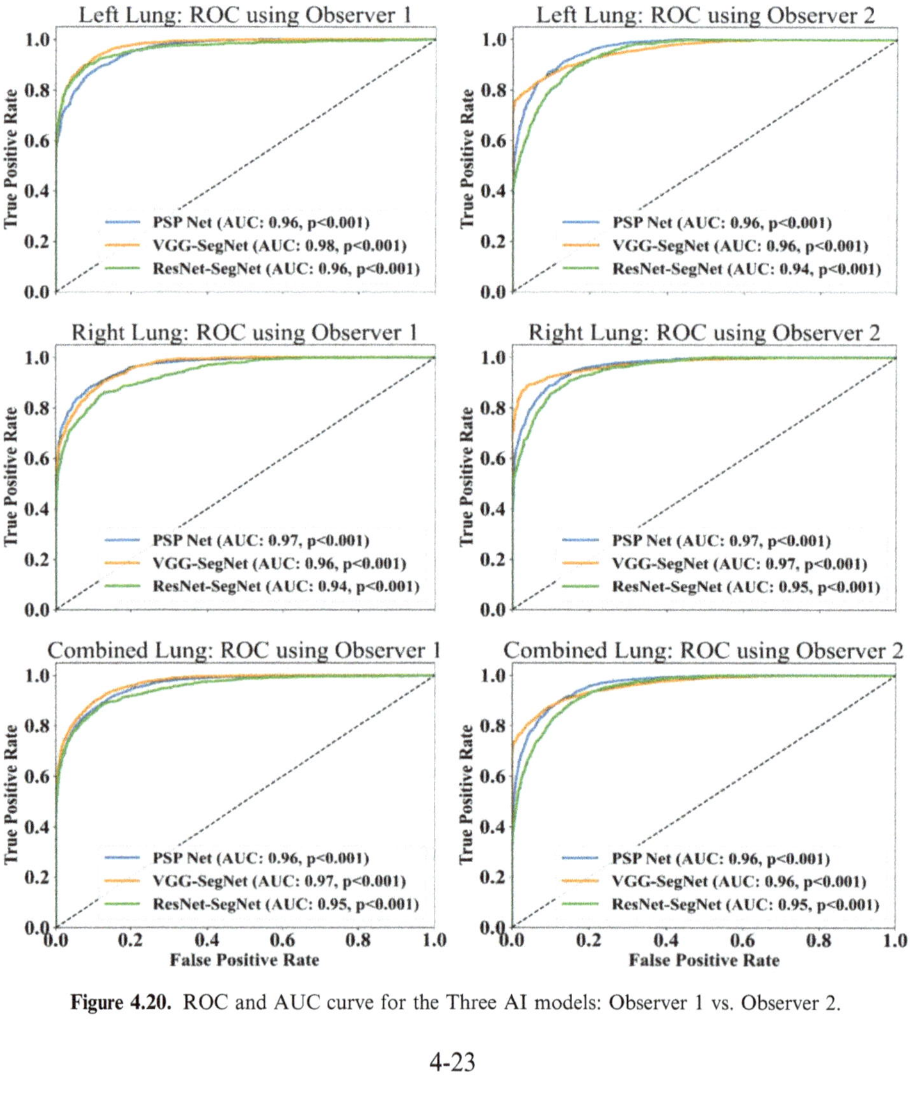

Figure 4.20. ROC and AUC curve for the Three AI models: Observer 1 vs. Observer 2.

and specificity and determine the optimal discrimination threshold for each AI model. The associated AUC values provide a quantitative measure of the model's performance, where a value closer to 1 indicates higher accuracy in differentiating between the two observers. The inclusion of the reference numbers and the corresponding sequence of the AI models ensures clear identification and connection with the provided information, facilitating a comprehensive understanding of the ROC analysis and its implications for the AI models' diagnostic performance.

4.4.2.3 Performance evaluation using lung long axis error

Lung long axis error cumulative frequency plot

Figures 4.21 and 4.22 present the cumulative frequency plots for the LLAE in the left and right lungs, corresponding to Observer 1 and Observer 2. These plots depict the distribution of LLAE values across the CT scans analyzed using the three AI models (PSP Net, VGG-SegNet, and ResNet-SegNet). For the left lung, using the 80% threshold, the LLAE values obtained for Observer 1 were 6.12 mm (PSP Net), 4.77 mm (VGG-SegNet), and 5.01 mm (ResNet-SegNet). On the other hand, for Observer 2, the LLAE values were 10.88 mm (PSP Net), 13.30 mm (VGG-SegNet), and 9.18 mm (ResNet-SegNet). These values represent the average discrepancy between the AI-generated long axis measurements and the ground truth measurements, indicating the accuracy of the AI models in estimating the lung long axis. By incorporating the reference numbers and specifying the corresponding AI models, it becomes easier to comprehend the LLAE analysis for each observer. The provided information allows for a comprehensive evaluation of the AI models' performance in estimating the lung long axis and the potential variations between different observers.

Correlation plot for lung long axis error

Figures 4.23 and 4.24 display the CC plots for the three AI models, focusing on the inter-observer variability analysis between Observers 1 and 2. These plots demonstrate the correlation between the AI-generated LLA measurements and the corresponding ground truth measurements for the left and right lungs. Table 4.2 provides the CC values for the right, left, and mean errors of the LLA. These values quantify the degree of correlation between the LLA estimates obtained from the AI models and the actual measurements, separately for each observer. Additionally, the table presents the percentage difference in CC values between Observers 1 and 2. The CC analysis results presented in table 4.2 support the assertion that there is a 5% difference in the outcomes obtained when comparing the two observers. This finding indicates that, within the specified inter-observer variability research scenario, the proposed system demonstrates clinical viability. The consistency of the AI models' performance across different observers is indicative of their reliability in estimating the lung long axis. By referring to figures 4.23 and 4.24 and analyzing the CC values in table 4.2, a comprehensive evaluation of inter-observer variability can be conducted. This analysis provides quantitative evidence of the AI models' performance and their ability to produce consistent results despite variations between observers.

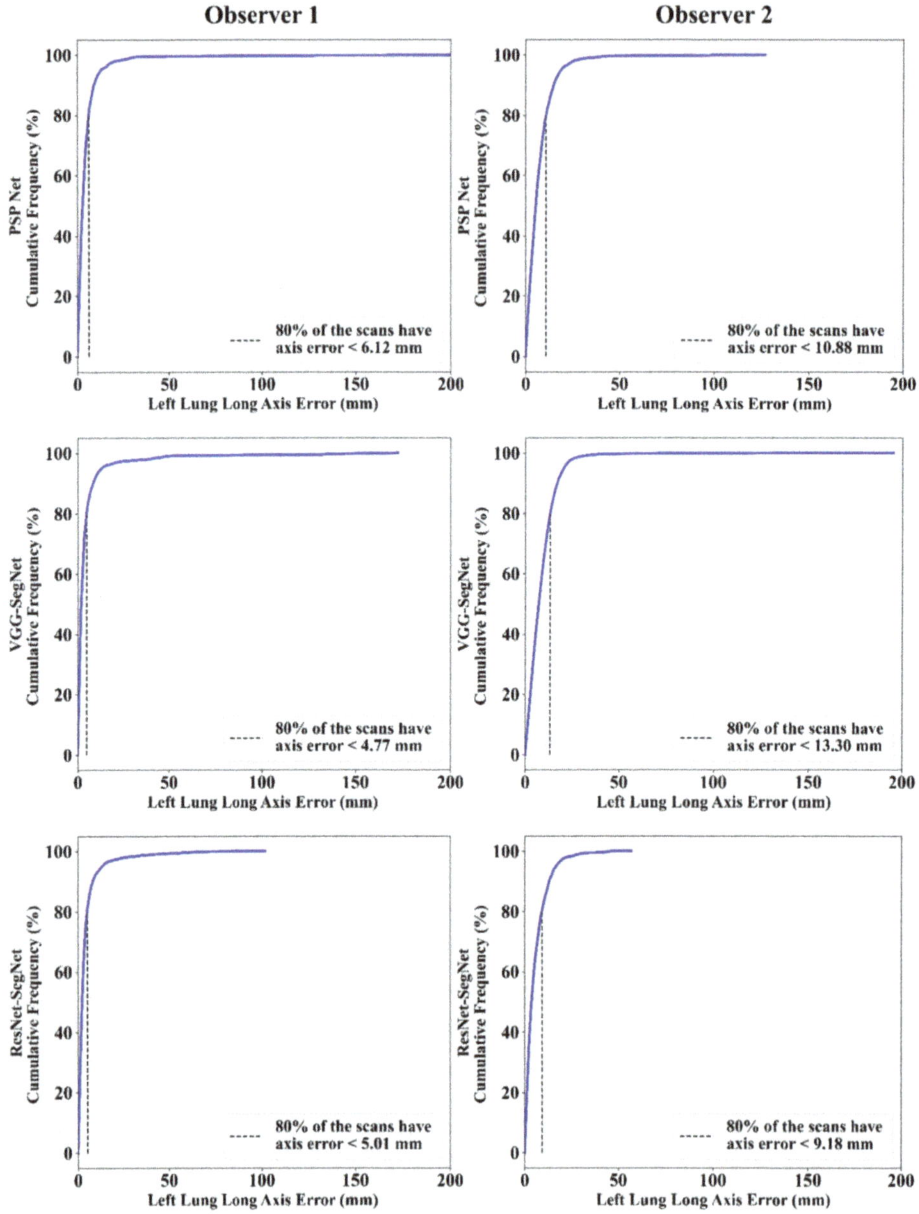

Figure 4.21. Cumulative frequency plot for left LLAE using three AI models: Observer 1 vs. Observer 2

Bland–Altman plots for lung long axis error

Figure 4.25 for the left lung and figure 4.26 for the right lung display the (I) standard deviation of the lung long axis corresponding to Observer 1 and Observer 2 for the three AI models and (ii) mean.

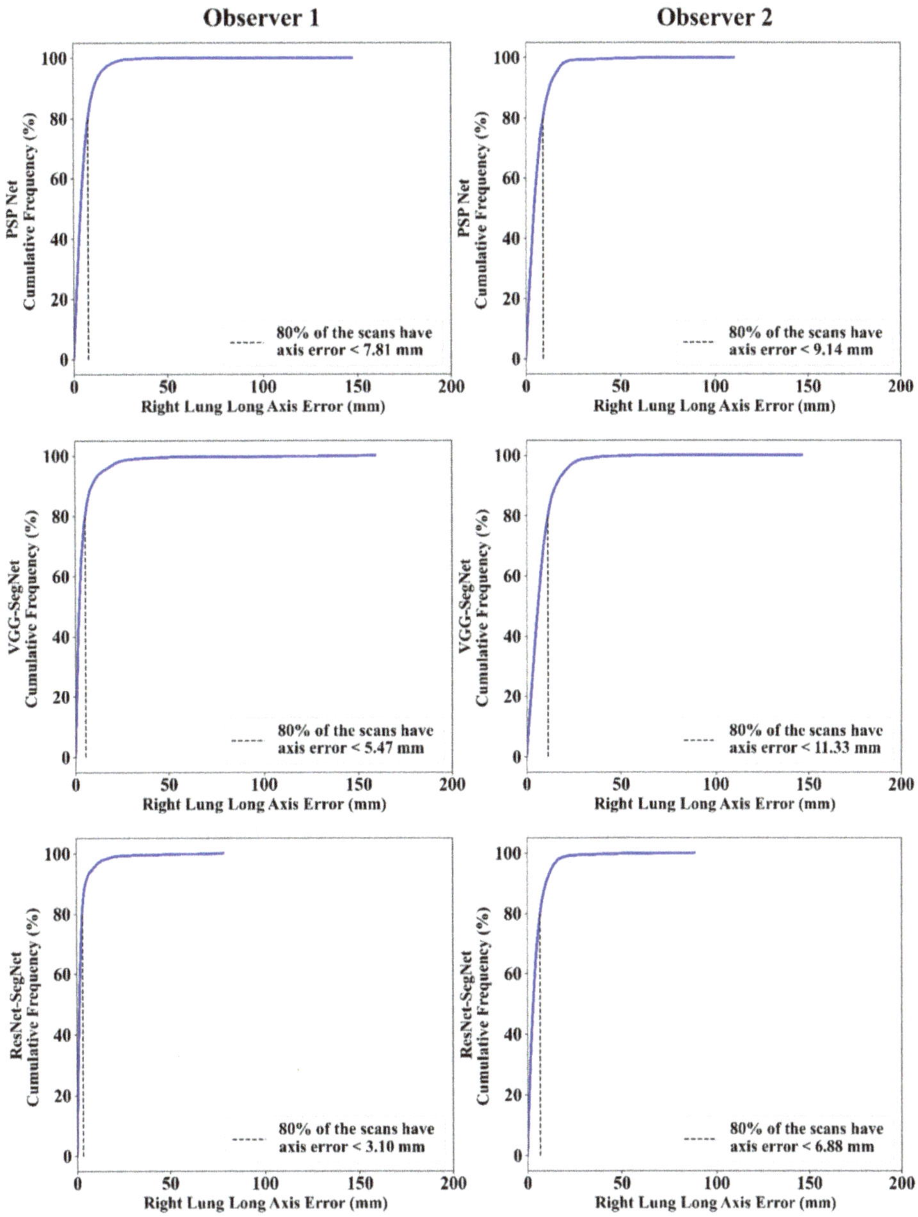

Figure 4.22. Cumulative frequency plot for right LLAE using three AI models: Observer 1 vs. Observer 2.

Statistical tests

The reliability and stability of the system were evaluated using a basic paired *t*-test, ANOVA, and Wilcoxon test. The paired *t*-test was used when the distribution was normal, while the Wilcoxon test was used when the distribution was not normal. ANOVA was employed to study the variance between group means of the input data. The statistical analysis was conducted using MedCalc software. In order to

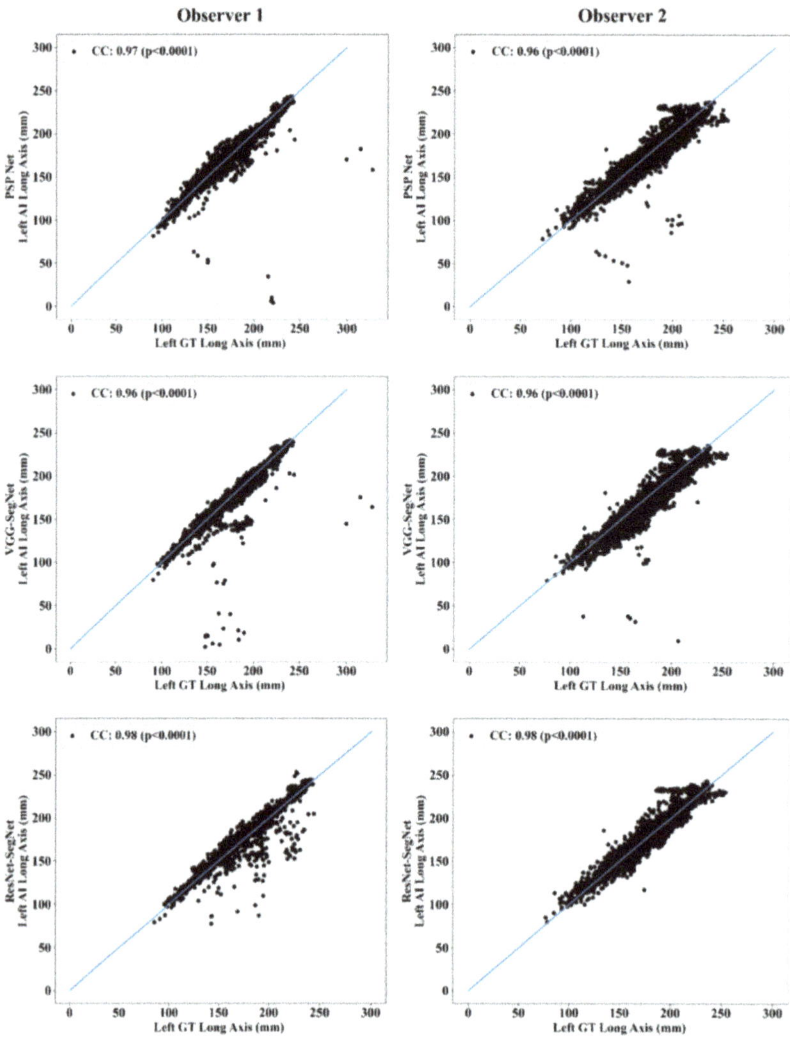

Figure 4.23. CC of left LL using three AI models: Observer 1 vs. Observer 2.

validate the system, all possible combinations of the three AI models between Observer 1 and Observer 2 were considered, resulting in a total of twelve combinations. The results of the paired ANOVA, Wilcoxon test, and *t*-test for these twelve combinations are presented in table 4.3.

Figure of merit

The figure of merit (FoM) is a metric that quantifies the probability of a system fault. In this study, we calculated FoM to demonstrate the acceptability of the hypothesis when the percent difference between the two observers is 5%. FoM was calculated for two parameters: (i) lung area and (ii) lung long axis. Table 4.4 presents the FoM

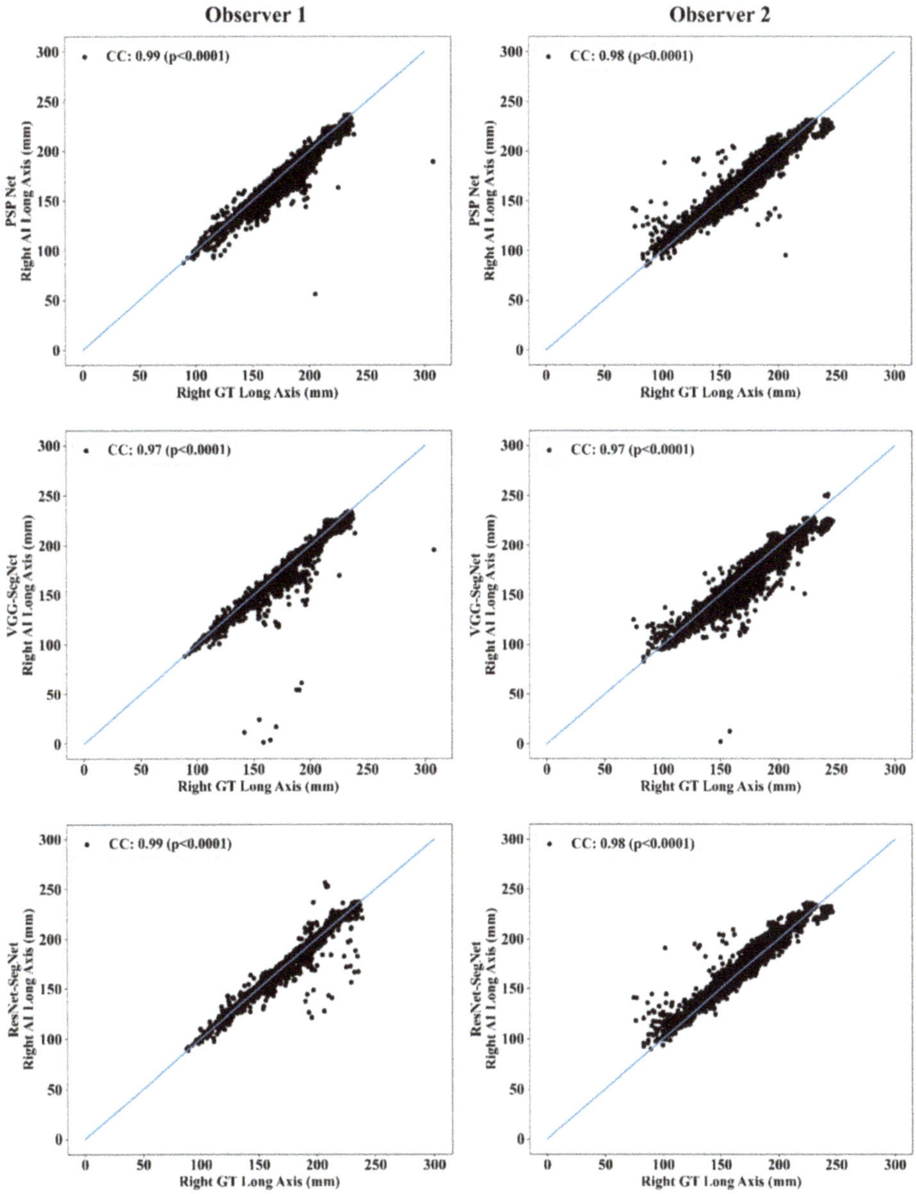

Figure 4.24. CC of right LLA using three AI models: Observer 1 vs. Observer 2.

values obtained using equation (4.5) along with the percentage difference between the three AI models and the two observers. Additionally, table 4.5 shows similar FoM values obtained using equation (4.6) and the corresponding percentage difference between the three AI models and the two observers. These tables provide insight into the acceptability of the hypothesis based on the calculated FoM values.

Table 4.2. Comparison of the CC values obtained between AI model lung long axis and the GT lung long axis corresponding to Observer 1 and Observer 2.

	Left	PSP Net Right	Mean	Left	VGG-SegNet Right	Mean	Left	ResNet-SegNet Right	Mean
Observer 1	0.97	0.99	0.98	0.96	0.97	0.97	0.98	0.99	0.99
Observer 2	0.96	0.98	0.97	0.96	0.97	0.97	0.98	0.98	0.98
% Difference	1.03	1.01	1.02	0.00	0.00	0.00	0.00	1.01	0.51

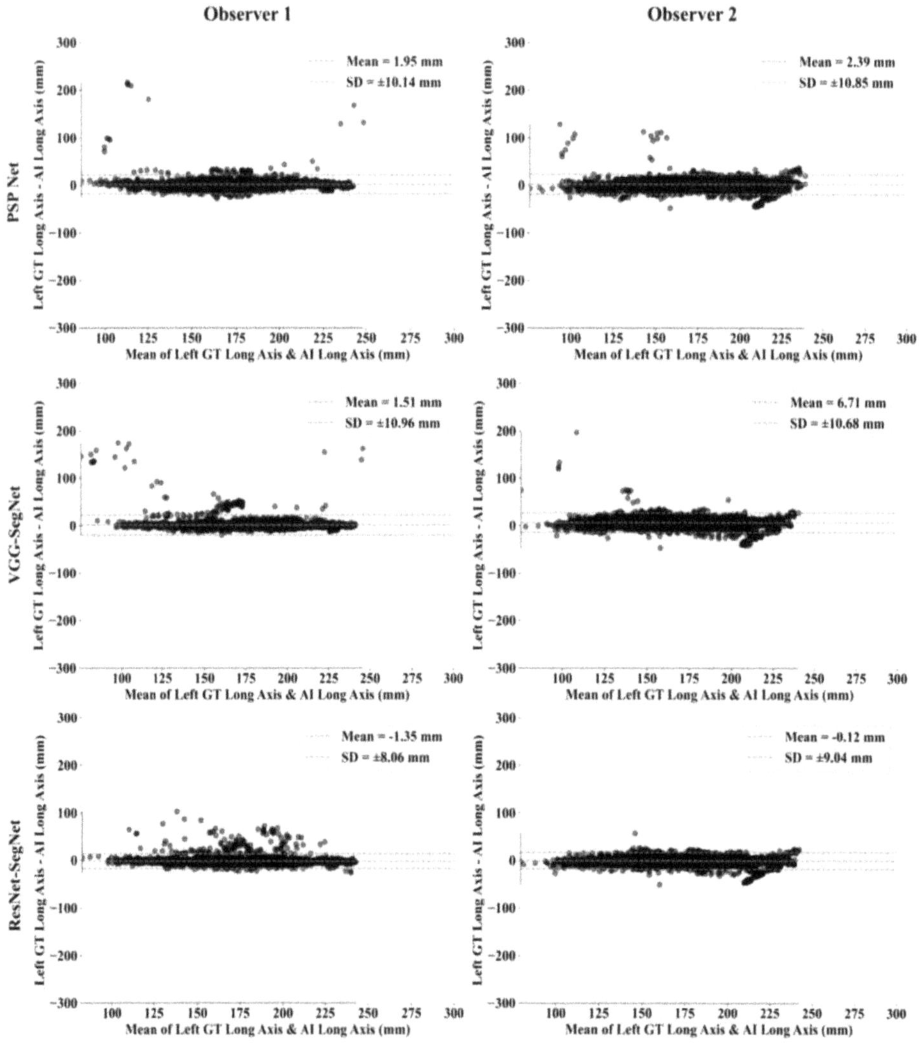

Figure 4.25. BA for the left LLA using the three: Observer 1 vs. Observer 2.

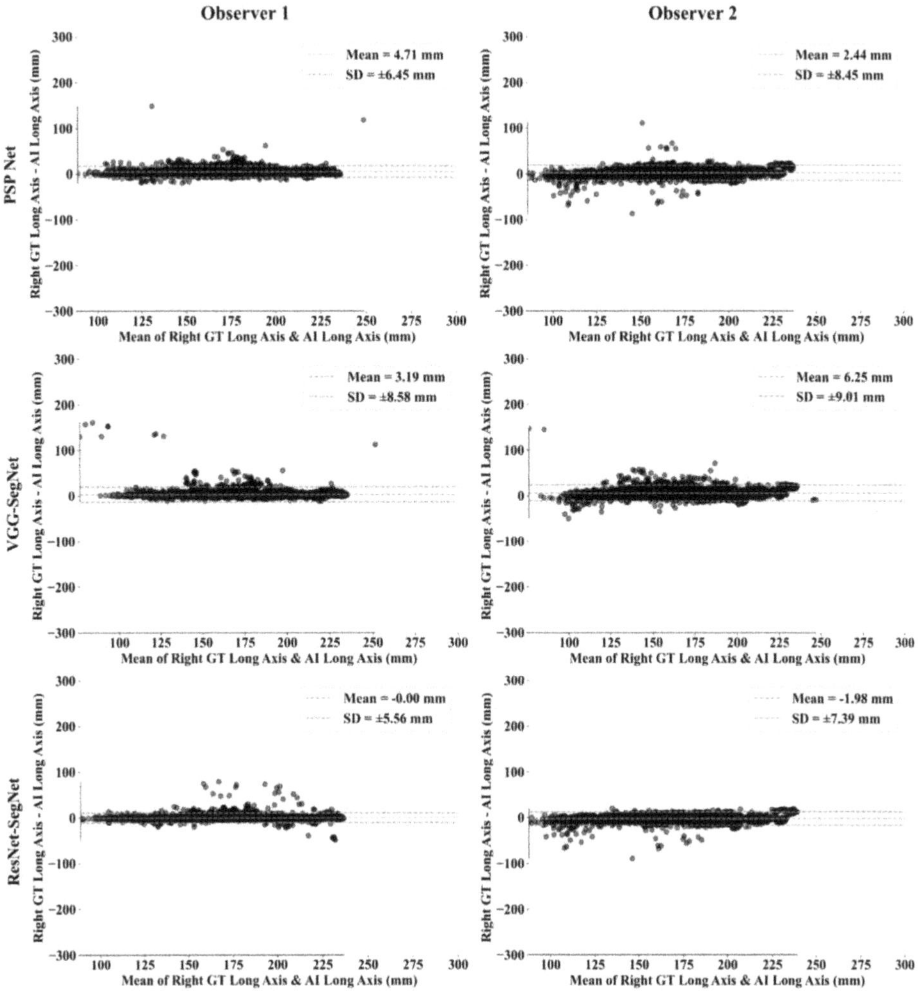

Figure 4.26. BA for the right LLA using the three AI models: Observer 1 vs. Observer 2.

$$FoM_A(m) = 100 - \left[\left(\frac{|\overline{A}_{ai}(m) - \overline{A}_{gt}|}{\overline{A}_{gt}}\right) \times 100\right],$$
$$FoM_{LA}(m) = 100 - \left[\left(\frac{|\overline{L}_{ai}(m) - \overline{L}_{gt}|}{\overline{L}_{gt}}\right) \times 100\right] \tag{4.5}$$

where $\overline{A}_{ai}(m) = \frac{\sum_{n=1}^{N} A_{ai}(m,n)}{N}$, $\overline{A}_{gt} = \frac{\sum_{n=1}^{N} A_{gt}(n)}{N}$,

Table 4.3. Paired t-test, Wilcoxon, ANOVA, and CC for LA and LLA for the 12 combinations.

		Lung area				Lung long axis			
SN	Combinations	Paired t-Test (p-Value)	Wilcoxon (p-Value)	ANOVA (p-Value)	CC [0–1]	Paired t-Test (p-Value)	Wilcoxon (p-Value)	ANOVA (p-Value)	CC [0–1]
1	P1 VS. V1	<0.0001	<0.0001	<0.001	0.9726	<0.0001	<0.0001	<0.001	0.9509
2	P1 vs. R1	<0.0001	<0.0001	<0.001	0.9514	<0.0001	<0.0001	<0.001	0.9506
3	P1 vs. P2	<0.0001	<0.0001	<0.001	0.9703	<0.0001	<0.0001	<0.001	0.9686
4	P1 vs. V2	<0.0001	<0.0001	<0.001	0.9446	<0.0001	<0.0001	<0.001	0.9445
5	P1 vs. R2	<0.0001	<0.0001	<0.001	0.9764	<0.0001	<0.0001	<0.001	0.9661
6	V1 vs. R1	<0.0001	<0.0001	<0.001	0.9663	<0.0001	<0.0001	<0.001	0.9561
7	V1 vs. P2	<0.0001	<0.0001	<0.001	0.9726	<0.0001	<0.0001	<0.001	0.9671
8	V1 vs. V2	<0.0001	<0.0001	<0.001	0.9766	<0.0001	<0.0001	<0.001	0.9638
9	V1 vs. R2	<0.0001	<0.0001	<0.001	0.9943	<0.0001	<0.0001	<0.001	0.9796
10	R1 VS. P2	<0.0001	<0.0001	<0.001	0.9549	<0.0001	<0.0001	<0.001	0.9617
11	R1 vs. V2	<0.0001	<0.0001	<0.001	0.9513	<0.0001	<0.0001	<0.001	0.9499
12	R1 vs. R2	<0.0001	<0.0001	<0.001	0.9690	<0.0001	<0.0001	<0.001	0.9726

Table 4.4. FoM for lung area.

	Observer 1			Observer 2			% Difference			Hypothesis (<5%)		
	Left	Right	Mean	Left	Right	Mean	Left	Right	Mean	Left	Right	Mean
PSP Net	95.07	95.11	95.09	97.37	97.49	97.43	2%	3%	2%	C	C	C
VGG-SegNet	96.73	97.40	97.04	97.74	97.27	97.52	1%	0%	0%	C	C	C
ResNet-SegNet	98.33	99.98	99.11	97.88	99.20	98.50	0%	1%	1%	C	C	C

$$\overline{LA}_{ai}(m) = \frac{\sum_{n=1}^{N} LA_{ai}(m, n)}{N} \ \& \ \overline{LA}_{gt} = \frac{\sum_{n=1}^{N} LA_{gt}(n)}{N} \tag{4.6}$$

4.5 Discussion

The study utilized three AI models, namely PSP Net, VGG-SegNet, and ResNet-SegNet, to analyze inter-observer variability in COVLIAS 1.0. These models have a focus on tissue characterization within the classification framework, allowing for better feature extraction to distinguish between ground and background [30, 37]. Our team has extensive expertise in tissue characterization techniques using various AI models and applications, including plaque, liver, thyroid, and breast, employing ML frameworks [21, 28, 30, 63–68], as well as DL frameworks [1, 36, 69, 70]. For

Table 4.5. FoM for lung long axis.

	Left	Observer 1 Right	Mean	Left	Observer 2 Right	Mean	Left	% Difference Right	Mean	Left	Hypothesis (<5%) Right	Mean
PSP Net	98.91	97.34	98.13	98.65	98.60	98.62	0%	1%	1%	C	C	C
VGG-SegNet	99.41	98.50	98.95	97.07	97.27	97.17	2%	1%	2%	C	C	C
ResNet-SegNet	99.73	99.37	99.83	99.51	98.75	99.13	0%	1%	1%	C	C	C

this study, the GT annotated data provided by the two observers were used to train the three AI models. The hypothesis was supported as the percentage difference between the outputs of the two AI models was found to be less than 5%. This demonstrates the effectiveness of the proposed system in addressing inter-observer variability.

4.5.1 A special note on three model behaviors with respect to the two observers

In the suggested inter-observer variability research, three AI models were employed, including PSP Net, which was used for COVID-19 lung segmentation for the first time. Benchmarking was conducted using the other models, namely VGG-SegNet and ResNet-SegNet. The AUC for the mean lung area exceeded 0.95 for all three AI models, indicating strong performance. Table 4.6 presents the findings, encompassing various metrics and the inter-observer variability research. The results indicated consistent behavior of all models when utilizing the two different observers. ResNet-SegNet exhibited the best performance across all performance evaluation measures. The percentage difference between the two observers for PSP Net, VGG-SegNet, and ResNet-SegNet was 0.4%, 3.7%, and 0.4%, respectively. These results further support the hypothesis and maintain an error threshold of less than 5% for each AI model. While all three AI models performed successfully, VGG-SegNet was found to be the least effective. This can be attributed to the architectural differences, as ResNet-SegNet and PSP Net both have 51 layers (including the encoder section), while VGG-SegNet has only 19 layers. Considering the findings from both observers, the order of model performance was determined as ResNet-SegNet > PSP Net > VGG-SegNet. Additionally, it can be concluded that the HDL models

Table 4.6. Comparison of PE metrics for Observer 1 and Observer 2 and their mean.

	Observer 1			Observer 2			Mean Obs. 1 & Obs. 2		
Attributes	PSP Net	VGG-SegNet	ResNet-SegNet	PSP Net	VGG-SegNet	ResNet-SegNet	PSP Net	VGG-SegNet	ResNet-SegNet
DS	0.96	0.98	0.98	0.96	0.95	0.97	0.96	0.97	0.98
JI	0.93	0.96	0.97	0.92	0.9	0.94	0.93	0.93	0.96
CC Left LA	0.98	0.98	0.98	0.98	0.98	1	0.98	0.98	0.99
CC Right LA	0.98	0.99	0.98	0.98	0.98	1	0.98	0.99	0.99
CC Left LLA	0.97	0.96	0.98	0.96	0.96	0.98	0.97	0.96	0.98
CC Right LLA	0.99	0.97	0.99	0.98	0.97	0.98	0.99	0.97	0.99
CF Left LA < 10%	0.83	0.85	0.90	0.81	0.75	0.89	0.82	0.80	0.89
CF Right LA < 10%	0.78	0.85	0.90	0.80	0.75	0.88	0.79	0.80	0.89
Aggregate Score	7.42	7.54	7.67	7.39	7.24	7.64	7.40	7.39	7.66

outperformed the shallow deep learning (SDL) model (PSP Net). Aggregate scores were obtained by taking the mean of all models for Observer 1, Observer 2, and the combined mean of the two observers. Although the performance of all models was comparable, ResNet-SegNet, VGG-SegNet, and PSP Net ranked in that order when closely examining Observer 1's performance.

4.5.2 Benchmarking

There have been relatively fewer investigations conducted specifically on COVID-19 using DL for lung segmentation compared to other applications [71–74], and even fewer studies have included variability analysis. Table 4.7 presents a benchmarking comparison with three variability studies that were evaluated alongside the suggested study by Suri *et al* These studies include Saba *et al* [48], Jeremy *et al* [75], and Joskowicz *et al* [76]. Saba *et al* utilized three observers and a dataset of 96 patients for their tracings, but the study did not include ROC curves. Jeremy *et al* demonstrated variability analysis with five different observers using the area error as the measure. However, there was no discussion of boundary error, ROC curves, JI, or DS. Joskowicz *et al* annotated a dataset of 480 images using 11 observers, but there were no boundary or area errors reported. Similar to the other studies, ROC curves, JI, and DS for the tracings were not presented. It is important to note that all three studies [48, 75, 76] used manual annotation exclusively, without employing AI techniques for automatic boundary delineation. In comparison, the suggested study provides a unique contribution by offering both HDL and shallow deep learning (SDL) solutions, along with an inter-observer variability analysis. This study

Table 4.7. Benchmarking table.

Attributes/Author	Saba *et al* [49]	Jeremy *et al* [77]	Joskowicz *et al* [78]	Suri *et al* (Proposed)
# of patients	96	33	18	72
# of Images	NA	NA	490	5000
# of Observers	3	5	11	2
Dataset	Non-COVID	Non-COVID	Non-COVID	COVID
Image Size	512	NA	512	768
# of tests/PE	5	0	2	13
CC	0.98	NA	NA	0.98
Boundary estimation	Manual	Manual	Manual	Manual & automatic
AI Models	NA	NA	NA	3
Modality	CT	CT	CT	CT
Area Error	C	C	C	C
Boundary Error	C	C	C	C
ROC	C	C	C	C
JI	C	C	C	C
DS	C	C	C	C

demonstrates that the error between the AI models trained using the two observers is less than 5%, supporting the claim made by the authors.

4.5.3 Strengths, weakness, and extensions

The suggested investigation successfully validated the hypothesis for inter-observer variability in COVID-19 lung segmentation using both shallow deep learning (SDL) and HDL models. However, the study had certain limitations, including the use of only two observers due to factors such as cost, time, and the availability of radiologists. To address the segmentation of denser pulmonary opacities like consolidation or mixed opacities, the imaging analysis component could be expanded. In the extension of the study, the HDL models, which combine ML and DL, or two separate DL models, could be further developed. Conventional techniques for lung segmentation [77, 78] could be employed along with denoising techniques [79] and compared with AI models. The system could be expanded to include unseen data by training the models on data from one clinical site and testing them on data from a different clinical location. It would also be interesting to investigate how the AI models trained on COVID-19 patients perform in segmenting lungs in healthy individuals. Other neural network methods, such as transfer learning and different loss schemes [38, 44, 81], or generative adversarial networks (GANs) [80], could be explored and modified as well [82].

4.6 Conclusions

The proposed study represents the first attempt to assess the influence of GT tracings on AI lung segmentation models specifically designed for COVID-19 CT scans. The study employs three distinct AI models: PSP Net, VGG-SegNet, and ResNet-SegNet, all of which are utilized for the purpose of lung segmentation. By using these different AI models, the study aims to explore and compare their performance in accurately segmenting lung regions in COVID-19 CT scans. Each model brings its own unique architectural design and approach to the task of lung segmentation, allowing for a comprehensive evaluation of their capabilities and effectiveness in this specific context. This study holds significance as it provides insights into the potential of AI models for COVID-19 lung segmentation, shedding light on their performance when trained with GT tracings. The results obtained from this study could have implications for the development and improvement of AI-assisted diagnostic tools for COVID-19 based on CT scans. Two independent observers annotated 5000 CT lung slices from 72 COVID-19 patients. This led to the creation of six AI training models by combining the three AI models with the annotations from the two observers. The performance of these six AI models was evaluated using a K5 cross-validation methodology and assessed using ten different criteria. To validate the hypothesis and ensure adherence to the 5% error threshold, the error metrics obtained from the two observers were compared for each AI model. The results showed that the differences in these errors were 0%, 0.51%, and 2.04% for the three AI models, respectively, all within the specified threshold, thus confirming the hypothesis. Statistical analysis, including basic paired *t*-test, ANOVA, and

Wilcoxon test, was conducted to support the proposed system's theory. Overall, the pilot investigation using the inter-variability COVLIAS 1.0 demonstrated clinically valid and statistically stable results. This indicates that the system is adaptable and can be applied effectively in clinical settings.

References

[1] Agarwal M, Saba L, Gupta S K, Johri A M, Khanna N N, Mavrogeni S, Laird J R, Pareek G, Miner M and Sfikakis P P 2021 Wilson disease tissue classification and characterization using seven artificial intelligence models embedded with 3D optimization paradigm on a weak training brain magnetic resonance imaging datasets: a supercomputer application *Med. Biol. Eng. Comput.* **59** 511–33

[2] Cau R, Pacielli A, Fatemeh H, Vaudano P, Arru C, Crivelli P, Stranieri G, Suri J S, Mannelli L and Conti M *et al* 2021 Complications in COVID-19 patients: characteristics of pulmonary embolism *Clin. Imaging* **77** 244–9

[3] Saba L, Gerosa C, Fanni D, Marongiu F, La Nasa G, Caocci G, Barcellona D, Balestrieri A, Coghe F and Orru G *et al* 2020 Molecular pathways triggered by COVID-19 in different organs: ACE2 receptor-expressing cells under attack? A review *Eur. Rev. Med. Pharmacol. Sci.* **24** 12609–22

[4] Cau R, Bassareo P P, Mannelli L, Suri J S and Saba L 2021 Imaging in COVID-19-related myocardial injury *Int. J. Cardiovasc. Imaging* **37** 1349–60

[5] Viswanathan V, Viswanathan V, Puvvula A, Jamthikar A D, Saba L, Johri A M, Kotsis V, Khanna N N, Dhanjil S K and Majhail M *et al* 2021 Bidirectional link between diabetes mellitus and coronavirus disease 2019 leading to cardiovascular disease: a narrative review *World J. Diabetes* **12** 215–37

[6] Suri J S, Agarwal S, Gupta S K, Puvvula A, Biswas M, Saba L, Bit A, Tandel G S, Agarwal M and Patrick A *et al* 2021 A narrative review on characterization of acute respiratory distress syndrome in COVID-19-infected lungs using artificial intelligence *Comput. Biol. Med.* **130** 104210

[7] Cau R, Falaschi Z, Paschè A, Danna P, Arioli R, Arru C D, Zagaria D, Tricca S, Suri J S and Karla M K *et al* 2021 Computed tomography findings of COVID-19 pneumonia in Intensive Care Unit-patients *J. Public Health Res.* **10** 2270

[8] Emery S L, Erdman D D, Bowen M D, Newton B R, Winchell J M, Meyer R F, Tong S, Cook B T, Holloway B P and McCaustland K A *et al* 2004 Real-time reverse transcription–polymerase chain reaction assay for SARS-associated coronavirus *Emerg. Infect. Dis.* **10** 311–6

[9] Wu X, Hui H, Niu M, Li L, Wang L, He B, Yang X, Li L, Li H and Tian J *et al* 2020 Deep learning-based multi-view fusion model for screening 2019 novel coronavirus pneumonia: a multicentre study *Eur. J. Radiol.* **128** 109041

[10] Pathak Y, Shukla P K, Tiwari A, Stalin S and Singh S 2020 Deep transfer learning based classification model for COVID-19 disease *IRBM* **43** 87–92

[11] Saba L and Suri J S 2013 *Multi-Detector CT Imaging: Principles, Head, Neck, and Vascular Systems* vol 1 (Boca Raton, FL: CRC Press)

[12] Gozes O, Frid-Adar M, Greenspan H, Browning P D, Zhang H, Ji W, Bernheim A and Siegel E 2020 Rapid ai development cycle for the coronavirus (COVID-19) pandemic: initial results for automated detection & patient monitoring using deep learning ct image analysis *arXiv* arXiv:05037

[13] Shalbaf A and Vafaeezadeh M 2021 Automated detection of COVID-19 using ensemble of transfer learning with deep convolutional neural network based on CT scans *Int. J. Comput. Assist. Radiol. Surg.* **16** 115–23
[14] Yang X, He X, Zhao J, Zhang Y, Zhang S and Xie P 2020 COVID-CT-dataset: a CT scan dataset about COVID-19 *arXiv* arXiv:13865
[15] Alqudah A M, Qazan S, Alquran H, Qasmieh I A and Alqudah A 2020 COVID-2019 Detection using X-ray images and artificial intelligence hybrid systems *Phys. Sci.* **2** 1
[16] Aslan M F, Unlersen M F, Sabanci K and Durdu A 2021 CNN-based transfer learning–BiLSTM network: a novel approach for COVID-19 infection detection *Appl. Soft Comput.* **98** 10691
[17] Wu Y H, Gao S H, Mei J, Xu J, Fan D P, Zhang R G and Cheng M M 2021 Jcs: an explainable COVID-19 diagnosis system by joint classification and segmentation *IEEE Trans. Image Process.* **30** 3113–26
[18] Xu X, Jiang X, Ma C, Du P, Li X, Lv S, Yu L, Ni Q, Chen Y and Su J *et al* 2020 A deep learning system to screen novel coronavirus disease 2019 pneumonia *Engineering* **6** 1122–9
[19] El-Baz A and Suri J 2021 *Machine Learning in Medicine* (Boca Raton, FL: CRC Press)
[20] Suri J S and Rangayyan R M 2006 *Recent Advances in Breast Imaging, Mammography, and Computer-Aided Diagnosis of Breast Cancer* (Bellingham, WA: SPIE Publications)
[21] Biswas M, Kuppili V, Edla D R, Suri H S, Saba L, Marinhoe R T, Sanches J M and Suri J S 2018 Symtosis: a liver ultrasound tissue characterization and risk stratification in optimized deep learning paradigm *Comput. Methods Programs Biomed.* **155** 165–77
[22] Acharya U R, Sree S V, Ribeiro R, Krishnamurthi G, Marinho R T, Sanches J and Suri J S 2012 Data mining framework for fatty liver disease classification in ultrasound: a hybrid feature extraction paradigm *Med. Phys.* **39** 4255–64
[23] Acharya U R, Sree S V, Krishnan M M R, Molinari F, Garberoglio R and Suri J S 2012 Non-invasive automated 3D thyroid lesion classification in ultrasound: a class of ThyroScan™ systems *Ultrasonics* **52** 508–20
[24] Acharya U R, Swapna G, Sree S V, Molinari F, Gupta S, Bardales R H, Witkowska A and Suri J S 2014 A review on ultrasound based thyroid cancer tissue characterization and automated classification *Technol. Cancer Res. Treat.* **13** 289–301
[25] Molinari F, Mantovani A, Deandrea M, Limone P, Garberoglio R and Suri J S 2010 Characterization of single thyroid nodules by contrast-enhanced 3-D ultrasound *Ultrasound Med. Biol.* **36** 1616–25
[26] Shrivastava V K, Londhe N D, Sonawane R S and Suri J S 2016 Computer-aided diagnosis of psoriasis skin images with HOS, texture and color features: a first comparative study of its kind *Comput. Methods Programs Biomed.* **126** 98–109
[27] Shrivastava V K, Londhe N D, Sonawane R S and Suri J S 2015 Reliable and accurate psoriasis disease classification in dermatology images using comprehensive feature space in machine learning paradigm *Expert Syst. Appl.* **42** 6184–95
[28] Pareek G, Acharya U R, Sree S V, Swapna G, Yantri R, Martis R J, Saba L, Krishnamurthi G, Mallarini G and El-Baz A *et al* 2013 Prostate tissue characterization/classification in 144 patient population using wavelet and higher order spectra features from transrectal ultrasound images *Technol. Cancer Res. Treat.* **12** 545–57
[29] McClure P, Elnakib A, El-Ghar M A, Khalifa F, Soliman A, El-Diasty T, Suri J S, Elmaghraby A and El-Baz A 2014 In-vitro and in-vivo diagnostic techniques for prostate cancer: a review *J. Biomed. Nanotechnol.* **10** 2747–77

[30] Mookiah M R K, Acharya U R, Martis R J, Chua C K, Lim C M, Ng E Y K and Laude A 2013 Evolutionary algorithm based classifier parameter tuning for automatic diabetic retinopathy grading: a hybrid feature extraction approach *Knowl.-Based Syst.* **39** 9–22

[31] Than J C, Saba L, Noor N M, Rijal O M, Kassim R M, Yunus A, Suri H S, Porcu M and Suri J S 2017 Lung disease stratification using amalgamation of Riesz and Gabor transforms in machine learning framework *Comput. Biol. Med.* **89** 197–211

[32] El-Baz A, Jiang X and Suri J S 2016 *Biomedical Image Segmentation: Advances and Trends* (Boca Raton, FL: CRC Press)

[33] Than J C, Saba L, Noor N M, Rijal O M, Kassim R M, Yunus A, Suri H S, Porcu M and Suri J S 2002 Shape recovery algorithms using level sets in 2-D/3-D medical imagery: a state-of-the-art review *IEEE Trans. Inf. Technol. Biomed.* **6** 8–28

[34] El-Baz A S, Acharya R, Mirmehdi M and Suri J S 2011 *Multi Modality State-of-the-Art Medical Image Segmentation and Registration Methodologies* vol 2 (Berlin: Springer Science & Business Media)

[35] El-Baz A and Suri J S 2019 *Level Set Method in Medical Imaging Segmentation* (Boca Raton, FL: CRC Press)

[36] Saba L *et al* 2021 Multimodality carotid plaque tissue characterization and classification in the artificial intelligence paradigm: a narrative review for stroke application *Ann. Transl. Med.* **9** 1206

[37] Acharya U R, Sree S V, Krishnan M M R, Krishnananda N, Ranjan S, Umesh P and Suri J S 2013 Automated classification of patients with coronary artery disease using grayscale features from left ventricle echocardiographic images *Comput. Methods Programs Biomed.* **112** 624–32

[38] Agarwal M, Saba L, Gupta S K, Carriero A, Falaschi Z, Paschè A, Danna P, El-Baz A, Naidu S and Suri J S 2021 A novel block imaging technique using nine artificial intelligence models for COVID-19 disease classification, characterization and severity measurement in lung computed tomography scans on an Italian cohort *J. Med. Syst.* **45** 1–30

[39] Saba L, Agarwal M, Patrick A, Puvvula A, Gupta S K, Carriero A, Laird J R, Kitas G D, Johri A M and Balestrieri A *et al* 2021 Six artificial intelligence paradigms for tissue characterisation and classification of non-COVID-19 pneumonia against COVID-19 pneumonia in computed tomography lungs *Int. J. Comput. Assist. Radiol. Surg.* **16** 423–34

[40] Skandha S S, Gupta S K, Saba L, Koppula V K, Johri A M, Khanna N N, Mavrogeni S, Laird J R, Pareek G and Miner M *et al* 2020 3-D optimized classification and characterization artificial intelligence paradigm for cardiovascular/stroke risk stratificationusing carotid ultrasound-based delineated plaque: Atheromatic™ 2.0 *Comput. Biol. Med.* **125** 103958

[41] Tandel G S, Balestrieri A, Jujaray T, Khanna N N, Saba L and Suri J S 2020 Multiclass magnetic resonance imaging brain tumor classification using artificial intelligence paradigm *Comput. Biol. Med.* **122** 103804

[42] Sarker M M K, Makhlouf Y, Banu S F, Chambon S, Radeva P and Puig D 2020 Web-based efficient dual attention networks to detect COVID-19 from x-ray images *Electron. Lett.* **56** 1298–301

[43] Sarker M M K, Makhlouf Y, Craig S G, Humphries M P, Loughrey M, James J A, Salto-Tellez M, O'Reilly P and Maxwell P 2021 A means of assessing deep learning-based detection of ICOS protein expression in colon cancer *Cancers* **13** 3825

[44] Jain P K, Sharma N, Giannopoulos A A, Saba L, Nicolaides A and Suri J S 2021 Hybrid deep learning segmentation models for atherosclerotic plaque in internal carotid artery B-mode ultrasound *Comput. Biol. Med.* **136** 104721

[45] Jena B, Saxena S, Nayak G K, Saba L, Sharma N and Suri J S 2021 Artificial intelligence-based hybrid deep learning models for image classification: the first narrative review *Comput. Biol. Med.* **137** 104803

[46] Suri J, Agarwal S, Gupta S K, Puvvula A, Viskovic K, Suri N, Alizad A, El-Baz A, Saba L and Fatemi M *et al* 2021 Systematic review of artificial intelligence in acute respiratory distress syndrome for COVID-19 lung patients: a biomedical imaging perspective *IEEE J. Biomed. Health Inform* **25** 4128–39

[47] Saba L, Banchhor S K, Araki T, Viskovic K, Londhe N D, Laird J R, Suri H S and Suri J S 2018 Intra- and inter-operator reproducibility of automated cloud-based carotid lumen diameter ultrasound measurement *Indian Heart J.* **70** 649–64

[48] Saba L, Than J C, Noor N M, Rijal O M, Kassim R M, Yunus A, Ng C R and Suri J S 2016 Inter-observer variability analysis of automatic lung delineation in normal and disease patients *J. Med. Syst.* **40** 142

[49] Zhang S, Suri J S, Salvado O, Chen Y, Wacker F K, Wilson D L, Duerk J L and Lewin J S 2005 Inter-and intra-observer variability assessment of *in vivo* carotid plaque burden quantification using multi-contrast dark blood MR images *Stud. Health Technol.Inform.* **113** 384–93

[50] Aggarwal D and Saini V 2020 Factors limiting the utility of bronchoalveolar lavage in the diagnosis of COVID-19 *Eur. Respir. J.* **56** 2003116

[51] Saba L, Banchhor S K, Suri H S, Londhe N D, Araki T, Ikeda N, Viskovic K, Shafique S, Laird J R and Gupta A *et al* 2016 Accurate cloud-based smart IMT measurement, its validation and stroke risk stratification in carotid ultrasound: a web-based point-of-care tool for multicenter clinical trial *Comput. Biol. Med.* **75** 217–34

[52] Zhao H, Shi J, Qi X, Wang X and Jia J 2017 Pyramid scene parsing network *Proc. of the IEEE Conf. on Computer Vision and Pattern Recognition (Honolulu, HI, USA, 21–26 July 2017)* pp 2881–90

[53] Simonyan K and Zisserman A 2014 Very deep convolutional networks for large-scale image recognition *arXiv* arXiv:1409.1556

[54] Suri J S, Agarwal S, Pathak R, Ketireddy V, Columbu M, Saba L, Gupta S K, Faa G, Singh I M and Turk M *et al* 2021 COVLIAS 1.0: lung segmentation in COVID-19 computed tomography scans using hybrid deep learning artificial intelligence models *Diagnostics* **11** 1405

[55] He K, Zhang X, Ren S and Sun J 2016 Deep residual learning for image recognition *Proc. of the IEEE Conf. on Computer Vision and Pattern Recognition (Las Vegas, NV, 27–30 June 2016)* (Piscataway, NJ: IEEE) pp 770–8

[56] Acharya U R, Faust O, Sree S V, Molinari F, Saba L, Nicolaides A and Suri J S 2012 An accurate and generalized approach to plaque characterization in 346 carotid ultrasound scans *IEEE Trans. Instrum. Meas.* **61** 1045–53

[57] Acharya U R, Saba L, Molinari F, Guerriero S and Suri J S 2012 Ovarian tumor characterization and classification: a class of GyneScan™ systems *Proc. of the 2012 Annual Int. Conf. of the IEEE Engineering in Medicine and Biology Society (San Diego, CA, 28 August–1 September 2012)* (Piscataway, NJ: IEEE)

[58] Araki T, Ikeda N, Dey N, Acharjee S, Molinari F, Saba L, Godia E C, Nicolaides A and Suri J S 2015 Shape-based approach for coronary calcium lesion volume measurement on intravascular ultrasound imaging and its association with carotid intima-media thickness *J. Ultrasound Med.* **34** 469–82

[59] Barqawi A B, Li L, Crawford E D, Fenster A, Werahera P N, Kumar D, Miller S and Suri J S 2007 Three different strategies for real-time prostate capsule volume computation from 3-D end-fire transrectal ultrasound *Proc. of the 2007 29th Annual Int. Conf. of the IEEE Engineering in Medicine and Biology Society (Lyon, France, 22–26 August 2007)* (Piscataway, NJ: IEEE)

[60] Suri J S, Haralick R M and Sheehan F H 1997 Left ventricle longitudinal axis fitting and its apex estimation using a robust algorithm and its performance: a parametric apex model *Proc. of the Int. Conf. on Image Processing (Santa Barbara, CA, 14–17 July 1997)* (Piscataway, NJ: IEEE)

[61] Singh B K, Verma K, Thoke A S and Suri J S 2017 Risk stratification of 2D ultrasound-based breast lesions using hybrid feature selection in machine learning paradigm *Measurement* **105** 146–57

[62] Riffenburgh R H and Gillen D L 2020 Contents *Statistics in Medicine* (Cambridge, MA: Academic) pp ix–xvi

[63] Acharya R U, Faust O, Alvin A P C, Sree S V, Molinari F, Saba L, Nicolaides A and Suri J S 2012 Symptomatic vs. asymptomatic plaque classification in carotid ultrasound *J. Med. Syst.* **36** 1861–71

[64] Acharya U R, Vinitha Sree S, Mookiah M R K, Yantri R, Molinari F, Ziele´znik W, Małyszek-Tumidajewicz J, St˛epie ´ n B, Bardales R H and Witkowska A *et al* 2013 Diagnosis of Hashimoto's thyroiditis in ultrasound using tissue characterization and pixel classification *Proc. Inst. Mech. Eng. Part H J. Eng. Med.* **227** 788–98

[65] Acharya U R, Faust O, Alvin A P C, Krishnamurthi G, Seabra J C, Sanches J and Suri J S 2013 Understanding symptomatology of atherosclerotic plaque by image-based tissue characterization *Comput. Methods Programs Biomed.* **110** 66–75

[66] Acharya U R, Faust O, Sree S V, Alvin A P C, Krishnamurthi G, Sanches J and Suri J S 2011 Atheromatic™: symptomatic vs. asymptomatic classification of carotid ultrasound plaque using a combination of HOS, DWT & texture *Proc. of the 2011 Annual Int. Conf. of the IEEE Engineering in Medicine and Biology Society (Boston, MA, 3 August–3 September 2011)* (Piscataway, NJ: IEEE)

[67] Acharya U R, Mookiah M R K, Sree S V, Afonso D, Sanches J, Shafique S, Nicolaides A, Pedro L M, Fernandes J F E and Suri J S 2013 Atherosclerotic plaque tissue characterization in 2D ultrasound longitudinal carotid scans for automated classification: a paradigm for stroke risk assessment *Med. Biol. Eng. Comput.* **51** 513–23

[68] Molinari F, Liboni W, Pavanelli E, Giustetto P, Badalamenti S and Suri J S 2007 Accurate and automatic carotid plaque characterization in contrast enhanced 2-D ultrasound images *Proc. of the 29th Annual Int. Conf. of the IEEE Engineering in Medicine and Biology Society (Lyon, France, 22–26 August 2007)* (Piscataway, NJ: IEEE)

[69] Saba L, Biswas M, Suri H S, Viskovic K, Laird J R, Cuadrado-Godia E, Nicolaides A, Khanna N N, Viswanathan V and Suri J S 2019 Ultrasound-based carotid stenosis measurement and risk stratification in diabetic cohort: a deep learning paradigm *Cardiovasc. Diagn. Ther.* **9** 439–61

[70] Biswas M, Kuppili V, Saba L, Edla D R, Suri H S, Sharma A, Cuadrado-Godia E, Laird J R, Nicolaides A and Suri J S 2019 Deep learning fully convolution network for lumen characterization in diabetic patients using carotid ultrasound: a tool for stroke risk *Med. Biol. Eng. Comput.* **57** 543–64

[71] Chaddad A, Hassan L and Desrosiers C 2021 Deep CNN models for predicting COVID-19 in CT and x-ray images *J. Med. Imaging* **8** 014502

[72] Gunraj H, Wang L and Wong A 2020 COVIDNet-CT: a tailored deep convolutional neural network design for detection of COVID-19 cases from chest CT images *Front. Med.* **7** 608525

[73] Iyer T J, Raj A N J, Ghildiyal S and Nersisson R 2021 Performance analysis of lightweight CNN models to segment infectious lung tissues of COVID-19 cases from tomographic images *PeerJ Comput. Sci.* **7** e368

[74] Ranjbarzadeh R, Jafarzadeh Ghoushchi S, Bendechache M, Amirabadi A, Ab Rahman M N, Baseri Saadi S, Aghamohammadi A and Kooshki Forooshani M 2021 Lung infection segmentation for COVID-19 pneumonia based on a cascade convolutional network from CT images *BioMed Res. Int.* **2021** 5544742

[75] Erasmus J J, Gladish G W, Broemeling L, Sabloff B S, Truong M T, Herbst R S and Munden R F 2003 Interobserver and intraobserver variability in measurement of non–small-cell carcinoma lung lesions: implications for assessment of tumor response *J. Clin. Oncol.* **21** 2574–82

[76] Joskowicz L, Cohen D, Caplan N and Sosna J 2019 Inter-observer variability of manual contour delineation of structures in CT *Eur. Radiol.* **29** 1391–9

[77] El-Baz A and Suri J 2019 *Lung Imaging and CADx* (Boca Raton, FL: CRC Press)

[78] El-Baz A and Suri J S 2011 *Lung Imaging and Computer Aided Diagnosis* (Boca Raton, FL: CRC Press)

[79] Sudeep P V, Palanisamy P, Rajan J, Baradaran H, Saba L, Gupta A and Suri J S 2016 Speckle reduction in medical ultrasound images using an unbiased non-local means method *Biomed. Signal Process. Control* **28** 1–8

[80] Sarker M M K, Rashwan H A, Akram F, Singh V K, Banu S F, Chowdhury F U, Choudhury K A, Chambon S, Radeva P and Puig D *et al* 2021 SLSNet: skin lesion segmentation using a lightweight generative adversarial network *Expert Syst. Appl.* **183** 115433

[81] Saba L, Agarwal M, Sanagala S S, Gupta S K, Sinha G R, Johri A M, Khanna N N, Mavrogeni S, Laird J R and Pareek G *et al* 2020 Brain MRI-based Wilson disease tissue classification: an optimised deep transfer learning approach *Electron. Lett.* **56** 1395–8

[82] El-Baz A and Suri J S 2019 *Big Data in Multimodal Medical Imaging* (Boca Raton, FL: CRC Press)

IOP Publishing

Multimodality Imaging, Volume 2
Heart, lungs and peripheral organs
Mainak Biswas and Jasjit S Suri

Chapter 5

A comparative analysis of tuberculosis-infected lung x-ray image segmentation: U-Net vs. U-Net++

Radhakrishn Birla, Gautam Chugh, Swastika Bishnoi, Riddhika Shringi, Piyush Kumar and Mainak Biswas

Tuberculosis (TB) is a significant global health issue, causing a high number of deaths each year, with 1.6 million people dying from TB in 2021. Lung segmentation is a crucial step in TB detection and other lung disorder diagnoses. Deep learning, particularly convolutional neural networks (CNNs), has shown promise in accurately segmenting lungs in medical images like chest x-rays or CT scans. U-Net is an encoder-decoder CNN-based model with a U-shaped architecture used for image segmentation. It introduces skip pathways layer-wise between encoder-decoder pathways, for reducing sematic loss of information that happens while convolving the images in the encoder pathway. The U-Net++ on the other hand uses reengineered skip connections to further reduce the semantic information gap between all the layers of encoder and decoder. In this work, both U-Net and U-Net++ has been used for segmenting tuberculosis annonated lung images. The dataset contained 566 x-ray images. The results clearly showed that U-Net++ performed better than U-Net in terms of Dice index (14%), pixel accuracy (4.9%), and specificity (6.5%) while interestingly, U-Net fared better at specificity (3.2%).

5.1 Introduction

In 2021, 1.6 million people died from TB, including 187 000 individuals with HIV [1, 2]. TB is currently the 13th leading cause of death worldwide and the second leading infectious killer, surpassing HIV and AIDS. Around 10.6 million people fell ill with TB in 2021. This includes 6 million men, 3.4 million women, and 1.2 million children. TB affects people of all ages and is present in all countries. TB is a curable and preventable disease. Ending the TB epidemic by 2030 is one of the health targets outlined in the United Nations Sustainable Development Goals (SDGs). The SDGs

doi:10.1088/978-0-7503-2352-9ch5

aim to improve global health and well-being by addressing various issues, including infectious diseases like TB. These facts emphasize the importance of continued efforts to combat TB, including increased funding, improved access to diagnosis and treatment, and the development of innovative strategies to prevent the spread of the disease. Achieving the goal of ending the TB epidemic by 2030 requires a concerted global effort.

X-ray lung image segmentation is the significant step for tuberculosis detection and characterization. Lung segmentation using deep learning [3–5] is an approach that utilizes advanced neural network architectures to automatically identify and delineate the boundaries of the lungs in medical images, such as chest x-rays or CT scans. Deep learning, particularly CNNs [6], has shown great promise in medical image analysis tasks, including lung segmentation [7]. It enables more precise identification and diagnosis of lung disorders such as pneumonia, lung cancer, and TB, lung segmentation is a crucial stage in the interpretation of chest x-rays. Accurate lung segmentation can be helpful in determining how lung disorders develop over time and how well therapies are working. Lung segmentation may be done using a variety of methods, including thresholding, region expanding, and machine learning-based methods. These techniques use several algorithms to recognise the characteristics of the lung regions and separate them from other x-ray picture components like the ribs and the heart. One such deep learning technique, U-Net, is an encoder-decoder deep learning model that uses two paths to process images [8–10]. The first path applies contraction operations to capture context, while the second path applies expansive operations to localize pixels and produce an output. U-Net is an extension of the fully convolutional network (FCN) architecture [11], which aims to improve segmentation accuracy with fewer training examples. The network's effectiveness relies on skip connections between layers that share the same level of abstraction. However, U-Net differs from FCN by incorporating more channels in the up-sampling layers to transfer more information from low to high resolution layers. U-Net++ is an upgraded version of the U-Net model that aims to overcome the semantic gap between the encoder and decoder blocks by introducing nested convolutional blocks and more skip connections [12]. The main goal is to make the intermediate feature maps between the encoder and decoder more similar, which is believed to make the optimization problem easier. The redesign of the architecture results in better performance, as demonstrated in experiments on medical image segmentation tasks. U-Net++ is also known as nested U-Net. U-Net is a CNN architecture used for accurate and efficient image segmentation. Its shape resembles a U-shape and is made up of three sections, each containing convolutional layers. The contraction section has two 3×3 convolutional layers and a 2×2 max pooling layer. The bottleneck section has two 3×3 convolutional layers followed by a 2×2 up-convolutional layer. The expansion section has two 3×3 convolutional layers followed by a 2×2 up-sampling layer. The U-Net++ is an updated version of U-Net. U-Net++ addresses the limitations of U-Net, such as the unknown optimal depth and restrictive skip connection design, by adding dense blocks and convolutional layers between the encoder and decoder. This has led to improved segmentation accuracy in medical image segmentation tasks. In the next few sections application of both U-Net and U-Net++ for lung image segmentation is studied and their performances are compared. In section 5.2, the dataset and methodology is discussed, results are given in section 5.3 and a conclusion is given in section 5.4.

5.2 Dataset and methodology

5.2.1 Dataset

The Shenzhen Hospital (SH) dataset of chest X-ray (CXR) images was acquired from Shenzhen No. 3 People's Hospital in Shenzhen, China [13, 14]. It contains normal and abnormal CXR images with marks of tuberculosis. The dataset consists of 566 chest x-ray images and their corresponding mask images. A chest x-ray image and its corresponding mask is shown in figure 5.1.

5.2.2 How does semantic segmentation using U-Net and U-Net++ work?

The concept of semantic segmentation is based on the concept of classification, but at the pixel level. General deep learning approaches such as CNN classifies the full image instead of at pixel level. The image classification by CNN follows extraction of features using convolution filters and down-sampling them using pooling at every layer. It results in a very rich set of features at higher level layers, which are flattened at the highest level into a one-dimension matrix from a two-dimension feature map. These features pertain only to minimum required details for classification by fully-connected layers appended at the end of the CNN. In contrast, semantic segmentation requires classification at pixel level. One of the essential characteristics of semantic segmentation is preparation of ground images with manually drawn borders separating regions of interest (ROIs) from other parts. These ROIs are color-coded and used as ground truth. For semantic segmentation, the flattening and fully-connected layers of the CNN are removed. Instead, the reverse process of up-sampling or convolution transpose is applied to the output feature map, preferably equal to the number of convolution-pooling layers. The contraction and expansion

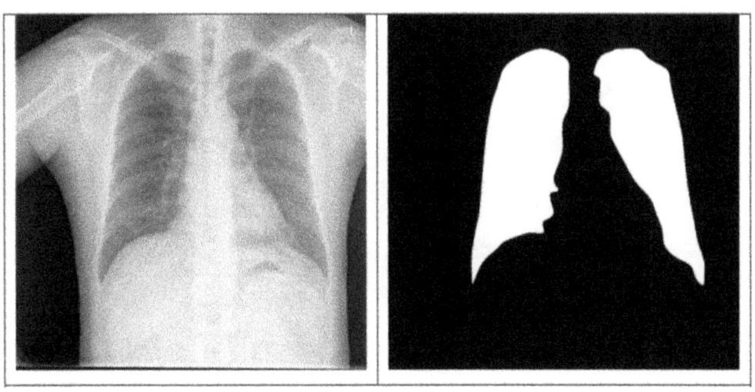

Figure 5.1. A x-ray chest image and its corresponding mask (image courtesy: [14], this data is distributed under a Creative Commons Attribution license).

path are known as the encoder and decoder path, respectively. The CNN weights are trained to adapt to different color-coded regions of the ground truth images during training. Additionally, merge operations are applied by combining outputs from intermediate layers of down-sampling with intermediate layers of up-sampling. This helps in prevention of semantic information loss that happens during down-sampling.

In this sub-section we discuss the architecture of U-Net and later its modification in the form of U-Net++ in detail. As other segmentation models of deep learning, the U-Net also has the encoder (contraction) and decoder (expansion) path. The U-Net diagram is shown figure 5.2. Each of the encoder and decoder consists of four layers. There is one intermediate layer between encoder and decoder. Each encoder layer has the following configuration: inputàconv1àconv2àmax-poolàoutput (E4, E3, E2, and E1 in figure 5.2). The decoder layer has the following configuration: inputàup-sampleàmergeàconv1àconv2àoutput (D1, D2, D3, and D4 in figure 5.2). The intermediate layer has the following configuration: inputàconv1àconv2àdropoutàoutput. In the encoding stage the number of filters applied in each layer is 64, 128, 256, and 512, respectively. The inverse is applied for the number of layers of the decoder. The intermediate layer uses 1024 filters. The size of each kernel filter for all the layers is 3-by-3. If an image of size (say) n-by-n is input to the a layer of the decoder, then in each layer it reduces to respectively. The reverse happens for the decoder, but up-sampling is applied earlier before feeding to the convolution pairs for each layer. The final output is a semantically segmented image. The skip connections in U-Net allow the feature maps at encoder layers to be received directly at decoder layers. An advanced version of U-Net++ was developed using reengineered skip connections, which is described next (shown in figure 5.3).

The idea behind U-Net++ is to minimize semantic information loss between each layer of the encoder and decoder. In this regard, the skip pathways in U-Net are replaced by a pyramid of convolutions, skipping and up-sampling operations in U-Net++. For the purpose of understanding, let's call them dense layers

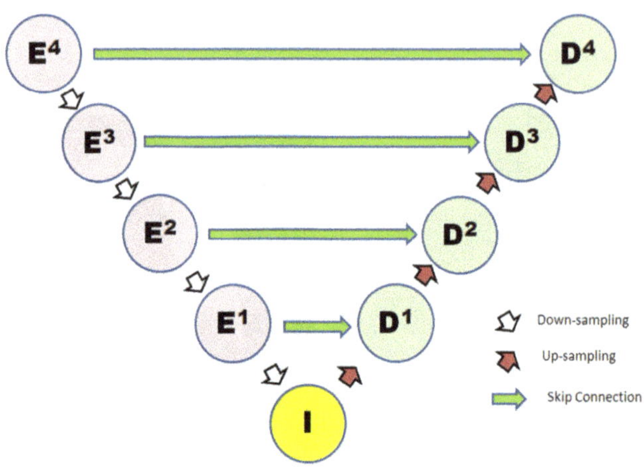

Figure 5.2. Conceptual diagram of U-Net.

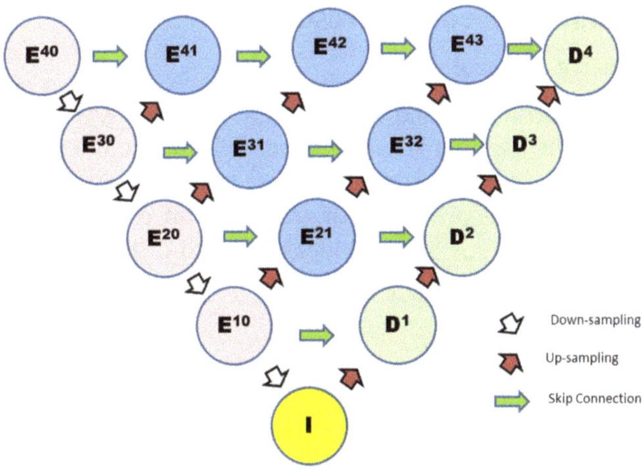

Figure 5.3. Conceptual diagram of U-Net++.

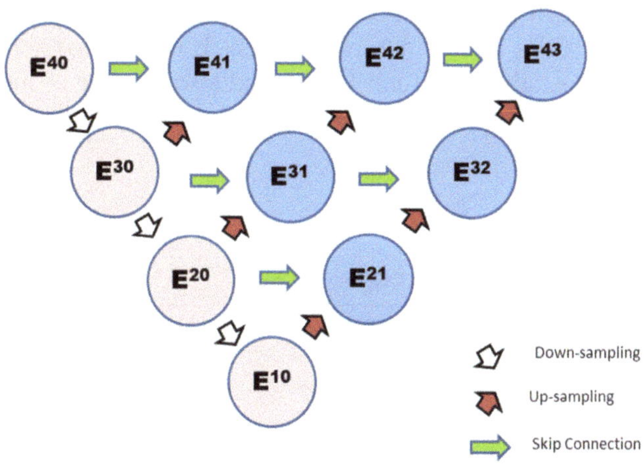

Figure 5.4. Dense layers of U-Net++.

(including encoder layers). Mathematically, each dense layer (shown in figure 5.4) can be described by:

$$E^{i,j} = \begin{cases} Conv\ (E^{i+1,j}) \text{ if } j = 0 \\ Conv\left([E^{i,\ k}]_{k=0}^{j-1}.\ \ Up(E^{i-1,\ j-1})\right) \text{ if } j > 0 \end{cases} \tag{5.1}$$

where [ ] represents concatenation, *Conv* represents convolution operation, and *Up* represents up-sample operation. For example, for E^{32}, the operations are merging outputs from E31 and E30 and up-sampled E^{21} i.e., $E^{32} = Conv(E^{31}.\ E^{30}.\ \ Up(E^{21}))$. Here, the up-sampling increases semantic similarity between encoder layers and decoder feature maps. In this work, both U-Net and U-Net++ have been used for segmenting lung x-ray images discussed in the previous section.

5.3 Experimental protocol and results

5.3.1 Cross-validation protocol

The cross-validation protocol used in this experiment is K10. In the K10 cross-validation protocol, the training size comprises 90% of the data and the test size is 10% of the data. For this purpose, the images are randomly divided into 10 blocks of equal sizes, and nine blocks are used for training and one block for testing. This same procedure is repeated for all the blocks until all of them are tested.

5.3.2 Hyper parameters

The hyper parameters used for both U-Net and U-Net++ are given in table 5.1. The activation function used is ReLu, the loss function is cross entropy, the batch size is 32 and the optimizer is Adam.

5.3.3 Training accuracy and loss values for U-Net and U-Net++

The comparison tables for training accuracy and loss of U-Net and U-Net++ for all epochs are given in tables 5.2 and 5.3, respectively. It is seen that accuracy values in table 5.2 for U-Net++ are better than U-Net and also the rate of increase is higher for every epoch. With regards to the loss values, the U-Net loss is static and higher than U-Net++.

Table 5.1. Hyper parameters for U-Net and U-Net++.

S. No.	Hyper parameters	Value
1	Activation function	ReLu
2	Loss function	Cross entropy
3	Batch size	32
4	Optimizer	Adam
5	Number of epochs	3
6	Steps per epoch	300

Table 5.2. Accuracy table.

S. No.	Model	Epoch 1	Epoch 2	Epoch 3
1	U-Net	0.744	0.744	0.745
2	U-Net++	0.817	0.821	0.827

Table 5.3. Loss table.

S. No.	Model	Epoch 1	Epoch 2	Epoch 3
1	U-Net	0.690	0.682	0.676
2	U-Net++	0.224	0.113	0.093

Table 5.4. Performance comparison of U-Net and U-Net++.

Sl	Metric	U-Net	U-Net++
01	Dice Similarity	0.790	**0.930**
02	Pixel accuracy	0.920	**0.969**
03	Sensitivity	**0.992**	0.960
04	Specificity	0.907	**0.972**

5.3.4 Comparison of segmentation performance of U-Net and U-Net++

The Dice similarity, pixel accuracy, sensitivity and specificity values for both U-Net and U-Net++ are given in table 5.4. A higher Dice index indicates better overlap between the predicted segmentation mask and the ground truth. Pixel accuracy measures the percentage of correctly classified pixels in the segmentation output. Specificity is a metric that measures the true negative rate, i.e., the ability of the model to correctly identify background or non-target regions. Sensitivity, also known as recall or true positive rate, measures the ability of the model to correctly identify the target regions. It is seen that except sensitivity, the performance attributes of U-Net++ are better than U-Net. The Dice index for U-Net++ is 14% better than U-Net, pixel accuracy for U-Net++ is 4.9% better than U-Net, and specificity is 6.5% better than U-Net. In the case of sensitivity only, the performance of U-Net is 3.2% better than U-Net++. The resultant segmented images are shown in figure 5.5, which clearly show that U-Net++ performance is far better than U-Net.

5.4 Conclusion

In this work, we have used two deep learning models, U-Net and U-Net++, for lung image segmentation. We have shown that reengineered skip connections of U-Net++ are better in terms of reducing sematic information loss than simple one-to-one skip connections in U-Net. The skip connections used in U-Net++ are not only used as one-to-one but up-sampled in all the layers and merged with corresponding layers of the decoder. This achieves a lot less information loss than in U-Net. In the future soft computing models as well as attention-based models [15] will be used for the purpose of segmentation.

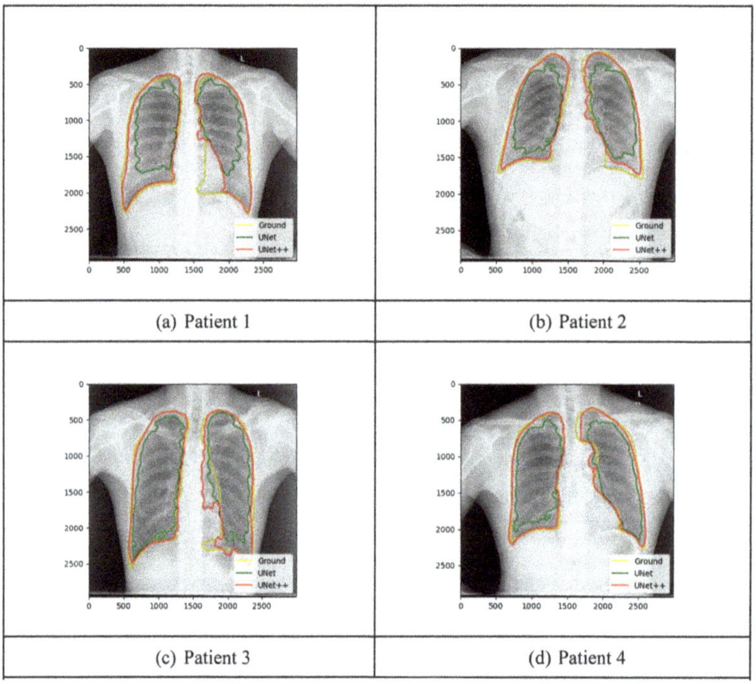

(a) Patient 1 | (b) Patient 2 | (c) Patient 3 | (d) Patient 4

Figure 5.5. Segmentation results of four patients using manual (yellow), U-Net (green), and U-Net++ (red)

References

[1] Chakaya J *et al* 2022 The WHO global tuberculosis 2021 report—not so good news and turning the tide back to End TB *Int. J. Infect. Dis.* **124** S26–9

[2] Corbett E L *et al* 2003 The growing burden of tuberculosis: global trends and interactions with the HIV epidemic *Arch. Intern. Med.* **163** 1009

[3] LeCun Y, Bengio Y and Hinton G 2015 Deep learning *Nature* **521** 436–44

[4] Saba L *et al* 2019 The present and future of deep learning in radiology *Eur. J. Radiol.* **114** 14–24

[5] Tandel G S *et al* 2019 A review on a deep learning perspective in brain cancer classification *Cancers* **11** 111

[6] Albawi S, Mohammed T A and Al-Zawi S 2017 Understanding of a convolutional neural network *2017 Int. Conf. on Engineering and Technology (ICET) (Antalya)* (IEEE) pp 1–6

[7] Suri J S 2019 State-of-the-art review on deep learning in medical imaging *Front. Biosci.* **24** 392–426

[8] Ronneberger O, Fischer P and Brox T 2015 U-Net: convolutional networks for biomedical image segmentation arXiv:1505.04597 (accessed 20 May 2023)

[9] Biswas M *et al* 2018 Deep learning strategy for accurate carotid intima-media thickness measurement: an ultrasound study on Japanese diabetic cohort *Comput. Biol. Med.* **98** 100–17

[10] Saba L *et al* 2019 Ultrasound-based carotid stenosis measurement and risk stratification in diabetic cohort: a deep learning paradigm *Cardiovasc. Diagn. Ther.* **9** 439–61

[11] Fu X and Qu H 2018 Research on semantic segmentation of high-resolution remote sensing image based on full convolutional neural network *2018 12th Int. Symp. on Antennas, Propagation and EM Theory (ISAPE) (Hangzhou, China)* (IEEE) pp 1–4
[12] Zhou Z, Siddiquee M M R, Tajbakhsh N and Liang J 2018 U-Net++: a nested U-Net architecture for medical image segmentation arXiv:1807.10165
[13] Stirenko S *et al* 2018 Chest x-ray analysis of tuberculosis by deep learning with segmentation and augmentation *2018 IEEE 38th Int. Conf. on Electronics and Nanotechnology (ELNANO) (Kiev)* (IEEE) pp 422–8
[14] Yang F *et al* 2022 Annotations of lung abnormalities in the shenzhen chest x-ray dataset for computer-aided screening of pulmonary diseases *Data* **7** 95
[15] Bahdanau D, Cho K and Bengio Y 2016 Neural machine translation by jointly learning to align and translate arXiv:1409.0473 (accessed 20 May 2023)

Part II

An overview of deep learning and its applications in cardiovascular diseases

Chapter 6

Reviewing the role of artificial intelligence in heart and vascular ultrasound: advancements and applications

Mayank Singhal, Radhakrishn Birla, Mainak Biswas, Sujoy Datta and Jasjit S Suri

The leading cause of death globally is cardiovascular disease (CVD). One primary cause of CVD is an artery inflammation disease known as atherosclerosis, which can result in endothelial dysfunction. This leads to formation of lipids, a necrotic core within the arterial wall leading to myocardial infarction or stroke. The existence of therapeutic medication greatly decreases the need for invasive examination. However, such examination is only possible post diagnosis of CVD. Medical imaging is one of the few key areas which allows doctors to have a look inside the body without performing any investigative surgery. Expert advice on the imaging can lead to faster and more effective treatment. However, there is a dearth of expert professionals in hospitals to deal with such a huge volume of images generated every day.

In such a situation, artificial intelligence (AI) can come to the rescue of hospitals by providing some initial advice regarding the condition of patients. There are over 7700 papers of AI in radiology, and more than 185 papers in cardiology. Most of the AI modules discussed in these articles have been trained and tested by studying patterns of CVD in a particular imaging modality. This study deals with some of the latest groundbreaking case studies of application of AI in medical imaging, focusing mainly on blood flow, echocardiography, plaque burden measurement, plaque characterization, left ventricle segmentation, and intravascular ultrasound imaging. The results from these papers clearly hold merit and conclusively prove that AI can definitely help in generating some preliminary diagnosis. Also, these results have substantial clinical significance as well as the ability for large-scale implementation at no significant cost and time. However, such widespread implementation should be done with due consideration to the safety and legality aspects of AI. This review also

doi:10.1088/978-0-7503-2352-9ch6

discusses some of these issues with some suggested solutions. Largely, this study allows scientists, medical students, engineers, and researchers to have a closer look into cardiac and cardiovascular applications of AI.

6.1 Introduction

The prevalence of cardiovascular diseases (CVD) among adults in the USA is 48.0% (121.5 million) and progressively increases with advancing age. Globally, CVDs were attributed to 17.6 million deaths in 2016, 14.5% higher than in 2006 [1]. Tobacco use, physical inactivity, malnutrition, high blood, cholesterol, high blood pressure, obesity, hyperglycemia, hyperlipidemia, etc. are the main causes of fatalities from CVDs [2, 3]. Most CVD events, which are linked to the inflammatory disease of the arteries called atherosclerosis, include peripheral artery disease, stroke and myocardial infarction (MI). The endothelium throughout coronary and carotid arteries becomes dysfunctional in light of atherosclerosis. Age-related fibrous tissue, fat, and lipid buildup within artery walls can result in plaque covered in a fine fibrous cap or more complex lesions. The fibrous cap may rupture if the plaque is asymptomatic, unstable, and not treated in a timely manner, resulting in thrombosis in the lumen and eventually stroke [4, 5]. However, if the plaque is stable and symptomatic, it can be treated with proper medication. The monitoring of pathophysiological progression of atherosclerosis, blood flow, left ventricle ejection fraction (LVEF), electrocardiogram (ECG) and heart rate variability [6] are key to patients' treatment and risk calculation. Surgical treatment of CVDs leads to post-surgery complications and a four to five-fold increase in incidents of stroke and death [7, 8]. A wide range of cardiovascular imaging techniques for patient monitoring have been developed thanks to technological advancements and improved therapeutic alternatives for treating CVDs instead of surgery [9]. Cardiac computed tomography (CT) [10], positron emission tomography (PET), single-photon emission computed tomography (SPECT) imaging, perfusion echocardiography, cardiac magnetic resonance imaging (MRI), ultrasound imaging (USI) [11], etc, are some of the key imaging tools [12]. Ultrasound-based imaging tools are more popular since they do not use ionized radiation and are therefore safer. Also, ultrasound devices are portable and comparatively less expensive than other imaging modalities. As ultrasound waves cannot penetrate bones, intravascular ultrasound (IVUS) techniques are used for inspection of heart and coronary arteries.

The risk calculators are convenient tools for predicting stroke/CVD events. The conventional risk calculators use laboratory-based biomarkers (LBBM) like Hb1Ac levels, high-density cholesterol, glucose levels, low-density lipoprotein cholesterol, etc, or office-based biomarkers (OBBM) like age, BMI, and blood pressure. Some risk calculators are ASCVD [13], Framingham [14], Joint British Societies (JBS3) [15], QRISK2 [16], Reynold Risk Score (RRS) [17], United Kingdom Prospective Diabetes Study (UKPDS56) [18],UKPDS60 [19], etc. However, because the plaque burden is not taken into account in the calculation, these risk calculators are less precise. AECRS 2.0 [20, 21] and AtheroEdge Composite Risk Score (AECRS 1.0) are new age risk calculators that use an autonomous morphological-based CVD risk prediction tool which integrates imaging-based phenotypes and LBBM, OBBM like plaque area (PA) and carotid intima-media thickness (cIMT). The carotid segment scans serve as the basis for the image-based biomarkers, for instance an internal carotid artery, bulb, or common carotid artery, obtained by 2D B-mode ultrasound [22–24].

Manual segmentation of ultrasound images to measure plaque volume and characterize plaque is tedious, error prone and slow. Many semi-automated and automated techniques have been developed to segment the plaque area and characterize plaque from ultrasound images. Similarly, diagnosis of CVDs from blood flow, LVEF and ECG has been performed using automated tools. Regarding measurement of plaque volume, the research has progressed for many decades. The automated segmentation of plaque volume is performed in two stages: ROI detection and boundary delineation. In the initial days of computing, ROI extraction was manual, whereas, wall delineation was based on threshold. In the subsequent years, several automated models came up for both stages such as deformable models, dynamic programming, signal processing based models, etc [25–28]. AI and its variants have seen considerable increase in adoption as the basis of automated CVD risk estimation and diagnosis models. One of the main reasons being that AI models are more robust and can handle noise better using experience and cognitive ability. In this review a strong emphasis has been made on the working of different AI models and their applications for estimating different image phenotypes.

In section 6.2, the working of different AI models like deep learning (DL), reinforcement learning (RL) as well as machine learning (ML) are discussed. Section 6.3 discusses AI application across several CVD risk models. Section 6.4 offers the discussion, and section 6.5 brings the paper to a conclusion.

6.2 Background literature survey on different AI models

AI refers to broad range of computing models capable of showing intelligent behavior in complex dynamic environments. AI models can be categorized into two classes in terms of their functionality: supervised and unsupervised [29, 30]. Supervised learning follows a training/testing paradigm where labelled instances are used to train a given AI model to characterize unlabeled instances during testing. Unsupervised learning on the other hand requires the given AI model to infer meaningful information from unknown data instances. AI models can also be categorized into three categories based on their architecture: machine learning (ML), deep learning (DL) [31] and reinforcement learning (RL) [32, 33].

ML refers to a host of tree-, rule-, distance-, and probability-based statistical models that learn complex patterns of information existing amongst data instances. ML algorithms are generally two-stage processes where in the first stage handmade features are extracted from instances, whereas in the second stage the ML model is applied on the extracted features during training (shown in figure 6.1(a)). Several popular ML models are artificial neural networks (ANNs) [34], K-nearest neighbors (KNN) [35], Bayesian classifiers [36], support vector machines (SVMs) [37], extreme learning machines (ELMs) [38, 39], etc. KNN characterizes instances based on the frequent classes of its neighbors. Bayesian classifiers use conditional probability for

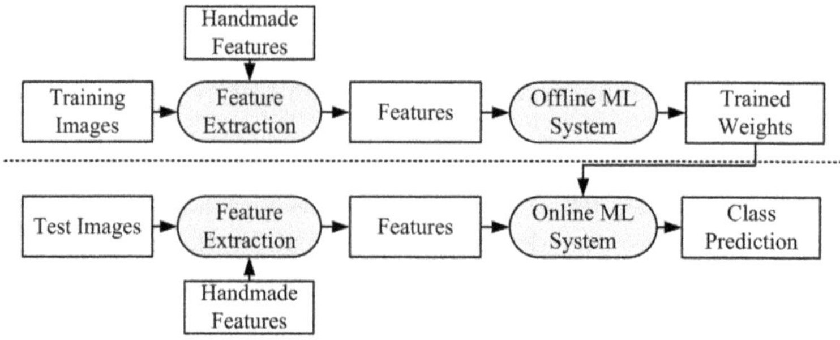

(a) ML architecture.

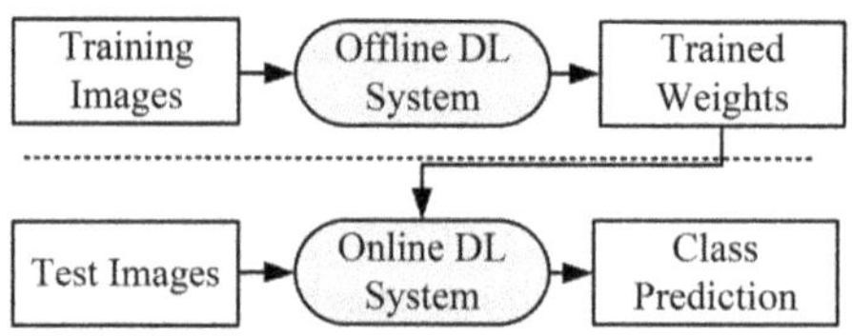

(b) DL architecture.

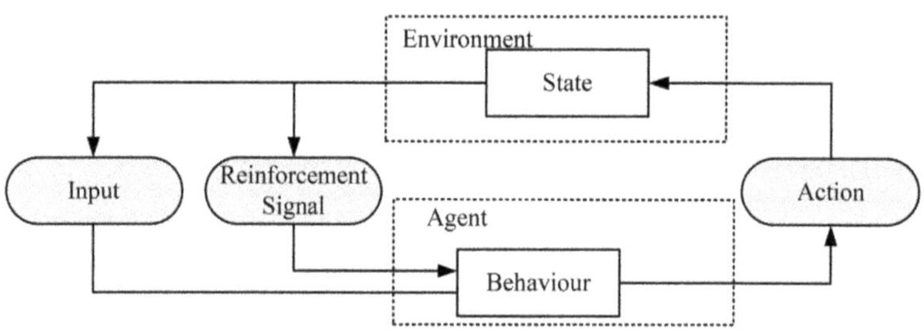

(c) RL architecture.

Figure 6.1. Architecture of (a) ML, (b) DL and (c) RL models.

labelling instances. SVMs draw optimal hyper-planes between instances of different classes to characterize them. ANNs are multiple layer neurons for characterizing instances. Single-layer feed-forward networks, often known as ELMs, use the Moore–Penrose matrix inverse to estimate the relationship between input features and instance class. DL refers to multiple layers of neural network mimicking the brain's visual cortex for characterizing and segmenting images [40]. Unlike ML, DL is a single-stage integrated-feature extraction and characterization protocol. The lower layers of the DL network detect essential characteristics like a boundary or edges, which are passed to higher layers. The higher layers perceive a more complex combination of the basic features to form a general understanding of shape, size, and other distinct geometrical properties of the instance (shown in figure 6.1(b)). This knowledge is the key to understanding distinct features of a class. Convolutional neural network (CNN), deep belief network (DBN), residual neural network (RNN), and fully convolutional networks (FCN) autoencoders are the most used DL networks. CNNs use a sequence of convolution, pooling, and fully connected network for characterization purposes. FCNs use a sequence of convolution and up-sampling layers for segmentation. Autoencoders are neural networks where the quantity of output neurons equals the quantity of input neurons. Reinforcement learning refers to models independent from supervised and unsupervised forms of learning. In RL, an agent must practice new behaviors through encounters in a dynamic environment. RL works by looking over the space of behaviors to identify the best one for the given situation (shown in figure 6.1(c)). RL is a relatively new but evolving field of study at the time of writing this chapter. We will see several uses of AI models in cardiovascular imaging in the chapter after this.

6.3 Applications of ML, DL and RL in cardiovascular imaging

AI is steadily replacing conventional automated models that have been traditionally used to diagnose cardiovascular issues. Similarly, it is changing the paradigm of traditional risk calculators to predict 10-year risks [20, 21]. The main reason is that AI is robust to small noises in data, which in normal situations would influence the results negatively. Also, AI models when trained efficiently give more accurate and quicker results. In the subsequent subsections we will see different applications of AI in cardiovascular imaging and related fields.

6.3.1 AI in blood flow

The cardiovascular system's performance is directly characterized by pressure wave propagation, artery wall mechanics and blood flow. Arterial velocity as well as wall displacement is key in monitoring health of the patients. Computational fluid dynamics model and non-invasive assessment, including 4D flow MRI flow provide a cheap and faster way to characterize propagation of blood's pressure and velocity waves in the carotid and coronary arteries. Coenen *et al* [41] used ML-based coronary CT fractional flow reserve (ML-FFR) due to which coronary artery disease diagnosis accuracy was increased. Data was collected from 351 patients. Three models were applied i.e., invasive-FFR (Inv-FFR), ML-FFR and computation fluid dynamics-based FFR (CFD-FFR). Regarding accuracy per vessel, CFD-FFR and ML-FFR both (78%) performed equally better than Inv-FFR. In the case of per-person basis, accuracy diagnosis of ML-based FFR increased substantially to 91%.

In another work by Kissas *et al* [42] a DL network is used to parameterize the differential equations pertaining to the system satisfying all the physical laws. The clinical MR common carotid artery images data (shown in figure 6.2(i)) were obtained from a healthy female volunteer utilizing a 1.5T Avanto scanner (weight = 51 kg, height = 160 cm, age = 27). The DL network predicted area, velocity correlated well with the actual values at aorta#3, while the pressure values matched with a healthy person as described in the literature, as shown in figure 6.2(ii). Kusunose *et al* [43] used DL-CNN to 300 echocardiographic images of the left anterior descending branch (LAD), left circumflex branch (LCX) and

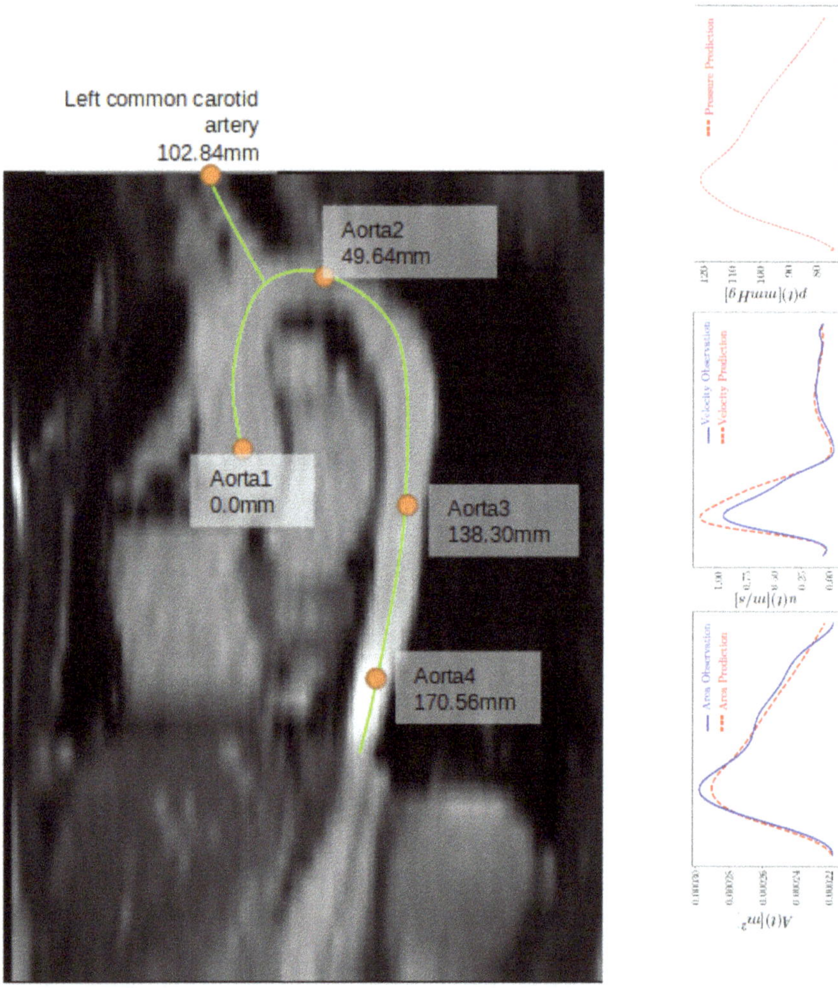

(i) (ii)

Figure 6.2. (i) Acquire CCA image, (ii) comparative correlation plots of area, velocity and pressure (image courtesy [42]).

right coronary artery (RCA) in order to identify any regional wall motion abnormality (RWMA). Ten cardiologists and ten resident observers were used to create the gold standard. Using AUC evaluation, the diagnostic effectiveness of the existence of RWMA was assessed. The DL-CNN network showed almost equal AUC (0.99 vs. 0.98) with respect to cardiologists and higher AUC (0.99 vs. 0.90) with respect to resident observers. Table 6.1 presents the findings.

6.3.2 AI in classification echocardiography ultrasound images

Echocardiography (ECG) captures high-resolution temporal and spatial images of the heart and its surroundings using ultrasound. One common method of managing

Table 6.1. AI in blood flow measurement.

S. No.	First Author (Year)	# Instances/ images/Patients	Imaging modality	Model	Performance parameter	Performance
1	Coenen *et al,* [41] (2018)	351	CT	ML	Diagnosis Accuracy	Per-vessel Acc: 78%, Per-person Acc:91%
2	Kissas *et al* [42] (2020)	One	MR	DL-network	NA	NA
3	Kusunose *et al* [43] (2020)	300	Ultrasound	DL-CNN	AUC	AUC: 0.99

and diagnosing CVDs is by this method. Interpretation of ECG images requires highly subjective expertise. AI is used as an automated diagnostic tool for interpretation of high-volume ECG images and high accuracy. Narula *et al* [44] used an ensemble of three ML models i.e., Physiological Hypertrophy found in Athletes (ATH) is contrasted with Hypertrophic cardiomyopathy (HCM) using SVM, ANN, and random forests. Data was collected from 139 male subjects (62 HCM, 77 ATH). A total of 20 LV speckle tracking variables over 6 different systolic time points were extracted from the ECG images for a total of 120 features. Some of them are maximum LV diameter, average velocity, LV volume, segmental velocity, longitudinal strain, minimum LV diameter, etc. The model performed showing 77% specificity and sensitivity of 96%. Yuan *et al* [45] applied ML protocol to classify LV-end systolic frame from end diastolic frames from 99 ECG images of ultrasound. A non-negative matrix factorization (NMF) was applied to all the frames to estimate cardiac cycle length and heart rate variability. Results from NMF were more closely correlated exhibiting a correlation coefficient as 0.9196.

In another work, a deep learning CNN model has been applied for characterization of ECG images of ultrasound to detect the existence of a pacemaker, a widened left atrium, an expanded left ventricle, and increased left ventricle end systolic and diastolic volumes [46]. Data in the form of 2.6 million echocardiogram images provided by 2850 patients was used for the experiment. The images were preprocessed, cropped and used for training the DL model as shown in figure 6.3. AUC of 0.75 indicated left ventricular hypertrophy, AUC of 0.85 indicated a severely swollen left atrium and AUC of 0.89 indicated the presence of the pacemaker. The predicted and actual systolic and diastolic volumes at the left ventricular end had a high CC of 0.74 and 0.70, respectively. All the results related to ECG are given in table 6.2.

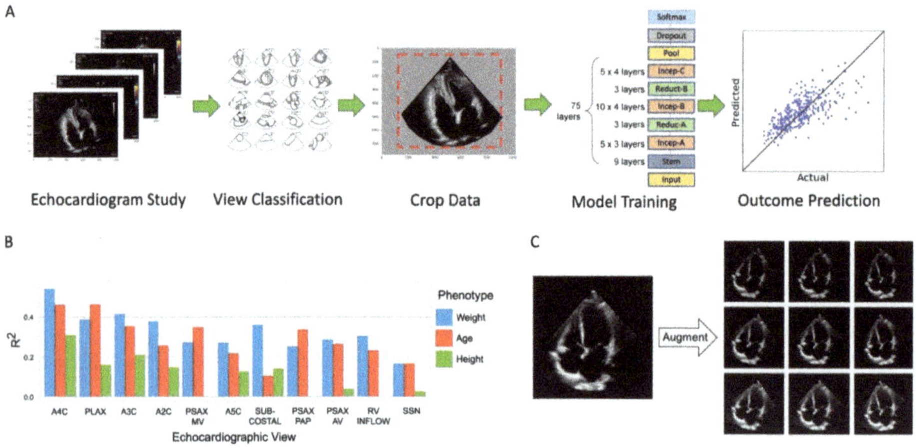

Figure 6.3. ECG characterization using DL network (image courtesy [46]).

Table 6.2. Characterization of ECG images.

S. No.	First Author (Year)	# Instances/ images/ Patients	Imaging modality	Model	Performance parameter	Performance
1	Narula *et al* [44] (2016)	139	Ultrasound	SVM, ANN, Ensemble Forest	Classification of HCM and ATH	Sensitivity: 96%, Specificity: 77%
2	Yuan *et al* [45] (2017)	99	Ultrasound	ML-NMF	Classification of systolic and diastolic frame	CC: 0.9196
3	Ghorbani *et al* [46] (2020)	2850	Ultrasound	DL-CNN	Characterization of a pacemaker, a widened left atrium, a thickened left ventricle, and elevated left ventricular end-systolic and end-diastolic volumes	LV hypertrophy AUC: 0.75, Left atrium AUC: 0.85, Pacemaker AUC: 0.89, Diastolic CC: 0.70 and Systolic CC: 0.74

6.3.3 AI for coronary and carotid plaque characterization

As discussed earlier, asymptomatic plaque is vulnerable and rupture could lead to myocardial infarction or stroke. Acharya *et al* [47] applied AdaBoost and SVM on ultrasound 346 carotid scans (150 asymptomatic, 196 symptomatic) to describe plaques that are symptomatic and those that are not. The SVM classifier performed with the highest specificity of 82.1%, sensitivity of 82.9% and accuracy of 82.4%. In a different study, Acharya *et al* [48] utilized the same cohort using SVM through a polynomial kernel of order 2 to reach 83.7% accuracy. In 2013, Acharya *et al* [49] used SVM, Gaussian mixture modelling, KNN, decision tree (DT), Bayes classifier, radial basis probabilistic neural network (RBPNN) and fuzzy classifier on two separate carotid ultrasound populations. One consists the earlier 346 scans, and another 146 (102 asymptomatic, 44 symptomatic) carotid scans from a different cohort. The SVM using radial basis function (RBF) obtained the highest accuracy of 82.3% for 346 scans, while the fuzzy classifier obtained an accuracy of 93.1% for 146 scans.

A deep learning CNN model was employed by Lekadir *et al* [50] to characterize the plaque morphology from 56 carotid ultrasound data. A patching model was applied and pixel-to-pixel characterization into three classes of lipids, fibrous tissue and calcium was performed by the DL network. Cross-validation findings revealed a correlation of roughly 0.90. Figure 6.4 displays the colored examples of the expected and actual plaque components, and table 6.3 provides the results.

6.3.4 AI for LV segmentation

It is crucial to segment the left ventricle (LV) of the heart in order to produce clinical indices for cardiac monitoring. Carneiro *et al* [51] used DBN network to track the LV from 496 ultrasound sequences. The DBN system was trained for different priors (non-LV, diastole, and systole) to tract LV contour. The system performed with an average Jaccard distance of 0.83. Avendi *et al* [52] used a combination of CNN and autoencoders to segment the LV region from 45 cardiac MR scans as shown in figure 6.5(a).

To establish region of interest (ROI) from MR images, MRI scans were entered into the CNN framework. The ROI was used to determine the first LV shape using a

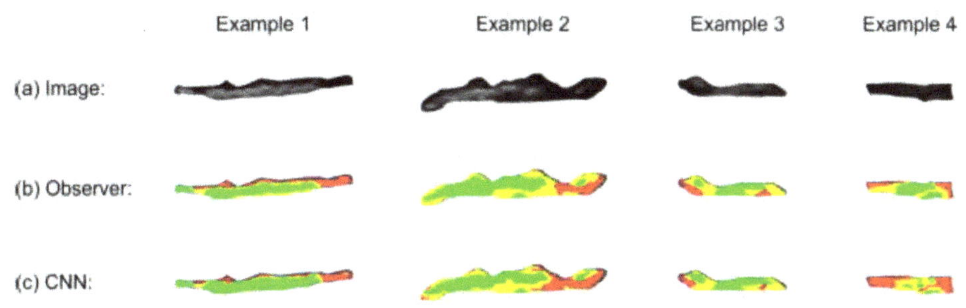

Figure 6.4. (a) Plaque sample, (b) ground truth and (c) CNN predicted plaque constituents (image reuse requested [50]).

Table 6.3. Characterization of plaque.

S. No.	First author (year)	# Instances/ images/ patients	Imaging modality	Model	Performance parameter	Performance
1	Acharya *et al* [47]	346	Ultrasound carotid scans	AdaBoost, SVM	Classification of symptomatic and asymptomatic plaque	SVM Acc: 82.4%, SVM specificity: 82.1%, SVM sensitivity: 82.9%
2	Acharya *et al* [48]	346	Ultrasound carotid scans	SVM with polynomial kernel	Classification of symptomatic and asymptomatic plaque	Acc: 83.7%
3	Acharya *et al* [49]	346 + 146	Ultrasound carotid scans	SVM, Gaussian mixture model, RBPNN, DT, KNN, Bayes classifier, fuzzy classifier	Classification of symptomatic and asymptomatic plaque	SVM-RBF (346) Acc: 82.3%, fuzzy classifier (146) Acc: 93.1%
4	Lekadir *et al* [50]	56	Ultrasound carotid scans	CNN	Characterization of plaque morphology	CC: 0.90

stacked autoencoder. Finally, the use of deformable models to generate the 3D LV segment. The Dice metric (DM), conformity metrics and average perpendicular distance (APD) were found to be 0.94, 0.86 and 1.81 mm, respectively. In the work by Romaguera *et al* [53] CNN network for semantic segmentation was used on whole MR images from 45 patients. Two separate optimization algorithms (OA1 and OA2) were used and two sets of performance indices were collected. The results are shown in figure 6.5(b). The system performed with Jaccard index of 0.97 and 0.97, specificity of 0.99 and 0.99, sensitivity of 0.92 and 0.90, Hausdorff distance (HD) of 4.48 and 5.43 and Dice score of 0.92 and 0.90, respectively. Table 6.4 displays the total results.

6.3.5 AI for measurement of plaque burden (cIMT and PA)

Measurement of plaque volume is critical for measurement of CVD risk. Plaque area and Carotid intima media thickness are used to calculate plaque volume. The

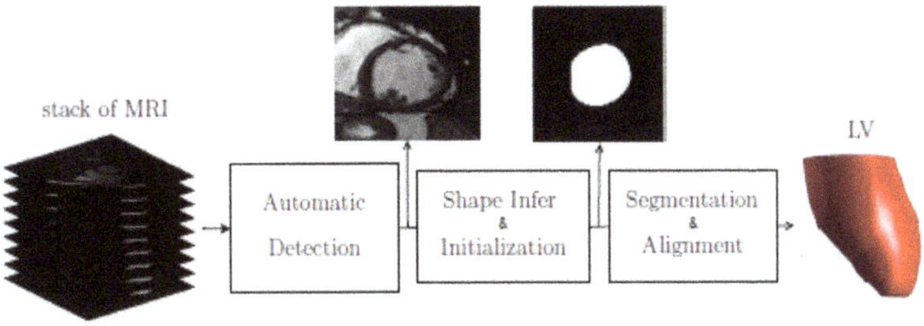

(a)

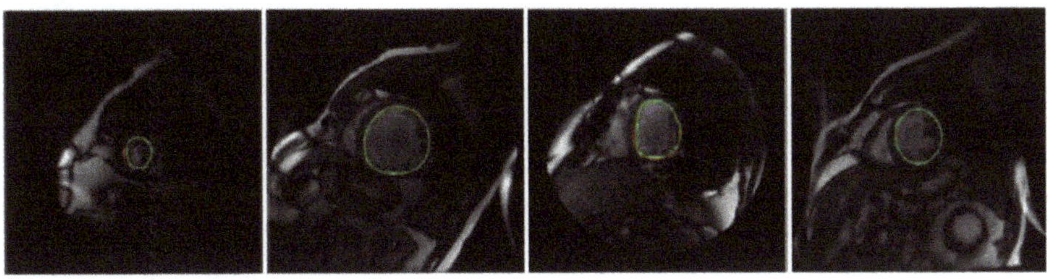

(b)

Figure 6.5. (a) LV segmentation using CNN and autoencoder (image reuse requested [52]), (b) LV segmentation from whole using CNN network (image reuse requested [53]).

plaque volume is usually segmented by delineating media adventitia (MA) and lumen intima (LI) walls automatically from the carotid ultrasound images using automated techniques. Over the years multiple ML and DL models have been used for estimating plaque volume either in computing cIMT or PA or both. Rosa *et al* [54] used ensemble of ANNs to estimate the cIMT from 60 CCA images of ultrasound. All the real-world tracings were performed by radiologists. The model had a cIMT error as 0.036 70 ± 0.024 29 mm. In the year 2014, Rosa *et al* [55] used radial basis neural network (RBNN) to 25 CCA images of ultrasound to calculate the cIMT. RBNN like ELM is single layer feed forward ANN which uses a radial basis function for characterization of pixels. System performance was recorded with a cIMT error of 0.065 ± 0.046 mm. Molinari *et al* [56] analyzed 200 ultrasound CCA images and using fuzzy K-means clustering (FKMC) for identifying pixels in the MA and LI wall. The FKMC model performed with acIMT error of 0.054 ± 0.035 mm.

Rosa *et al* [57] analyzed a combination of DL and ML in two stages for segmenting 55 CCA ultrasound images. In the first stage, two autoencoders were used to extract features from the MA and LI walls. In the following stage, ANNs were used to characterize the MA and LI wall pixels. Biswas *et al* [58] used fully convolutional network (FCN) for segmenting plaque from 396 CCA images of ultrasound using two real-world tracings from two different experts for training the

Table 6.4. Performance of LV segmentation using AI.

S. No.	First author (Year)	# Instances/ images/ patients	Imaging modality	Model	Performance parameter	Performance
1	Carneiro *et al* [51]	496	Ultrasound	DBN	Jaccard distance (JD)	JD: 0.83
2	Avendi *et al* [52]	45	MR	CNN + Autoencoder	DM, APD and conformity metrics	DM: 0.94 Conformity metrics: 0.86 APD: 1.81 mm
3	Romaguera *et al* [53]	45	MR	CNN-OA1, CNN-OA2	DM. HD, JD, sensitivity, specificity	DM (OA1): 0.92, DM(OA2): 0.90, HD (OA1): 4.48, HD (OA2): 4.43, JD (OA1): 0.97, JD (OA2): 0.97, Sens (OA1):0.92 Sens (OA2): 0.90 Spec (OA1): 0.99 Spec (OA2):0.99

DL network. The performance for the two real-world tracings were 0.124 ± 0.10 and 0.126 ± 0.134 mm, respectively. Elisa *et al* [59] analyzed DL model and ground truth tracings to estimate the plaque area, which were 20.52 and 19.44 mm^2, respectively. In the year 2020, Biswas *et al* [60] derived a two-step DL network for segmenting the plaque volume from 250 ultrasound CCA images. The images were divided into patches initially. In the initial step, a CNN was used to characterize the patches to identify the ROI consisting the plaque volume. The next stage was using an FCN to separate and define the MA and LI margins out of the ROI. The PA, PA error and cIMT error were measured to be 21.5794 ± 7.9975, 2.7939 ± 2.3702 and 0.0935 ± 0.0637 mm^2, respectively. Table 6.5 depicts the results.

6.3.6 AI in IVUS imaging

IVUS is a powerful invasive imaging tool for detection of stenosis inside coronary arteries since non-invasive ultrasound cannot penetrate through bones. Many different AI modules have been developed in the past decades to measure stenosis or plaque volume inside coronary arteries. Su *et al* [62] used a combination of sparse

Table 6.5. Measurement of plaque burden.

S. No.	First author (year)	# Instances/ images/ patients	Imaging modality	Model	Performance parameter	Performance
1	Rosa *et al* [54]	60	Ultrasound	ANN	cIMT error	cIMT error: 0.03670 ± 0.02429 mm
2	Rosa *et al* [55]	25	Ultrasound	RBNN	cIMT error	cIMT error: 0.046 ± 0.065 mm
3	Molinari *et al* [56]	200	Ultrasound	FKMC	cIMT error	cIMT error: 0.035 ± 0.054 mm
4	Rosa *et al* [57]	55	Ultrasound	ANN + Autoencoder	cIMT error	cIMT error: 0.0498 ± 0.0499 mm
5	Biswas *et al* [58]	396	Ultrasound	FCN	cIMT error	cIMT error: 0.134 ± 0.126 mm (DL1) cIMT error: 0.10 ± 0.124 mm (DL2)
6	Elisa *et al* [59]	396	Ultrasound	FCN	PA	PA1: 20.52 mm^2 (DL1). PA2: 19.44 mm^2 (DL2)
7	Biswas *et al* [61]	250	Ultrasound	CNN + FCN	cIMT error, PA and PA error	PA error: 2.7939 ± 2.3702 mm^2, cIMT error: 0.0935 ± 0.0637 mm, PA: 21.5794 ± 7.9975 mm^2

autoencoders for determination of plaque volume from 461 IVUS images. The ensemble autoencoders were used separately for detecting the MA and LI borders. The first autoencoder detects the MA/LI wall while the second autoencoder smoothens the borders of the LI/MA wall. The model's results showed that the Lumen and MA Jaccard Index were, respectively, 91.828 and 91.353. Lumen and MA have Hausdorff distances of 0.22431 and 0.32186, respectively. The output of the model for the intermediate stages is shown in figure 6.6.

Sofian *et al* [63] in 2018 used a DL residual neural network for characterization of calcification present in IVUS images. The dataset consisted of 2175 IVUS (530 calcified, 1645 non-calcified) images. The model performed with a sensitivity of 0.9981, highest accuracy of 0.9995 and specificity of 1.0000. In 2020, Olender *et al* [64]

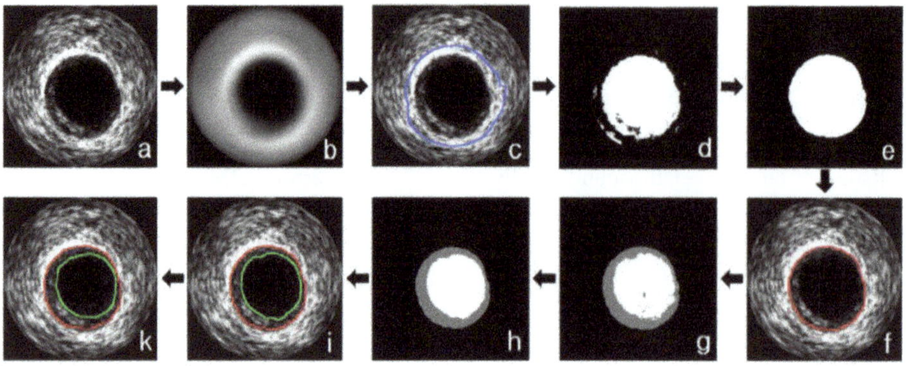

Figure 6.6. LI/MA boundary detection using two-stage autoencoder model (image reuse requested [62]).

Table 6.6. Characterization and segmentation in IVUS imaging.

S. No.	First author (Year)	# Instances/ images/ patients	Imaging modality	Model	Performance parameter	Performance
1	Su *et al* [62]	461	Ultrasound	Autoencoder 1 (Lumen) + autoencoder 2 (MA)	JD, HD	JD (Lumen): 91.828, JD (MA): 91.353, HD (Lumen): 0.224 31, HD (MA): 0.321 86
2	Sofian *et al* [63]	2175	Ultrasound	RNN	Classification between calcified and non-calcified	Acc: 0.9995 Sens: 0.9981 Spec: 1.0000
3	Olender *et al* [64]	553	Ultrasound	CNN	Classification Accuracy	Acc: 93.5%

used CNN for 553 IVUS frames for characterization of diseased vessel walls. The system performed with an accuracy of 93.5%. All results related to IVUS imaging is given in table 6.6.

6.4 Discussion

In this paper we have discussed AI applications in various areas of cardiovascular imaging such as blood flow, characterization of ECG images, plaque characterization, LV segmentation, plaque burden measurement and IVUS imaging. The AI applications' performance have shown considerable clinical implications. The characterization accuracy in some cases are better than human diagnosis. However, there are several safety, legal, psychological, social obstacles for the generalized implementation of AI in medical science. The safety requirement entails that the AI healthcare system should perform without failure since it assumes primary responsibility in diagnosis of vulnerable patients. The legality constraint of liability ownership in case of catastrophic failure. It could be the joint responsibility of medical doctor and system designer. The biggest constraint is the final acceptance of AI systems by the public who are highly suspicious regarding its general use due to overall antagonistic representation of AI in media. Nevertheless, the benefits of AI far outrun the apprehensions. One of the major benefits of usage of AI is better economy-of-scale. There are only very few radiologists available compared to the volume of radiology images produced. The AI models can be scaled up to handle a large number of medical images with an investment of a few thousand dollars. AI can be teamed up with telemedicine to serve people and hospitals in far flung areas. Third world countries in Asia and Africa can benefit immensely from the large scale implementation of AI, as they can be served in real time.

Another area where AI is making the mark is in the area of CVD risk calculators. As already discussed earlier, the AECRS1.0 and AECRS2.0 are composite risk scores combining OBBM, LBBBM and imaging phenotypes such as PA and cIMT. The AI has shown considerable success in robust measurement of these phenotypes as discussed in subsection 3.4, 3.5 and 3.6. Therefore, the risk calculators can take these measurements and predict CVD events. Also the risk calculators can themselves be modified using ML models of regression to predict automated future events if all the data is available. Overall, AI can be stipulated as a daily use system just like desktop computers.

6.5 Conclusion

In this work, we have covered AI applications in different areas of CVD imaging and how this can be helpful for future medical professionals. AI in combination with telemedicine can be put to good use where there is a dearth of imaging specialists. AI can be scaled up to serve millions of patients worldwide and can help in their future treatment. This review also focuses on different legal, safety, social issues related to AI. Also this review talks about the larger benefits of AI to mankind as general. It is foreseen that AI along with human intelligence can help patient treatment in better way.

References

[1] Benjamin E J, Muntner P, Alonso A, Bittencourt M S, Callaway C W, Carson A P, Chamberlain A M, Chang A R, Cheng S and Das S R 2019 Heart disease and stroke Statistics-2019 update a report from the American heart association *Circulation* **139** e56–528

[2] Mozaffarian D, Benjamin E J, Go A S, Arnett D K, Blaha M J, Cushman M, De Ferranti S, Després J-P, Fullerton H J and Howard V J 2015 Executive summary: heart disease and stroke statistics—2015 update: a report from the American Heart Association *Circulation* **131** 434–41

[3] Mozaffarian D, Benjamin E J, Go A S, Arnett D K, Blaha M J, Cushman M, Das S R, De Ferranti S, Després J-P and Fullerton H J 2016 Executive summary: heart disease and stroke statistics—2016 update: a report from the American Heart Association *Circulation* **133** 447–54

[4] Suri J S, Kathuria C and Molinari F 2010 *Atherosclerosis Disease Management* (Springer Science & Business Media)

[5] Libby P, Ridker P M and Maseri A 2002 Inflammation and atherosclerosis *Circulation* **105** 1135–43

[6] Acharya U R, Joseph K P, Kannathal N, Lim C M and Suri J S 2006 Heart rate variability: a review *Med. Biol. Eng. Comput.* **44** 1031–51

[7] Greenstein A J, Chassin M R, Wang J, Rockman C B, Riles T S, Tuhrim S and Halm E A 2007 Association between minor and major surgical complications after carotid endarterectomy: results of the New York Carotid Artery Surgery study *J. Vasc. Surg.* **46** 1138–46

[8] Naylor A, Payne D, London N, Thompson M, Dennis M, Sayers R and Bell P 2002 Prosthetic patch infection after carotid endarterectomy *Eur. J. Vasc. Endovasc. Surg.* **23** 11–6

[9] Wolk M J, Allen J M and Raskin I E 2005 ACCF proposed method for evaluating the appropriateness of cardiovascular imaging *J. Am. Coll. Cardiol.* **46**

[10] Budoff M J, Achenbach S, Blumenthal R S, Carr J J, Goldin J G, Greenland P, Guerci A D, Lima J A, Rader D J and Rubin G D 2006 Assessment of coronary artery disease by cardiac computed tomography: a scientific statement from the American heart association committee on cardiovascular imaging and intervention, council on cardiovascular radiology and intervention, and committee on cardiac imaging, council on clinical cardiology *Circulation* **114** 1761–91

[11] Sanches J M, Laine A F and Suri J S 2012 *Ultrasound Imaging* (Berlin: Springer)

[12] Suri J S, Wilson D and Laxminarayan S 2005 *Handbook of Biomedical Image Analysis* vol 2 (Springer Science & Business Media)

[13] Preiss D and Kristensen S L 2015 The new pooled cohort equations risk calculator *Can. J. Cardiol.* **31** 613–9

[14] Lloyd-Jones D M, Wilson P W, Larson M G, Beiser A, Leip E P, D'Agostino R B and Levy D 2004 Framingham risk score and prediction of lifetime risk for coronary heart disease *The Am. J. Cardiol.* **94** 20–4

[15] Board J B S 2014 Joint British Societies' consensus recommendations for the prevention of cardiovascular disease (JBS3) *Heart* **100** ii1–ii67

[16] Tillin T, Hughes A D, Whincup P, Mayet J, Sattar N, McKeigue P M, Chaturvedi N and Group S S 2014 Ethnicity and prediction of cardiovascular disease: performance of QRISK2 and Framingham scores in a UK tri-ethnic prospective cohort study (SABRE—Southall And Brent REvisited) *Heart* **100** 60–7

[17] Ridker P M, Buring J E, Rifai N and Cook N R 2007 Development and validation of improved algorithms for the assessment of global cardiovascular risk in women: the Reynolds risk score *JAMA* **297** 611–9

[18] Stevens R J, Kothari V, Adler A I, Stratton I M and Holman R RGroup UKPDS 2001 The UKPDS risk engine: a model for the risk of coronary heart disease in Type II diabetes (UKPDS 56) *Clin. Sci.* **101** 671–9

[19] Kothari V, Stevens R J, Adler A I, Stratton I M, Manley S E, Neil H A and Holman R R 2002 UKPDS 60: risk of stroke in type 2 diabetes estimated by the UK Prospective Diabetes Study risk engine *Stroke* **33** 1776–81

[20] Khanna N N, Jamthikar A D, Gupta D, Piga M, Saba L, Carcassi C, Giannopoulos A A, Nicolaides A, Laird J R and Suri H S 2019 Rheumatoid arthritis: atherosclerosis imaging and cardiovascular risk assessment using machine and deep learning–based tissue characterization *Curr. Atheroscler. Rep.* **21** 7
[21] Viswanathan V, Jamthikar A D, Gupta D, Puvvula A, Khanna N N, Saba L, Viskovic K, Mavrogeni S, Turk M and Laird J R 2020 Integration of eGFR biomarker in image-based CV/stroke risk calculator: a south Asian-Indian diabetes cohort with moderate chronic kidney disease *Int. Angiol.* **39** 290–306
[22] Saba L, Sanches J M, Pedro L M and Suri J S 2014 *Multi-Modality Atherosclerosis Imaging and Diagnosis* (Berlin: Springer)
[23] Seabra J and Sanches J 2012 *Ultrasound Imaging: Advances and Applications* (New York: Springer)
[24] Suri J S and Laxminarayan S 2003 *Angiography and Plaque Imaging: Advanced Segmentation Techniques* (Boca Raton, FL: CRC Press)
[25] Molinari F, Meiburger K M, Acharya U R, Zeng G, Rodrigues P S, Saba L, Nicolaides A and Suri J S 2011 CARES 3.0: a two stage system combining feature-based recognition and edge-based segmentation for CIMT measurement on a multi-institutional ultrasound database of 300 images *2011 Annual Int. Conf. of the IEEE Engineering in Medicine and Biology Society* (Piscataway, NJ: IEEE) pp 5149–52
[26] Wendelhag I, Liang Q, Gustavsson T and Wikstrand J 1997 A new automated computerized analyzing system simplifies readings and reduces the variability in ultrasound measurement of intima-media thickness *Stroke* **28** 2195–200
[27] Cheng D-c, Schmidt-Trucksäss A, Cheng K-s and Burkhardt H 2002 Using snakes to detect the intimal and adventitial layers of the common carotid artery wall in sonographic images *Comput. Methods Programs Biomed.* **67** 27–37
[28] Kumar P K, Araki T, Rajan J, Saba L, Lavra F, Ikeda N, Sharma A M, Shafique S, Nicolaides A and Laird J R 2017 Accurate lumen diameter measurement in curved vessels in carotid ultrasound: an iterative scale-space and spatial transformation approach *Med. Biol. Eng. Comput.* **55** 1415–34
[29] Bishop C M 2006 *Pattern Recognition and Machine Learning* (Berlin: Springer)
[30] Biswas M, Kuppili V, Saba L, Edla D R, Suri H S, Cuadrado-Godia E, Laird J R, Marinhoe R T, Sanches J M and Nicolaides A 2019 State-of-the-art review on deep learning in medical imaging *Front. Biosci. (Landmark edition)* **24** 392–426
[31] LeCun Y, Bengio Y and Hinton G 2015 Deep learning *Nature* **521** 436–44
[32] Sutton R S and Barto A G 2018 *Reinforcement Learning: An Introduction* (Cambridge, MA: MIT Press)
[33] Kaelbling L P, Littman M L and Moore A W 1996 Reinforcement learning: a survey *J. Artif. Intell. Res.* **4** 237–85
[34] Rosenblatt F 1958 The perceptron: a probabilistic model for information storage and organization in the brain *Psychol. Rev.* **65** 386
[35] Hart P 1968 The condensed nearest neighbor rule (Corresp.) *IEEE Trans. Inf. Theory* **14** 515–6
[36] Bayes T 1968 *Naive Bayes Classifier* Article Sources and Contributors 1–9
[37] Cortes C and Vapnik V 1995 Support-vector networks *Mach. Learn.* **20** 273–97
[38] Huang G-B, Wang D H and Lan Y 2011 Extreme learning machines: a survey *Int. J. Mach. Learn. Cybern.* **2** 107–22

[39] Kuppili V, Biswas M, Sreekumar A, Suri H S, Saba L, Edla D R, Marinhoe R T, Sanches J M and Suri J S 2017 Extreme learning machine framework for risk stratification of fatty liver disease using ultrasound tissue characterization *J. Med. Syst.* **41** 152

[40] Saba L, Biswas M, Kuppili V, Godia E C, Suri H S, Edla D R, Omerzu T, Laird J R, Khanna N N and Mavrogeni S 2019 The present and future of deep learning in radiology *Eur. J. Radiol.* **114** 14–24

[41] Coenen A, Kim Y-H, Kruk M, Tesche C, De Geer J, Kurata A, Lubbers M L, Daemen J, Itu L and Rapaka S 2018 Diagnostic accuracy of a machine-learning approach to coronary computed tomographic angiography-based fractional flow reserve: result from the MACHINE consortium *Circ: Cardiovasc. Imaging* **11** e007217

[42] Kissas G, Yang Y, Hwuang E, Witschey W R, Detre J A and Perdikaris P 2020 Machine learning in cardiovascular flows modeling: predicting arterial blood pressure from non-invasive 4D flow MRI data using physics-informed neural networks *Comput. Meth. Appl. Mech. Eng.* **358** 112623

[43] Kusunose K, Abe T, Haga A, Fukuda D, Yamada H, Harada M and Sata M 2020 A deep learning approach for assessment of regional wall motion abnormality from echocardiographic images *JACC: Cardiovascular Imaging* **13** 374–81

[44] Narula S, Shameer K, Omar A M S, Dudley J T and Sengupta P P 2016 Machine-learning algorithms to automate morphological and functional assessments in 2D echocardiography *J. Am. Coll. Cardiol.* **68** 2287–95

[45] Yuan B, Chitturi S R, Iyer G, Li N, Xu X, Zhan R, Llerena R, Yen J T and Bertozzi A L 2017 Machine learning for cardiac ultrasound time series data *Medical Imaging 2017: Biomedical Applications in Molecular, Structural, and Functional Imaging* (International Society for Optics and Photonics) pp 101372D

[46] Ghorbani A, Ouyang D, Abid A, He B, Chen J H, Harrington R A, Liang D H, Ashley E A and Zou J Y 2020 Deep learning interpretation of echocardiograms *NPJ Digit. Med.* **3** 1–10

[47] Acharya R U, Faust O, Alvin A P C, Sree S V, Molinari F, Saba L, Nicolaides A and Suri J S 2012 Symptomatic vs. asymptomatic plaque classification in carotid ultrasound *J. Med. Syst.* **36** 1861–71

[48] Acharya U R, Faust O, Sree S V, Molinari F, Saba L, Nicolaides A and Suri J S 2011 An accurate and generalized approach to plaque characterization in 346 carotid ultrasound scans *IEEE Trans. Instrum. Meas.* **61** 1045–53

[49] Acharya U R, Mookiah M R K, Sree S V, Afonso D, Sanches J, Shafique S, Nicolaides A, Pedro L M, Fernandes J F and Suri J S 2013 Atherosclerotic plaque tissue characterization in 2D ultrasound longitudinal carotid scans for automated classification: a paradigm for stroke risk assessment *Med. Biol. Eng. Comput.* **51** 513–23

[50] Lekadir K, Galimzianova A, Betriu À, del Mar Vila M, Igual L, Rubin D L, Fernández E, Radeva P and Napel S 2016 A convolutional neural network for automatic characterization of plaque composition in carotid ultrasound *IEEE J. Biomed. Health Inform* **21** 48–55

[51] Carneiro G and Nascimento J C 2013 Combining multiple dynamic models and deep learning architectures for tracking the left ventricle endocardium in ultrasound data *IEEE Trans. Pattern Anal. Mach. Intell.* **35** 2592–607

[52] Avendi M, Kheradvar A and Jafarkhani H 2016 A combined deep-learning and deformable-model approach to fully automatic segmentation of the left ventricle in cardiac MRI *Med. Image Anal.* **30** 108–19

[53] Romaguera L V, Costa M G F, Romero F P and Costa Filho C F F 2017 Left ventricle segmentation in cardiac MRI images using fully convolutional neural networks *Medical Imaging 2017: Computer-Aided Diagnosis* (International Society for Optics and Photonics) 101342Z

[54] Menchón-Lara R-M, Bastida-Jumilla M-C, Morales-Sánchez J and Sancho-Gómez J-L 2014 Automatic detection of the intima-media thickness in ultrasound images of the common carotid artery using neural networks *Med. Biol. Eng. Comput.* **52** 169–81

[55] Menchón-Lara R-M and Sancho-Gómez J-L 2014 Ultrasound image processing based on machine learning for the fully automatic evaluation of the Carotid Intima-Media Thickness *2014 12th Int. Workshop on Content-Based Multimedia Indexing (CBMI)* (Piscataway, NJ: IEEE) pp 1–4

[56] Molinari F, Zeng G and Suri J S 2010 Intima-media thickness: setting a standard for a completely automated method of ultrasound measurement *IEEE Trans. Ultrason. Ferroelectr. Freq. Control* **57** 1112–24

[57] Menchón-Lara R-M and Sancho-Gómez J-L 2015 Fully automatic segmentation of ultrasound common carotid artery images based on machine learning *Neurocomputing* **151** 161–7

[58] Biswas M, Kuppili V, Araki T, Edla D R, Godia E C, Saba L, Suri H S, Omerzu T, Laird J R and Khanna N N 2018 Deep learning strategy for accurate carotid intima-media thickness measurement: an ultrasound study on Japanese diabetic cohort *Comput. Biol. Med.* **98** 100–17

[59] Cuadrado-Godia E, Srivastava S K, Saba L, Araki T, Suri H S, Giannopolulos A, Omerzu T, Laird J, Khanna N N and Mavrogeni S 2018 Geometric total plaque area is an equally powerful phenotype compared with carotid intima-media thickness for stroke risk assessment: a deep learning approach *J. Vasc. Ultrasound* **42** 162–88

[60] Biswas M, Saba L, Chakrabartty S, Khanna N N, Song H, Suri H S, Sfikakis P P, Mavrogeni S, Viskovic K and Laird J R 2020 Two-stage artificial intelligence model for jointly measurement of atherosclerotic wall thickness and plaque burden in carotid ultrasound: a screening tool for cardiovascular/stroke risk assessment *Comput. Biol. Med.* **123** 103847

[61] Biswas M, Saba L and Chakrabartty S *et al* 2020 Two-stage artificial intelligence model for jointly measurement of atherosclerotic wall thickness and plaque burden in carotid ultrasound: a screening tool for cardiovascular/stroke risk assessment *Comput. Biol. Med.* **123** 103847

[62] Su S, Hu Z, Lin Q, Hau W K, Gao Z and Zhang H 2017 An artificial neural network method for lumen and media-adventitia border detection in IVUS *Comput. Med. Imaging Graph.* **57** 29–39

[63] Sofian H, Ming J T C, Mohamad S and Noor N M 2018 *Calcification Detection Using Deep Structured Learning in Intravascular Ultrasound Image for Coronary Artery Disease 2018 2nd Int. Conf. on BioSignal Analysis, Processing and Systems (ICBAPS) (Piscataway, NJ)* (IEEE) pp 47–52

[64] Olender M L, Athanasiou L S, Michalis L K, Fotiadis D I, Edelman E and Domain A 2020 Enriched deep learning approach to classify atherosclerosis using intravascular ultrasound imaging *IEEE J. Sel. Top. Signal Process* **14** 1210–20

IOP Publishing

Multimodality Imaging, Volume 2
Heart, lungs and peripheral organs
Mainak Biswas and Jasjit S Suri

Chapter 7

A deep learning perspective in internal carotid artery and bulb segmentation: review

Amita Banerjee, Pankaj Kumar Jain, Radhakrishn Birla, Mainak Biswas and Jasjit S Suri

In the present, we have reviewed the different causes of stroke and cardiovascular diseases (CVDs). Atherosclerosis is a disease of the arteries that causes stenosis, which can subsequently lead to stroke or other CVDs. Stroke and CVDs have a massive impact on people's health worldwide. Early detection of atherosclerosis may reduce the risk of death and diminish treatment costs. Lumen diameter and carotid intima-media thickness are the two parameters that are used as markers for atherosclerosis. In view of the rising number of strokes due to CVDs, we need a robust computer-aided diagnosis system. Many automated techniques, such as methods based on machine learning (ML), can diagnose CVDs at an early stage, but nowadays attention is shifting toward the development of diagnosis systems based on deep learning (DL). Therefore, the current review explores DL-based strategies in a competitive environment. The major drawback of ML strategies is their self-correction capability, which requires the system designer's input. DL strategies, on the other hand, provide a robust solution for the problem of self-correction.

ML strategies are based on extracting custom-built features from regions of interest in ultrasound images of the carotid artery and classifying using supervised algorithms, such as support vector machines, artificial neural networks, and AdaBoost classifiers. However, in modern DL techniques, convolutional features replaced the custom-built features. The depth of the neural network refines these convolutional features, a process that is dependent on CPU speed. Fortunately, the advent of modern graphics processing units has accelerated CPU operations. We have studied the different applications of ML and DL systems in medical image diagnosis of various diseases.

ML and DL strategies have their pros and cons, so the system designer may need to select the best method for a given application. Overall, we find that DL strategies prove to be more competent than ML strategies.

7.1 Introduction

In a report published by the WHO in 2016, cardiovascular diseases (CVDs) were identified as the leading contributor to fatalities in the United States, accounting for 43.2% of deaths, while stroke ranked second at 16.9% [1]. CVDs profoundly affect human life, resulting in approximately 17.6 million deaths globally each year. This figure is projected to increase to 23.6 million by the year 2030, emphasizing the significant impact of CVDs on mortality rates worldwide [2]. Low- and middle-income group countries are also affected by CVDs. For India, CVDs have gained more importance due to rising fatality rates [3–5]. CVDs lead to high costs in treatment, productivity, and mortality; therefore, the total direct expenditure in CVDs is projected to increase to $749 billion in 2035 [1, 6]. Prabhakaran *et al* [5] reported that the urban population of India is likely to be more susceptible to CVDs (1%–13.2%) than its rural population (1.6%–7.4%). According to the World Health Organization (WHO), the age-standardized mortality rate for cardiovascular diseases (CVDs) in India is estimated to range from 363 to 443 per 100 000 for males and from 181 to 281 per 100 000 for females [5].

Atherosclerosis in carotid arteries is the primary cause of stroke. Early detection of CVDs may reduce mortality rates and, subsequently, treatment expenditure. Atherosclerosis results in the constriction of arteries, impeding the circulation of blood within the cardiovascular and cerebrovascular systems [7–14]. Significant parts of the carotid arteries include the internal carotid artery (ICA), common carotid artery (CCA), external carotid artery (ECA), and bulb, which are all affected by atherosclerosis. The CCA has been studied extensively, but the ICA and the bulb have not. The ICA is deeply embedded and covered with flesh in the neck; hence, anterior to posterior manoeuvring is very difficult [15]. The ECA dispenses blood to the face and the scalp, whereas the ICA supplies blood to the brain. The plaque growth inside the ICA and the bulb has no pattern; however, the focal thickening around the lumen region can be viewed in a 3D cross-sectional slice of the carotid artery. The unidirectional plaque, as shown in figure 7.1(a), can be quantified using conventional carotid intima-media thickness (cIMT) measurement methods. The bidirectional stenotic plaques shown in figure 7.1(b) can be quantified using North American Symptomatic Carotid Endarterectomy Trial (NASCET) or European Carotid Surgery Trial (ECSET) criteria by utilizing various tools of lumen diameter (LD) measurements [16–18]. The cIMT- or LD-quantified measurements are conventionally used as gold standards for stroke risk assessments [17, 19–26].

Various imaging techniques, including magnetic resonance imaging (MRI), computed tomography (CT) angiography, and ultrasound (US) scans, can be utilized to detect and locate atherosclerosis within the arteries., as shown in figure 7.1 [7, 27–29]. B-mode US is preferred over the other imaging modalities as it does not use a contrast agent and is radiation-free and non-invasive. Therefore, the diagnosis

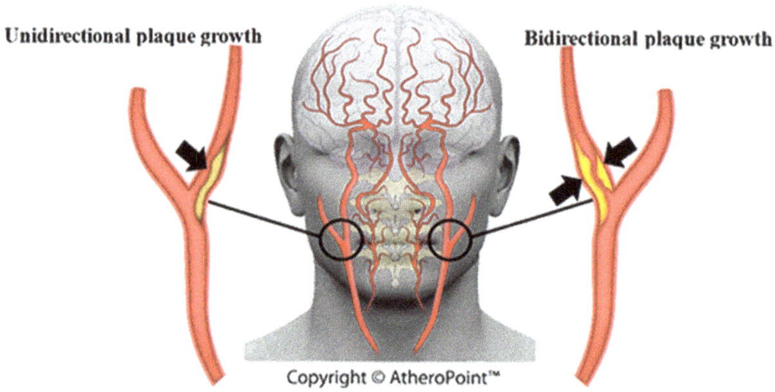

Figure 7.1. Anatomy of the human cerebrovascular system.

with US is better than the diagnoses based on MRI or CT scans. The internal carotid artery (ICA) is positioned behind and to the side of the external carotid artery (ECA) and has a slightly larger diameter compared to the ECA; therefore, ICA scanning is a complicated job. The sonographer can identify the ECA by tapping their finger on the ipsilateral temporal artery, which produces a serration-like artifact in the Doppler spectrum [30].

Since there are only very few radiologists compared with the huge number of medical images generated every day, the screening is slow and error-prone. Therefore, a computer-aided diagnosis (CAD) system or decision support system (DSS) is needed to overcome these drawbacks of manual measurements. Machine learning (ML) has proven to be a great tool for learning from experiences automatically [31]. The polyline distance metric method is a benchmark for segmentation and used in many studies [19, 25, 32–34] for region-of-interest (ROI) detection. Biswas *et al* used this method in combination with convolution neural networks (CNNs) for the automatic detection of LD [35]. The shape-based approach developed by Cootes *et al* [36] inspired researchers to develop methods for LD and cIMT measurements [24, 37–39]. ML tools for measuring plaques and stenosis rely upon accessing the texture from the ROI. These texture features contain spatial and temporal information about the images' ROI. Depending on system design and analysis, the extracted features may indicate statistically significant alterations, which sometimes enhance the precision and reliability of the system and sometimes collapses the estimations. ML-based risk assessment systems using LD and cIMT as biomarkers overcome the difficulties of manual methods [17, 18, 40, 41]. In this aspect, Saba *et al* [40] and Araki *et al* [17] show that the careful extraction and selection of features is an essential step for ML systems. Detailed investigation of the ML system designed by Araki *et al* [17, 18, 41] and Saba *et al* in [40] reveals various checkpoints, e.g., ROI detection, feature extraction, and feature selection, that can improve image classification and evaluation and, therefore, the overall performance of the system [17, 18, 40, 41].

DL-based techniques, such as CNNs, overcome these checkpoints and improve the system's ability, hence becoming a very attractive tool for the analysis of medical images. These brain-inspired deep neural networks use convolutional, pooling, and fully connected layers to extract features from images, and the depth of the network provides highly improved feature maps. However, CNN layers use a large stack of images; therefore, the depth of the network increases the computational costs; hence, computations performed on standard CPUs may slow down the whole operation. Modern GPUs are built to process a large amount of data in parallel and do not affect the speed of the CPUs; therefore, these units can speed up the operation and measure in almost real-time [35, 42–50]. The DL systems designed (by Biswas *et al* in [22] used encoder/decoder layers; 10-layer CNN by Sudha *et al* [23]; and the autoencoders used by Menchón-Lara *et al* [45]) have gained remarkable success in cIMT measurements compared to conventional ML methods. Biswas *et al* [22] showed an improvement of 20% in measuring cIMT values compared to sonographers using a two-stage system (a class of AtheroEdge™ system). Furthermore, with their DL system consisting of 13 encoder layers and 3 upsample decoder layers, Biswas *et al* [35] have accomplished significant improvements in lumen characterization based on US images. A similar perspective was carried out by Zreik *et al* [52] to identify coronary plaques and stenosis in coronary CT angiographic images. Apart from making contributions to implementing DL systems, many researchers have attempted to write reviews on DL in medical imaging. In their reviews, Tandel *et al* [42] wrote about brain cancer classification, Liu *et al* [53] focused on medical ultrasound, and Lundervold *et al* [47] concentrated on MRI images. Saba *et al* [54] and Meyer *et al* [48] discussed different aspects of the DL strategy in radiology. Based on the above studies, we can write an informative review of different segmentation techniques using ML and DL systems in vascular imaging. Further, in this review, we focused on automated detection of LD, cIMT, and the bulb in US images, using various ML and DL techniques. We also addressed techniques that are an amalgamation of conventional and DL techniques.

7.2 Article search strategy

Our work started with a rigorous search on academic websites, such as Google Scholar, PubMed, and Web of Science Arxiv, and non-academic websites, such as Github, Kaggle, and Stack Overflow. We confined our research only to articles published within the past ten years in high impact factor and peer-reviewed journals. Since we focused on the segmentation of ICA images based on a DL strategy, we used keywords associated to the title of this review. Figure 7.2(a) shows the bar chart of the articles studied for this review. Figure 7.2(b) shows the pie chart of the distribution of the articles based on the school of thought (SOT) studied in this review. Table 7.1 represents the sector-wise representation of SOTs. SOT1 is based on the biology of atherosclerosis and the acquisition of B-mode ultrasound (BUS) images of carotid arteries. SOT2 is based on the conventional methods of artery segmentation. SOT3 and SOT4 are motivated by ML- and DL-based methods of segmentation, respectively. In figure 7.2(a), the vertical axis shows the year of

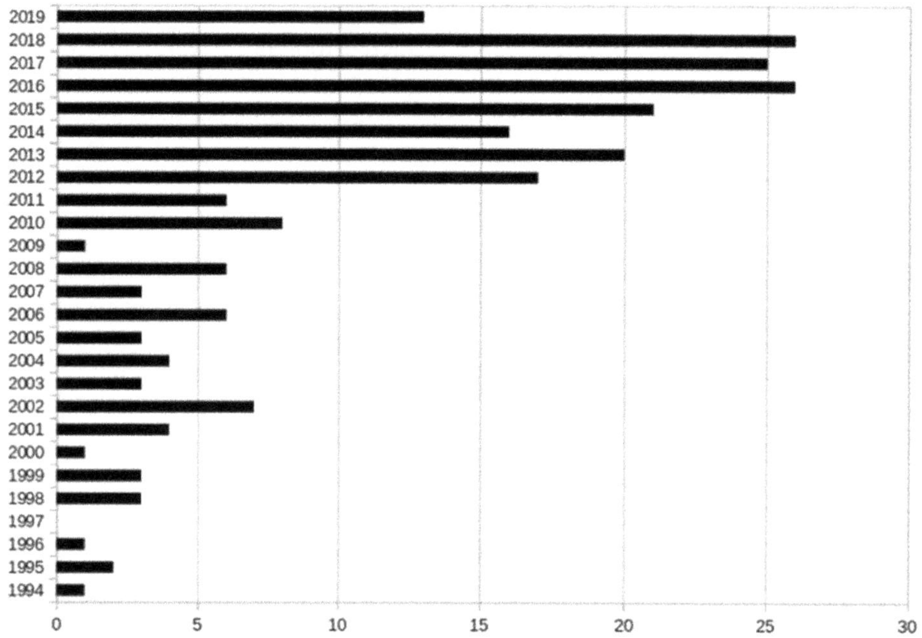

Figure 7.2(a). Number of articles from each year studied in this review.

Figure 7.2(b). Pie-chart of the distribution of articles based on schools of thoughts (SOT).

publication, and the horizontal axis shows the number of articles studied. A total of 226 articles were found after rigorous searching, and nearly half of them were selected based on their relevance. We employed the following search terms: carotid artery segmentation, ultrasound imaging of carotid arteries, carotid artery

Table 7.1. Sector wise subjects of the articles studied in this review.

S. No.	Sector	Subject
1	Sector 1	Biology of the atherosclerosis, US image reconstruction, plaque tissue characterization, cIMT and LD as biomarker (appendix A: table A.1)
2	Sector 2	Conventional methods of LD and cIMT segmentation (appendix A: table A.2)
3	Sector 3	ML methods of segmentation of LD and cIMT (appendix A: table A.3)
4	Sector 4	DL method of segmentation of LD and cIMT (appendix A: table A.4)

bifurcation, automated segmentation methods, segmentation using machine learning, and segmentation based on deep learning. Cardiologists and radiologists from our team reviewed the papers related to the anatomy of the cardiovascular system. Experts in the field of computer science and electrical engineering inspected whether the papers were worth including from a technical point of view. Finally, we prepared a consolidated list of references based on the advice of our team members. Articles were selected using these criteria; however, the selection was not limited to the above criteria.

7.3 Biology of atherosclerosis and plaque formation

Atherosclerosis, a progressive inflammatory condition, typically manifests as individuals age. The disease advances through distinct stages, beginning with endothelial dysfunction, followed by the formation of fatty streaks, and ultimately leading to the development of complex lesions [28, 55–58]. The chemical, mechanical, or immunological changes damage the endothelium and lead to endothelial dysfunction, which starts the atherosclerosis process. Endothelial permeability, leukocyte migration, endothelial adhesion, and leukocyte adhesion are responsible for endothelium dysfunction [55]. Under normal conditions, endothelial monolayers and blood flow prevent the adhesion of leukocytes. However, with endothelium dysfunction, leukocytes are recruited, and pro-inflammatory cytokines are expressed at the inflammation sites, which further leads to the early stage of atherogenesis. VCAM-1 is an adhesion leukocyte that binds monocytes and T lymphocytes at the site of the atheroma initiation. E and P selectin and intercellular adhesion molecule-1 are some other endothelial adhesion molecules, whereas L selectin, integrins, and platelet endothelial cell adhesion molecules are leucocyte adhesion molecules [13, 56–60].

In a state of normal physiological conditions, the endothelial layers effectively obstruct the infiltration of leukocytes into the intimal layers, functioning as a barrier between the blood and the intima. However, adverse conditions such as abnormal arterial structures (such as tortuosity, kinking, and coiling) create an environment of low shear stress and flow turbulence. Consequently, this facilitates the penetration of large molecules, like low-density lipoproteins (LDL), through the endothelial layers

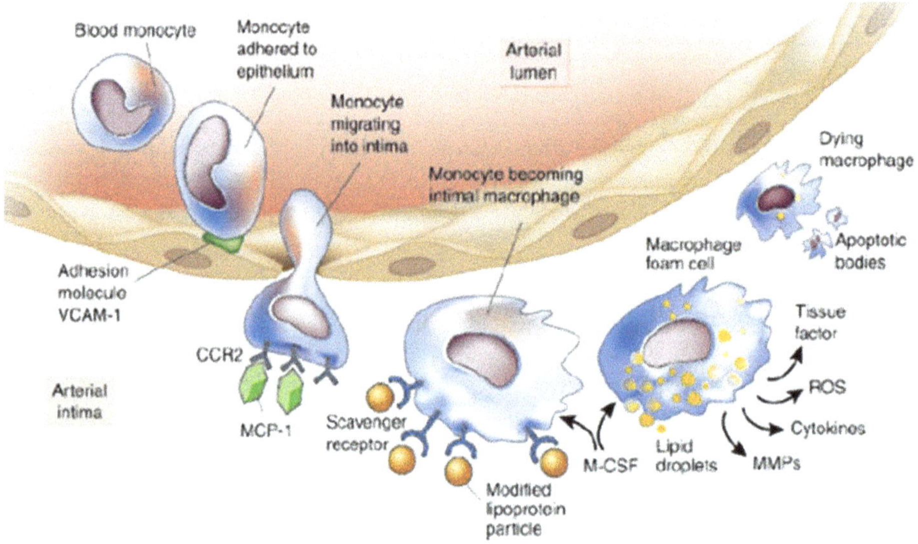

Figure 7.3. A schematic view of the atherogenesis courtesy of Libby *et al* [55].

and their accumulation beneath the intimal layer. This accumulation process, referred to as atheroma formation, intensifies in the presence of elevated levels of LDL cholesterol. Growth factors such as VCAM-1 and ICAM-1 (macrophage colony-stimulating factors) are responsible for atheroma formation. Phagocytosis by monocyte-derived macrophages oxidizes LDL, and foam cells appear, which is the first appearance of plaques in the form of fatty streaks [28, 29, 56–62]. Figure 7.3 shows the schematic diagram of the progression of the disease in arteries. Cardiologists and cardiovascular pathologists found that culprit plaques are responsible for cardiovascular events. In the clinical practice, these culprit plaques are also known as vulnerable, high-risk, or unstable plaques. Naghavi *et al* [56] have reviewed and explored various classifications of vulnerable plaques. Vulnerable plaques render people prone to thrombosis, and patients are at risk of cardiovascular events, stroke, or sudden death [28, 29, 62–65].

7.3.1 Plaque morphology

The presence of an atherosclerotic plaque poses a risk for thrombus formation, which can lead to the obstruction of carotid arteries and the development of CVDs such as myocardial infarction (MI) and unstable angina [1, 3, 5, 28, 29, 51, 65–69]. Additionally, when the carotid arteries become embolized, it can result in strokes and ischemic attacks. These conditions, including thrombosis and embolization, can have severe consequences such as death or long-term disability. Consequently, the atherosclerotic plaque is considered a high-risk or vulnerable plaque [70, 71]. As the plaque ages, it undergoes changes and develops various components, which can be visualized using different imaging modalities [27, 72–75]. Plaques can be categorized into three primary types based on their echolucency in US images: hyperechoic, hypoechoic, and moderately echoic plaques [18, 61, 76, 77]. However, it is important to note that US imaging is largely influenced by the operator's skills, making the reproducibility of images operator-dependent [9, 19, 27, 62, 78]. Furthermore, we classify plaques into low-risk (asymptomatic) and high-risk (symptomatic) categories. Previous research has explored various parameters related to carotid arteries, such as pulse wave velocity (PWV), total plaque area (TPA), total plaque volume (TPV), bifurcation angle, tortuosity, and bulb diameter [64, 69, 79, 80]. However, these parameters have not proven to be useful biomarkers for a coronary artery disease (CAD) system or a decision support system (DSS) [68].

7.3.2 Physics of the reconstruction of ultrasound images

US image reconstruction involves three steps: (1) the generation of US waves, (2) propagation and reception of the echo from inside the tissues, and (3) the reconstruction of images using the echo received. The generated frequency range lies between 1 and 18 MHz, and when the US waves propagate through the physiological system in a controlled manner, soft and hard tissues produce reverberations (echo) of the sound waves. The echo received is converted into electrical pulses and analyzed based on the amplitude and delay properties of the sound waves. Londhe *et al* [81] highlighted the application of superharmonic waves in the reconstruction of US images. Higher harmonics improve the resolution along with the optimized penetration. Naim *et al* [28] reported on the echolucency and stability of the atherosclerotic plaque in US images. US images have different shades between black and white, which means that the absorption and scattering of the sound waves are different in different tissue regions. Blood cells and plasma absorb almost all the sound waves; hence, no reverberation produces a black shade. Hard tissues such as bone maximally scatter the waves impeded upon them; hence, the waves received produce a white shade. Soft tissues partially absorb the sound waves; therefore, grey shades indicate a soft-tissue region with partially reflected waves [76, 82–84].

7.3.3 A non-invasive ultrasound scanning of ICA

The internal carotid artery (ICA) ascends through the neck and passes through the carotid canal, located in the petrous part of the temporal bone just above the jugular fossa. In the anterior portion of the canal, there is a thin layer of bone that separates the artery from both the cochlea and the trigeminal ganglion [85–88]. Figure 7.4 provides an illustration of the ICA's location from the neck to the brain. Understanding the distinction between the ICA and the external carotid artery (ECA) becomes crucial when one or both of them are occluded. US imaging allows for the visualization of the proximal segment of the common carotid artery (CCA), with the left CCA originating from the aorta. However, the imaging cannot capture the proximal phase of the left CCA. It is essential for the examiner to differentiate between the ICA and the ECA. Figure 7.5 demonstrates the acquisition of a US image of the carotid artery in a patient's neck. Furthermore, figures 7.6(a)–(c)

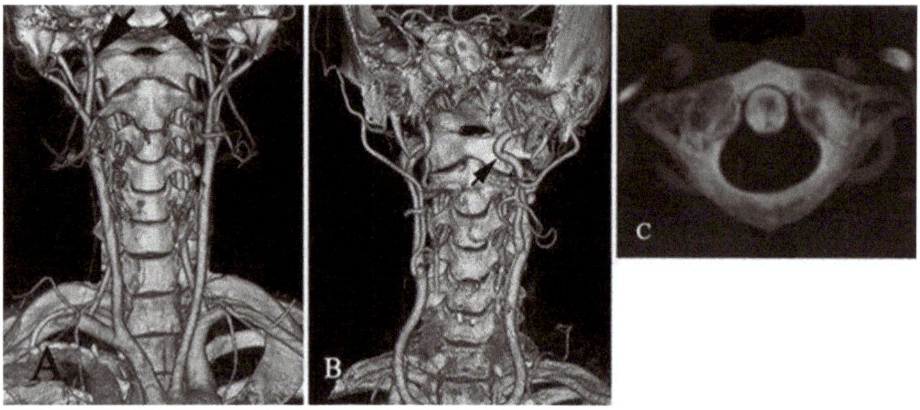

Figure 7.4. Location of the internal carotid artery through the neck [89].

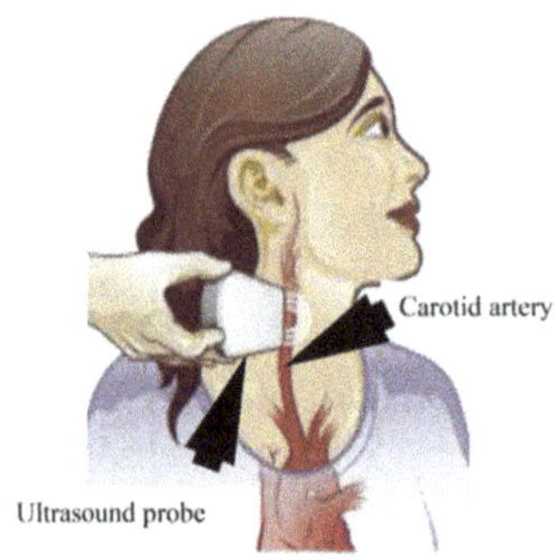

Figure 7.5. Acquisition of carotid B-mode ultrasound image from patient (Reprint from Banchhor *et al* Elsevier) [90].

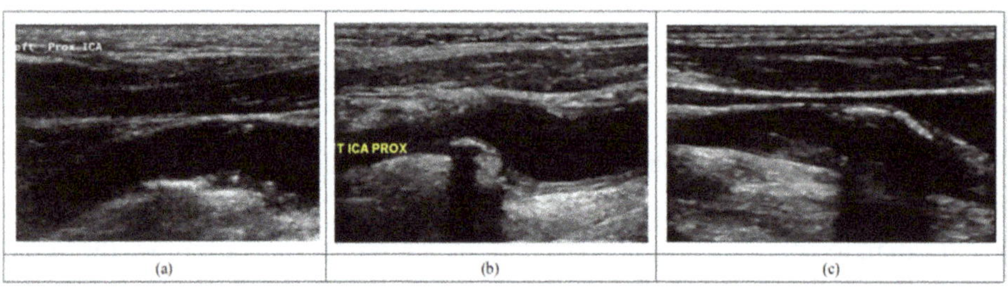

Figure 7.6. B-mode ultrasound images of the low risk, medium risk and high risk patients with 50%, 60% and 80% stenosis.

present US images of patients with stenosis at different levels, namely 50% (low-risk), 60% (medium-risk), and 85% (high-risk), respectively.

7.4 Machine learning paradigm

ML systems employ a diverse range of parameters for classification and validation, which vary based on data types, variables, and the number of classes. Figure 7.7

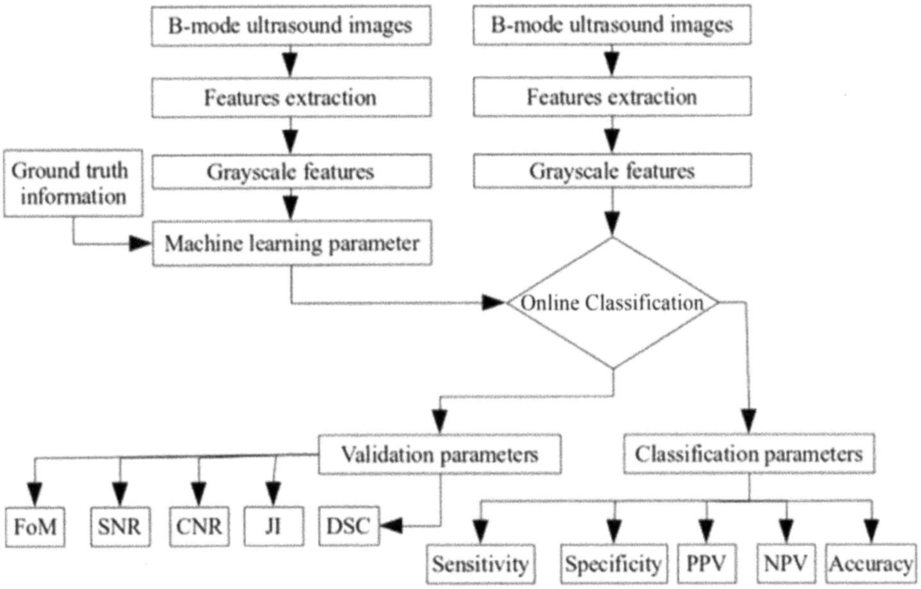

Figure 7.7. General block diagram of a ML based system for ultrasound image classification.

illustrates a general block diagram of an ML system. Various statistical tests are utilized in the analysis of the system, such as the Z-test, t-test, KS test, ANOVA test, chi-squared test, Mann-Whitney test, and Wilcoxon test. Additionally, a set of indices can be employed to assess the performance of the system when comparing ground-truth images with test images. These indices encompass the Jaccard similarity coefficient (JSC), signal-to-noise ratio (SNR), contrast-to-noise ratio (CNR), and Dice similarity coefficient (DSC) [19, 90]. Table 7.2 shows some parameters of the ML system used to design the system.

7.5 Deep learning architecture

In their recent review, Biswas *et al* [91] have discussed various algorithms that are used along with CNNs as significant components in a DL system. A DL system can be used for classification, clustering, and predictive analysis [92–94]. However, DL strategies are not restricted to such applications; researchers are developing many applications in biomedical engineering, e-commerce, surveillance, and automation. A discussion of all such applications goes beyond the scope of this paper; however, key components of DL systems are described below.

Convolutional layer: Convolution is the primary and fundamental operation of a CNN and constitutes the computational load of the network. The kernel or filter (mask) is a small matrix that plays a vital role in the convolution operation.

The size of the filter (WK×HK×DK) is usually smaller than the image size (WI×HI×DI), and the depth D_K is similar to the depth of the image, D_I, which refers to the RGB channel. Convolution is the dot product of these two matrices. During this operation, the center of the kernel is placed over the first element of the image

Table 7.2. Performance parameters of the machine learning system.

S. No.	ML parameters	Mathematical expression
1	Jaccard similarity coefficient (Jaccard index) is a statistical parameter to calculate intersection over union of the two sets of data. $0 \leqslant JI_{XY} \leqslant 1$	$JSC_{XY} = \frac{XUY}{XUY}$
2	F1 score, also known as Dice similarity coefficient, $0 \leqslant DSC_{XY} \leqslant 1$	$DSC_{XY} = 2\frac{X \cap Y}{X + Y}$
3	SNR is the ratio of signal and background noise	$SNR = \frac{S_A(i,j)S_B(k,l)}{\sqrt{2}\,\sigma(k,l)}$
4	Contrast-to-noise ratio determines the image quality	$CNR = \sqrt{\frac{\mu_A(i,j)\mu_B(k,l)}{\sigma_A(i,j) + \sigma_B(k,l)}}$
5	The central tendency of the error can be computed using FoM	$FoM \equiv 100 - [\frac{\overline{Auto} - \overline{Manual}}{\overline{Manual}}] * 100$ $\overline{Auto} = \frac{1}{N}\sum_{i=1}^{N} Auto(i)$ $\overline{Manual} = \frac{1}{N}\sum_{i=1}^{N} Manual(i)$
6	Sensitivity is the percentage of positive values which are correctly identify by the classifier.	$Sen = \frac{TP}{TP + TN} * 100$
7	Specificity or recall is the percentage of the negative values that are efficiently recognized by the classifier.	$Spec = \frac{TN}{TN + FP} * 100$
8	Accuracy is the percentage proportion of the correctly classified cases to the total number cases	$ACC = \frac{TP + TN}{TP + TN + FP + FN} * 100$
9	Precision (Positive predictive value)	$Precision = \frac{TP}{TP + FP} * 100$

X ∩ Y and X ∪ Y: intersection and unions of ground-truth (GT) and the test images
S_A (i, j) and S_B (k, l): mean signal strength in the ROI and background region
$\sigma_B(k, l)$: standard deviation of the background
$\mu_A(i, j)$ and $\mu_B(k, l)$: mean signal strength in the ROI with a lesion at the location (i, j) and the background ROI without a lesion at the location (k, l)
$\sigma_A(i, j)$ and $\sigma_B(k, l)$: standard deviation of the signal strength in the ROI with lesion at the location (i, j) and the background ROI without a lesion at location (k, l)
TP = True positive; TN = True negative; FP = False positive; FN = False Negative.

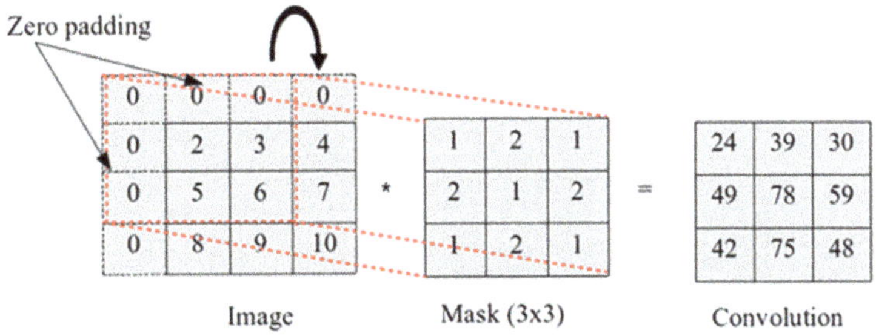

Figure 7.8. Convolution of two matrices with zero padding $P = 1$, stride $S = 1$.

matrix, and the overflow of the kernel beyond the image matrix is padded with zeros. After convolution, the filter is moved one element ahead of the previous image matrix, and this movement is known as the stride. A stride (S) of one retains the size of the input matrix after convolution; $S = 2$ and $S = 3$ reduce the image matrix size considerably. The width and height of the output signal are given by the following equations, with strides S and zero padding P. Figure 7.8 shows the convolution of the image and the kernel matrix with padding $P = 1$ and stride $S = 1$.

$$\mathrm{W}_o = \frac{W_I - W_K + 2P}{S + 1} \tag{7.1}$$

$$H_o = \frac{H_I - W_K + 2P}{S + 1} \tag{7.2}$$

Activation layer: The activation layer consists of a nonlinear activation function that imposes nonlinearity on the input signal received from the previous layer. A higher-degree nonlinear function maps the input signal to a high-dimensional, complicated, and complex output. The neural network, in a sense, is a universal function approximator; hence, learning from such a complex and nonlinear data set with a deep architecture is possible. Appendix B shows a list of the most used activation functions in DL systems.

Fully connected layer: In a fully connected layer, neurons establish connections with all the activations in the preceding layer, similar to traditional (non-convolutional) artificial neural networks. The activation of these neurons can be computed through an affine transformation, involving matrix multiplication followed by the addition of a bias offset (a vector addition of a learned or fixed bias term).

Pooling layer: A pooling layer is inserted between successive convolutional layers in a CNN architecture. The primary function of this layer is to reduce the computational complexity of the neural network.

$$W_o = \frac{W_I - W_K}{S + 1} \tag{7.3}$$

$$H_o = \frac{H_I - W_K}{S + 1} \tag{7.4}$$

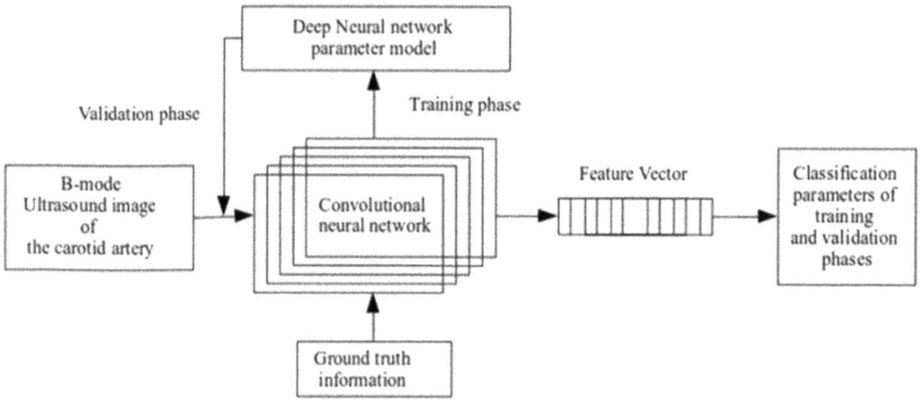

Figure 7.9. Block diagram of a general deep neural network for medical image classification.

$$D_o = D_I \tag{7.5}$$

The pooling layer commonly uses a 2 ×2 kernel size ($WI \times WI$) with stride $S = 2$ over the depth of the input. The pooling operation requires only two hyperparameters, stride S and kernel size W. Pooling in itself performs a convolution operation; however, it does so with the intent of reducing the size of the input. Stride $S = 2$ reduces the size of the input by as much as 25%.

Loss functions: The 'loss layer' is responsible for determining how training penalizes the disparity between the predicted output and the actual labels, typically serving as the final layer in a neural network. Different loss functions suitable for specific tasks can be employed. Examples of loss functions include cross-entropy, squared loss, multiclass, root-mean-square-error (RMSE), cross-entropy, and Poisson regression.

Batch size and batch normalization: A batch represents the complete data set, and its size is the total number of training examples in the available data set. The mini-batch size is the number of examples the learning algorithm processes in a single pass (forward and backward). A mini-batch is a small part of the data set of a given mini-batch size.

Iterations and epochs: In one epoch, the entire data set is passed forward and backward through the neural network only once. Since the data set is too big to feed to the computer at once, we divide it into several smaller batches, and the number of iterations is the number of batches needed to complete one epoch. Figure 7.9 shows a general block diagram of a DL-based system for the classification of US images.

7.5.1 Characterization of carotid plaque in ultrasound image

Considerable efforts have been dedicated to characterizing plaque tissue within the region of interest (ROI) in carotid and coronary arteries. Lal *et al* [95] employed computer-assisted duplex ultrasound (DU) images to conduct pixel distribution analysis (PDA) on the carotid arteries of 19 patients. They conducted histologic analysis of the plaque tissue post carotid endarterectomy and identified the grayscale intensity range of blood, lipid, fibromuscular tissue, and calcium. Additionally, they characterized symptomatic and asymptomatic plaques based on their echogenicity. Symptomatic plaque tissues exhibited lower calcium levels but higher intraplaque hemorrhage and lipid content. Conversely, asymptomatic plaque tissues displayed a higher calcium content [95]. Grayscale texture features have also been recognized for their potential, as demonstrated by Araki *et al* [18] and Saba *et al* [40]. They developed a computer-aided diagnosis (CAD) system and carefully selected 16 grayscale texture features from the plaque region of 407 images. These features were then classified using an ML system based on a support vector machine (SVM). An SVM is a maximum margin classifier that separates two classes by maximizing the distance between them.

Their work stands out for its innovative approach in using lumen diameter threshold values as biomarkers for the risk prediction system. They developed the system using five sets of kernel functions: linear functions, radial basis functions (RBFs) [96], and polynomials of different orders (1, 2, and 3). Among these, the polynomial functions of order 2 achieved an impressive accuracy of 98.80% in recognizing far-wall plaques [18, 40]. In a separate study, Acharya *et al* [70] conducted a classification of plaque tissue using 160 B-mode ultrasound (BUS) images of carotid arteries, distinguishing between symptomatic and asymptomatic plaques. They extracted a comprehensive set of 32 features, including local binary pattern (LBP), fuzzy grey level co-occurrence matrix (FGLCM), fuzzy run-length matrix (FRLM), trace transform, higher-order spectrum (HOS), and Fourier spectrum features. These features were then classified using an SVM classifier [97]. Additionally, they proposed and computed a plaque risk index (PRI) based on these features. Through a cross-validation procedure with a linear kernel function, their system achieved an accuracy of 90.66%, sensitivity of 83.33%, specificity of 95.39%, and positive predicted value (PPV) of 89.71%. Qian and Yang [98] introduced a classification method for atherosclerotic plaques using ML techniques. Their approach involved initial ROI segmentation from US images, followed by the application of a classification algorithm to generate a probability map for plaque tissue morphology within the ROI. They evaluated four classification algorithms: SVM with a linear kernel, SVM with an RBF kernel, AdaBoost, and a random forest algorithm. The AdaBoost classifier employed an adaptive boosting technique to prioritize the misclassified classes, while the random forest algorithm utilized an ensemble classification method with uncorrelated trees. The best results were achieved with a random forest algorithm using the auto-context method (sensitivity: 80.4% ± 8.4%; specificity: 96.5% ± 2.0%; Dice similarity: 81.0% ± 4.1%; area under the curve (AUC): 0.897) [98].

In a recent study, Lekadir *et al* [99] explored the application of DL techniques for characterizing plaque tissue in carotid US images. They conducted an extensive experiment using over 90 000 patches from 56 patient images to classify plaque data into lipid, calcium, and fibrous cap categories. Their deep neural network architecture consisted of four convolutional layers, three fully connected layers with L-ReLU activation, and fully connected layers followed by the softmax function. The proposed CNN model achieved a mean accuracy of 75% ± 16%, surpassing the mean single-scale SVM accuracy of 63% ± 14% and mean multiscale SVM accuracy of 69% ± 16%. While the network training time was a critical factor (1.5 h ± 0.6 h), they achieved nearly real-time testing with an average time of 52 ms ± 13 ms per single image.

Dong *et al* [73] employed a morphology-enhanced probabilistic plaque segmentation (MEPPS) method on MRI images. They successfully identified necrotic core, hemorrhage, calcification, and fibrous tissue, quantifying their distribution within the ROI in terms of average pixel count per location. Pedro *et al* [63] computed an enhanced activity index (EAI) using a symptomatic plaque classification system based on 146 carotid US images. The mathematical expression of the EAI is given by equation (7.6):

$$\mathrm{EAI} = G(\Gamma(X)) \tag{7.6}$$

where $\Gamma(X) = \frac{R_S(X)}{R_{AS}(X)}$ is the Bayes factor, $R_S(X) = p(X|\mu_S)$ is the likelihood of the symptomatic class and $R_{AS}(X) = p(X|\mu_{AS})$ is the likelihood of the asymptomatic class. $G(X) = \frac{100}{(1+\exp(1-x))}$ is the rescaling function to map the Bayes function in the range (0,100).

They compared the EAI with the degree of stenosis and the activity index (AI), with cut-offs of 55, 80, and 65. In a binary class system based on simple thresholding of EAI cut-offs, the accuracy is found to be 76.92% as compared to 72.85% and 74.66% for DS- and AI-based systems.

7.5.2 Lumen diameter and stenosis measurement

Arteries are composed of distinct layers, including the lumen, lumen intima, media, and adventitia. The space between the lumen-intima interfaces of the far wall and the near wall is referred to as the lumen region. Figure 7.10 illustrates the selection of the region of interest (ROI) within an artery, specifically showcasing the area between the media-adventitia interfaces (MAIs) of the far wall and the near wall. Two established methods, namely NASCET and ECST, can be employed to assess the presence of stenosis in the artery, providing measurements in percentage form. ECST era stenosis calculation is shown in Figure 7.11(a) and Nascet Stenosis calculation is shown in Figure 7.11(b) [100].

$$\%\,\mathit{Stenosis} = 100 \times (\mathit{Reference\ area} - \mathit{residual\ lumen\ area})/\mathit{reference\ area}$$

$$\%\,\mathit{Stenosis} = 100 \times (\mathit{Area\ of\ normal\ lumen} - \mathit{area\ of\ residual\ lumen})/\mathit{area\ of\ normal\ lumen}$$

The use of LD as a biomarker in clinical studies has gained recognition, leading to the exploration of numerous methods in this area. Over time, researchers have put forth various automated and semi-automated approaches for measuring LD, indicating an ongoing progression of work in this field.

7.5.3 Conventional methods of detection of LD

In a comprehensive review conducted by Kumar *et al* [16], various automated methods incorporating region-based and boundary-based approaches for LD/IAD

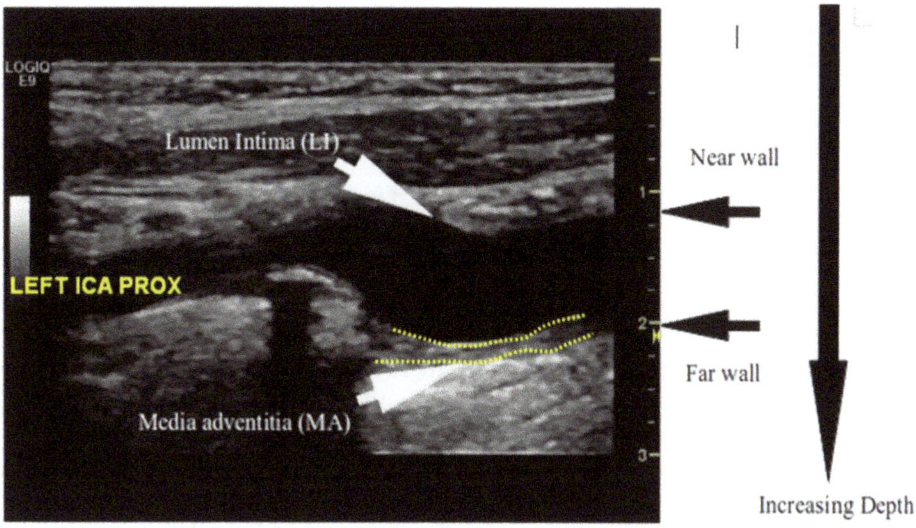

Figure 7.10. ICA images showing LII, MAI and lumen region.

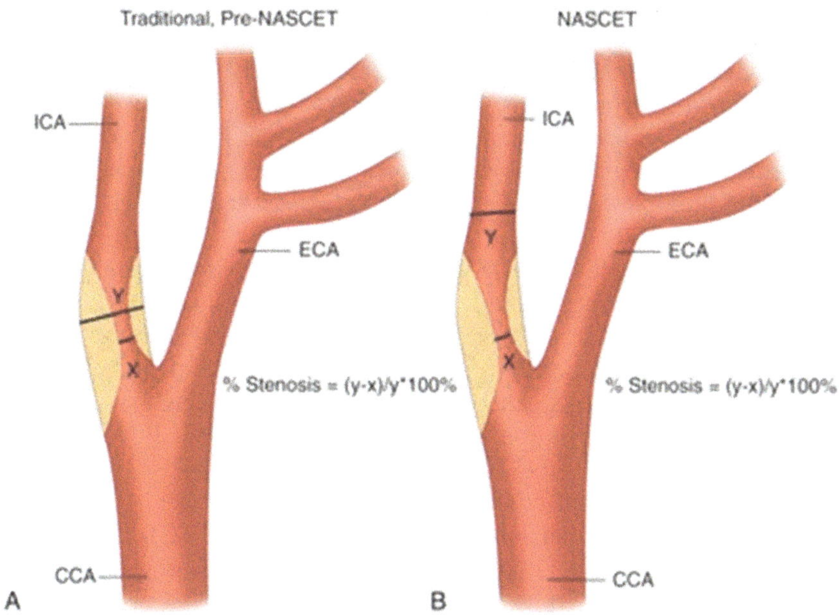

Figure 7.11. Stenosis measurement method (a) ECST criteria (b) NASCET criteria (Reprint from Vincent Ho *et al* 2010).

measurements were examined. The study emphasized the significance of LD and IAD as imaging biomarkers, offering reliable and accurate measurements. Region-based methods were found to be more suitable than boundary-based methods for LD/IAD segmentation. Utilizing a risk index scale of 1 to 5, the researchers identified 9 patients in the high-risk category and 27 patients in the medium-risk category out of a total of 203 patients. In a separate study, Saba *et al* proposed the stenosis severity index (SSI) [33]. The SSI is expressed as a percentage and can be calculated using equation (7.7):

$$SSI = \left(1 - \frac{LD_{narrow}}{LD_{normal}}\right) \times 100 \tag{7.7}$$

Most traditional methods for measuring LD are semi-automated, but there are a few methods that aim to provide automated responses. Ikeda *et al* [21] developed a conventional method to locate the carotid bulb in BUS images of the carotid artery. The carotid bulb serves as a reference area where the cIMT of the CCA and ICA can be measured. The system accurately detects and validates the reference point in approximately 13 seconds, achieving nearly 100% accuracy.

In a comparative study, Saba *et al* [32] demonstrated the superiority of carotid IAD over LD in relation to the plaque score. Analyzing a database of 404 patients, they found a correlation of 0.38 ($p < 0.0001$) between IAD and the plaque score, while the correlation between LD and polyline distance (PS) was 0.25 ($p < 0.0001$). The determination of LD and IAD involved a polyline distance algorithm followed by the application of higher-order Gaussian filters.

With advancements in technology and the need for more efficient techniques, AI-based automated methods have become increasingly relevant. ML techniques such as artificial neural networks, SVMs [101], and AdaBoost classifiers offer compact and time-efficient solutions. Shrivastava *et al* [102–105] utilized an SVM as part of a CAD system for psoriasis disease risk stratification. Their work involved classifying the disease into multiple categories and selecting features using principal component analysis (PCA).

Araki *et al* [17] employed an SVM classifier in a CAD system for plaque classification in coronary intra-vascular ultrasound (IVUS) images. Their innovative approach introduced cIMT as a risk biomarker in the binary classification system, revealing a significant relationship between coronary plaques and cIMT values. They utilized a comprehensive dataset of 4004 images from high-risk and low-risk patients, achieving a classification accuracy of 98.43% with a PCA-based polling strategy and a radial basis kernel function.

In another study by Araki *et al* [41], 56 texture features were extracted from a large dataset of high-risk and low-risk plaque frames in IVUS images. Various feature matrices and statistical methods were utilized for feature extraction, and plaque segmentation was performed using the ImgTracer™ software. The system was validated for both fixed and variable data sizes, achieving accuracies of 95.30% and 97.36%, respectively.

Saba *et al* [19] observed significant inter- and intra-observer variability in LD measurements for clinical use. To assess the differences between inter- and intra-observer measurements, they conducted an experiment using 200 BUS CCA images from 100 patients. The study employed an automated cloud-based software called AtheroCloud™ for LD measurement and evaluated the inter- and intra-observer variability between automated and manual measurements. The automated measurements achieved an average precision of merit (POM) of 94.27% ± 7.14% and a figure of merit (FOM) of 98.94%, with the lowest mean absolute error (MAE) of 5.42% ± 4.67%.

7.5.4 Deep learning based methods of LD measurement

As techniques continue to advance, there is a growing need for faster and more accurate methods. The emergence of modern GPUs has paved the way for the development of high-performance computing (HPC) methods. One notable example of such methods is the utilization of deep neural networks based on DL techniques [106]. Deep neural networks are constructed by interconnecting convolutional layers, activation layers, batch normalization, fully connected layers, and pooling layers, with the depth of these layers carefully determined [107]. Through specific refinements, each layer produces feature maps that contribute to the final output [108, 109]. Although various pre-tested deep neural networks are readily available, they can be fine-tuned to suit specific applications. Prominent examples include GoogleNet, ResNet, VGG, and LeNet-5. In the field of medical image diagnosis, CNNs have been increasingly applied, demonstrating their versatility [94]. For a historical overview of the development of these architectures, please refer to table 7.3.

Avendi *et al* [115] participated in the MICCAI LV 2009 challenge and introduced a deep neural network that surpassed existing methods in performance. Using a database of 45 MR images, they achieved an accuracy of 96.69%, outperforming the 79.2%–95.62% accuracy range achieved by other methods. Another application of deep neural networks (DNNs) in brain MR image segmentation was presented by Havaei *et al* [116], focusing on glioblastomas, a type of tumor. By utilizing local and contextual features extracted by a CNN, they successfully distinguished between low-grade and high-grade gliomas. Different architectures, including TwoPathCNN, MFCascadeCNN, and LocalCascadeCNN, were presented and compared based on evaluation metrics such as Dice coefficient, sensitivity, and specificity. The InputCascadeCNN architecture demonstrated the highest accuracy among the tested architectures, with a Dice coefficient of 0.88, a sensitivity of 0.87, and a specificity of 0.89 in the tumor region.

Table 7.3. Different CNN architectures developed by the researchers.

Network	No. of layers	Developed by
VGG 19	16 Conv layers, 3 FC	Simonyan and Zisserman [110]
ResNet	151 Conv layers, 1 FC	He *et al* [111]
AlexNet	5 Conv, 3 FC	Krizhevsky *et al* [112]
GoogLeNet	3 Conv, 18 inception, 1 FC	Szegedy *et al* [113]
LeNet-5	3 Conv, 2 FC	LeCunn *et al* [114]

In the work of Tajbaksha *et al* [117], the question of whether it is more advantageous to develop a CNN model from scratch or fine-tune pre-trained CNN models was explored. They conducted classification, detection, and segmentation tasks on three different imaging modalities (MR, US, and colonoscopy) using both approaches. Fine-tuning of the AlexNet architecture and training an AlexNet from scratch were employed for tasks such as polyp detection, colonoscopy frame classification, IMT boundary segmentation, and pulmonary embolism detection. The findings indicated that the fine-tuned pre-trained CNN yielded superior results compared to the CNN trained from scratch. Pre-trained models are particularly beneficial when the available training data is limited; however, the extent of refinement required for a pre-trained CNN architecture depends on the specific application. In a recent study by Biswas *et al* [35], a 13-CNN architecture consisting of 13 convolutional encoder layers and three upsample decoder layers was developed for lumen segmentation. Evaluation against ground-truth values revealed a mean LD of 6.07 ± 0.93 mm, closely matching the ground-truth value of 5.99 ± 0.95 mm. Furthermore, there have been other applications involving parameter fine-tuning of existing networks such as GoogleNet, AlexNet, ResNet, and VGG for various tasks [110, 113, 118–120].

7.5.5 Vision diagram for LD/cIMT measurement

The DL system architecture involves the processing of B-mode US images as input and the generation of ROI comprising the lumen region and the adventitia boundaries of both the near and far walls. The system comprises three main stages: (1) pre-processing, (2) DL-based artery area localization, and (3) LI/MA boundary detection using conventional methodologies.

During the initial pre-processing stage, various steps such as noise removal and image registration are performed on the US image. The subsequent stage incorporates a DL system employing a fast region convolutional neural network (RCNN) specifically designed for segmenting the tissue region surrounding the lumen area. The region proposal network (RPN) utilizes ground-truth coordinates to generate a bounding box around the lumen region. This proposed region is then provided to the spatial pyramidal network [121], which generates a bounding box encompassing the adventitial region of both the near and far walls. By utilizing this bounding box, the artery tissue region is effectively segmented. Finally, a conventional method, such as a snake-based or shape-based approach, is applied to accurately delineate the LI/MA boundary within the tissue region.

7.6 Discussion

The sole goal of this study was to examine the factors associated with cardiovascular and cerebrovascular diseases. Our investigation indicated that atherosclerosis plays a significant role in the development of these diseases. Atherosclerosis is closely linked to the increase in both intima-media thickness (IMT) and luminal diameter (LD). However, previous research has also identified other factors that can influence the elevation of LD and IMT.

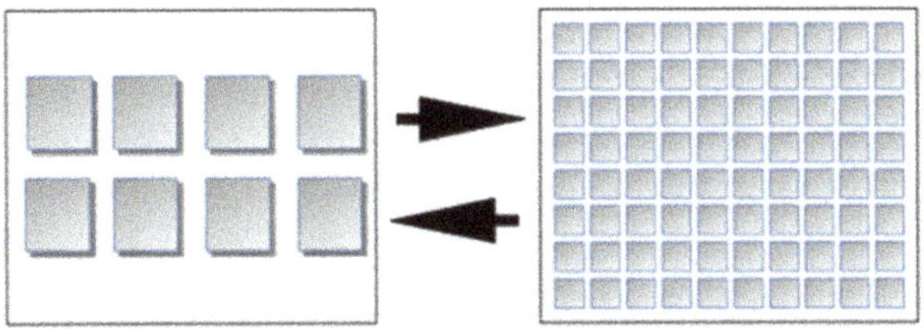

Figure 7.12. CPU and GPU schematic diagram with core comparison.

7.6.1 Hardware and software for DL system

One of the most important reasons for the development of DL systems in medical image segmentation is the widespread availability of open-source software packages. Some of these open-source ML frameworks include Tensorflow, Theano, Torch, CUDNN, and Caffe. Another contribution to the development of DL systems is the availability of GPU and GPU core-computing libraries. GPUs have a considerable number of smaller cores compared to a general-purpose CPU, as shown in figure 7.12. CUDA (Compute Unified Device Architecture) or OpenGL facilitates these GPUs with a large number of cores for accessing parallel computing. CUDA provides a parallel computing platform through an application programming interface (API) developed by NVIDIA™ for GPUs. All the above-mentioned GPU frameworks and GPU computing libraries are nowadays supported by Python, MATLAB, and C++.

7.6.2 Other biological aspects of enhanced LD and cIMT

Ozdemir *et al* [122] the researchers discovered that obesity, indicated by a high body mass index (BMI), was among the factors associated with elevated luminal diameter (LD) and intima-media thickness (IMT) in a group of 71 individuals. Specifically, individuals with a BMI of 23 exhibited an average RCCA LD of 6.90 ± 0.93 mm, while those with a BMI of 27.62 ± 1.72 displayed an increased LD of 7.40 ± 0.79 mm. These findings indicate a correlation between BMI and heightened LD. Llyod *et al* [123]. A similar study was conducted on postmenopausal women who were either obese or overweight. The researchers observed that women who were currently undergoing hormone therapy exhibited a luminal diameter (LD) of 5.31 mm and an intima-adventitia diameter (IAD) of 6.79 mm. On the other hand, women who were former users of hormone therapy showed an LD of 5.44 mm and an IAD of 6.94 mm. Another gender-based study by Kreja *et al* [124] on 500 patients found a difference in lumen diameter between men and women. The mean LD in 305 women was found to be 4.66 ± 0.78 mm, whereas in men, the mean was 5.11 ± 0.87 mm.

In addition to gender-based investigations, Ruan *et al* [125] examined the relationship between LD and cIMT in a cohort consisting of 1040 individuals, with 306 classified as Black and 734 as White. The mean LD for individuals of the

Caucasian was 5.72 ± 0.53, while for individuals of the people of colour it was 5.81 ± 0.65 [106]. Mirek *et al* [65] employed a multivariate logistic regression model incorporating LD, age, and myocardial infarction (MI) as parameters to explore the association between LD and three-vessel disease (3VD). Their model, with an AUC value of 0.8, revealed a significant correlation between LD and intima-media thickness (IMT) in the right carotid and left femoral arteries of patients with 3VD compared to those without 3VD [53]. This approach facilitates the identification of arterial abnormalities. Togay *et al* [10] observed a relationship between abnormalities and reduced LD in 345 patients. In patients with tortuosity, the LD at the origin of the internal carotid artery (ICA) and the associated carotid abnormality measured 6.45 mm and 4.55 mm, respectively. For patients with kinking, these values were 6.30 mm and 3.75 mm, while for patients with coiling, they were 6.08 and 2.80 mm. Additionally, patients with kinking and coiling exhibited higher maximum systolic velocity (MSV) values of 103 and 120 cm s^{-1}, respectively, compared to patients with tortuosity, whose MSV measured 84.5 cm s^{-1}. Table 7.4 presents some previous studies on ICA occlusion.

7.6.3 Rational for choosing the LI and MA boundary

The crucial purpose of carotid artery US is to measure whether stenosis is present. Polak and John Mancini *et al,* Cohn *et al*, Amato *et al,* and O'Leary *et al* [14, 126–129] have established that the cIMT is a surrogate for the subclinical result of atherosclerosis and can be used for risk assessment. We use the LD and cIMT as biomarkers to represent the narrowing of arteries, with stenosis being the area of increased thickness. Stenosis is measured between the LI layers of the neighbor wall and the farther wall. The manual delineation of the near wall and far wall borders for the LI and MA is a laborious process and susceptible to inaccuracies. Consequently, there is a pressing need for an automated method that can effectively and swiftly separate the LI and MA walls with minimal errors.

7.6.4 A note on DL strategy on carotid arterty imaging

Smistad and Lovstakken [130] successfully detected vessels (cross-section region) in US images using their proposed DNN. They were able to identify different regions of the femoral arteries, such as muscle, bone, shadows, and blood vessels, from 12804 sub-images from 15 patients and acquired a mean accuracy of 94.5%. Instead of fine-tuning the pre-trained network parameters, they first used AlexNet and then sequentially started reducing the convolutional-pooling block while consistently maintaining the accuracy. Most recently, Tajbaksh *et al* [117] have established that developing a CNN architecture from scratch is never a requirement, and higher accuracy can be achieved with a fine-tuned network. Recently, Shin *et al* [46] also developed a DL model from scratch for the detection of ROI and cIMT measurements in BUS videos of carotid arteries. Their database has a total of 92 videos from 23 patients, from which they used 48 videos (4456 frames) for training and 44 videos (3565 frames) for testing. They validated their experiment against previous findings

Table 7.4. Internal carotid artery literature study.

S. No.	Author (year)	Study-type (relationship examined)	Patients size(ethnicity)/ demographics (diabetic, coronary artery disease/carotid artery disease/events*)	ICA LD(mean) Thresh (mm) Stenosis(%)	Stratification/observations/ performance (HR—high risk LR—low risk)
1	Kane *et al* [137]	ICA diameter measurement in the absence of Ipsilateral or hypoplasia of A1 segment	$N = 104$ Women: 54 Men: 50 Age:54±16	Mean:4.65±0.70 mm (R); 4.59±0.66 mm (L) Ipsilateral: 3.63 ±0.41 mm Contralateral: 5.25 ±0.52 mm	Hypoplastic A1: 4.30 ±0.56 mm Asymmetric A1: 4.50 ±0.64 mm Isolated FP: 4.75±0.38 mm Balanced: 4.43±0.52 mm
2	Ascher *et al*[138]	Pseudo occlusion of ICA	$N = 17$ M/F: 12/5, Age:73±7 Stroke: 7 TIA:3; Hypertension:81% Diabetes:44% Endarterectomy:12 Death:1; CAD: 25%	Min: 0.7 mm Max: 3.6 mm Mean LD: 2.0±1.1mm (Post stenotic); 4.4±0.3 mm (post operative)	LR: 7 (41%)
3	Bartlett *et al* [139]	Quantification of carotid stenosis on CT angiography	$N = 466$ 268: ICA disease 76: Stroke 192: Carotid atherosclerotic disease	2.2 mm: 50% stenosis [65] 1.4–2.2 mm: 50%–70% [47] 1.3 mm: 70% stenosis [24]	Sensitivity =88.2% Specificity = 92.4%
4	Hyde *et al* [140]	ICA stenosis comparison between 3D CRA and DSA	42 Images of $n = 26$ (16M/10F) Age: 63±10	MRA: 4.62±0.68 mm DUS : 3.5±0.8 mm mean Distal ICA: 3.8 mm	Sensitivity:0.89 Specificity:0.935 interobserver variability between DSA and CRA: 9.1% and 9.4%

5	Baradaran *et al* [141]	Quantifying intracranial internal carotid artery stenosis on MR angiography	$N = 193$ (386 MRA) 91F Age: 71.9±14.0 White:173; Black:11; Other:9; Diabetes:55; Hypertension:55	Mean LD: 3.97±0.3 mm (247 w/o stenosis) <70% stenosis: 1.4–2.1 mm [126] >70% stenosis: 1.3 mm [12]	>70% stenosis AUC = 0.99; Sensitivity = 95.5%; Specificity = 100%;

and found a reduced thickness error of 2.8 ± 2.1 mm, compared to 13.8 ± 31.9 mm [46] 80 ± 40 mm [131], 30 ± 30 mm [132], 43 ± 93 mm [133].

7.6.5 Ml vs. DL system; which is best?

The advent of the DL strategy has raised a discussion among researchers about the selection of the best strategy. However, in a broad sense, DL is a part of ML, although there are differences between both strategies. An ML system progresses in a sequence of steps for classification and segmentation [143]. These steps include feature extraction, selection of the best feature from a large pool (if necessary), and applying an ML algorithm for drawing the boundary between the two classes. Several classifiers, such as AdaBoost, SVM, random forest, nearest neighbor, and artificial neural networks (ANNs), are used to design the ML system [25]. Based on the labeling of the data, ML and DL systems can be classified into supervised and unsupervised learning systems. The system designer provides a label for the data in supervised learning, whereas unsupervised learning deals with unlabeled data. In their review, Saba *et al* [54] reported significant differences between current DL and conventional ML strategies.

A DL system is, however, designed using multiple layers of ANNs, as described in section 7.5. A DL architecture is inspired by the biological neural networks in the human brain, which has self-learning capabilities. One of the major shortcomings of a DL-based proposed model is the need for high-performance computing resources, such as GPUs. The training of a DL system requires a large stack of the data set, and having a large number of layers in the CNN architecture enhances the operations. Moreover, multiple iterations enhance the time of training; therefore, a lack of high-performance computing resources may slow down the operations [54]. Many may think of a DL system as a black box, using multiple layers of any type and giving input to one side and the required output to the other side of the system. However, the scope of a DL system is beyond a black box because the final output is the result of complex interactions between multiple neurons. This interaction is so complex that the computational data acquire higher dimensions, which cannot be visualized by the human brain [134, 135]. In a paper by Lipton, understandable models are called transparent and incomprehensible models - black boxes [136]. However, the strategy of a DL system can be explored with the help of other methods. Further, many methods proposed in the literature are classified as semi-automated and automated based on their mode of operation.

7.6.6 Clinical evaluation and scientific validation of the system

A CAD system should be clinically evaluated before it is used in medical practice. The whole system passes through a series of processes, such as reviews, regulatory compliances, and sanctions, before it is finally approved by the authorities. Scientific validation of the system is also part of this evaluation, in which the system can be validated against the ground truth obtained from other sources. Furthermore, experts who are blind to the previous output can validate the system. Saba *et al* [19] presented a user-interactive, cloud-based system for IMT measurements named

AtheroCloud™ and its validation in routine mode and pharmaceutical mode. Attractive features of AtheroCloud™ are its anytime-anywhere access and support of many image formats, such as DIACOM, JPG, PNG, BMP, GIF, and TIFF. Physicians and cardiologists can upload BUS images from local servers and use the lucrative services of the cloud-based system. Various scientific tools and indices, such as the Framingham risk score, POM, receiver operating characteristics (ROC), sensitivity, and specificity, are used to validate the system against established results. The scientific validation is not a one-time process; the system designer is required to submit an update of the system and the validation process (if any). Saba *et al* [19] reported the update in the same AtheroCloud™ system and introduced intra- and interobserver variability and reproducibility analysis of the system for LD measurement.

7.7 Conclusion

In this paper, we have reviewed the applications of DL technology in the domain of medical imaging. In the past couple of years, DL has emerged as a free and extensive technology in the field of AI. Many researchers have deployed DL strategies in the fields of data science, medical science, and finance. Further, we have focused our study on LD segmentation and stenosis measurement in the ICA using US images. Previously, researchers have focused mainly on stenosis and LD of the CCA and applied ML and DL strategies to CCA images. Although a few researchers have shifted their attention to stenosis and LD measurements of ICA images, such studies are still in the minority. The reasons for this are a lack of availability of databases and a limited intent to focus on ICA images. Further, stenosis measurements in CCA and ICA images have been explored by ML and DL strategies, but clinically validated and scientifically approved methods that are accepted by medical experts are needed.

Appendix A

Table A.1. Summary of the articles of sector 1.

Author	Summary
Benjamin *et al*	A WHO report on statistics of the stroke and cardiovascular disease published in 2019
Kamalkannan *et al*	Statistics of stroke and cardiovascular disease in perspective of India
Chauhan *et al*	A survey of the rising incidence of strokes and CVDs in India
Prabhakaran *et al*	Current status and future aspects of stroke and CVDs in India
Benjamin *et al*	A WHO report on statistics of the stroke and cardiovascular disease published in 2018
Suri *et al*	Cause of atherosclerosis and preventive measures

(*Continued*)

Table A.1. (*Continued*)

Author	Summary
Park *et al*	Carotid plaque imaging using US
Togay Ishikay *et al*	Carotid artery abnormalities its association with stroke risk.
Tracqui *et al*	Study of mechanical properties of the atherosclerotic plaque via atomic force microscopy
Teng *et al*	Different properties of the atherosclerotic plaque in carotid arteries
Patel *et al*	Mechanical properties of the atherosclerotic plaque studied via US.
Rothwell *et al*	Study of ischemic stroke in patients with reduced ICA LD
Kim *et al*	US based study for atherosclerosis detection
Naim *et al*	Atherosclerotic plaque study based on US, MRI and CT scans
Naim *et al*	Plaque tissue characterization using US and MRI
Lee W	Principles and methods of carotid USG
Libby P	Biology of the atherosclerosic
Naghavi *et al*	Clinical trial of the patients with atherosclerotic plaque
Ross R.	Biology of atherosclerosis
Hopkins E.	Biology of the atherosclerosis
Mannarino PN.	Biology of the atherosclerosis
Lo and Plutzky	Biology of atherosclerosis in HIV patients
Mohebali *et al*	Plaque characterization using acoustic shadowing in US images
Chiu *et al*	Plaque characterization in 3-D US and MR images by using surface based algorithm for image registration
Pedro *et al*	Characterization of symptomatic and asymptomatic carotid plaque using enhance activity index (EAI) based system
Ho *et al*	A review of the carotid US in atherosclerosis diagnosis
Mirek *et al*	Peripheral artery disease LD biomarker relation with coronary atherosclerosis
Nambi *et al*	Role of cIMT and plaque in CVD prediction
Picano *et al*	Vulnerable Plaque tissue characterization in US images
Cuadrado-Godia *et al*	CSVD review using pathophysiology, biomarkers and ML strategies.
De Korte *et al*	Review of mechanical characteristics of the arteries and atherosclerotic plaque
Turan *et al*	Clinical trial for characterization of intracranial atherosclerotic plaque using MR images (CHIASM)
Gupta *et al*	Echolucency based Plaque characterization for stroke risk in US images
Hunt *et al*	ARIC study: Prediction of ischemic stroke using acoustic shadowing in BUS images of carotid arteries
Chu *et al*	Study of atherosclerotic plaque hemorrhage in MRI images
Kamanskeiy *et al*	Age and disease related structural remodeling in carotid arteries
Remington LA.	Clinical anatomy and physiology of ICA
Binning MJ	ICA anatomy study
Webb WG	Anatomy of the nervous system
Love and Biller	Anatomy of the neurovascular system
Londhe and Suri	Effect of superharmonic frequency in US image formation

Gronholdt *et al*	Echogenicity based characterization of carotid plaque and stroke risk prediction
Jashari *et al*	Echogenicity based plaque tissue characterization and cerebrovascular symptoms
Madore *et al*	Reconstruction algorithm for improving US image quality
Barnett *et al*	Carotid endarterectomy performed in patients with symptomatic moderate or severe stenosis
Ozdemir *et al*	Effect of overweight on LD, PSV and cIMT
Llyod *et al*	CCA diameter and CVD risk factor on obese postmenopausal women
Kreja *et al*	Sex based study of CCA LD
Polak *et al*	Study of cIMT biomarker for prediction of CVD risk
Ruan *et al*	Race based study of CCA LD
Mancini CBJ	Study of structural markers in relation with CVD
Cohn JN	Study of functional markers in relation with CVD
Amato *et al*	Relation of cIMT and coronary atherosclerosis.
Breiman L	
Lipton ZC	
Ascher *et al*	
Hong *et al*	Relation between age and ICA
Bartlett *et al*	Relation between carotid stenosis diameter and cross sectional areas
Hyde *et al*	ICA stenosis measurement in 3D CTA and digital subtraction angiography

Table A.2. Summary of the articles of sector 2.

Author	Summary
Molinari *et al*	Snake based method for measurement of cIMT measurement
Molinari *et al*	Edge based method for measurement of cIMT (CAUDLES-EF)
Molinari *et al*	Automated method of cIMT measurement using edge snapper (CAMES)
Molinari *et al*	Automated method of cIMT measurement using edge snapper (CARES 2.0)
Coots *et al*	Active shape-based method for detecting the shape
Yang *et al*	CCA segmentation using active shape-based model
Araki *et al*	Shape based method applied for calcium volume detection in coronary IVUS images
Loizou *et al*	Shape based method applied for cIMT measurement in CCA images
Araki *et al*	Calcium volume measurement in IVUS videos using FCM, K-means and HMM
Lal *et al*	Plaque characterization in BUS images using PDA and histologic features
Bastida-Zumilla *et al*	cIMT measurement in CCA images using active contour based method

Table A.3. Summary of the articles of sector 3.

Authors	Summary
Krishna Kumar *et al*	Review of the LD and adventitial border measurement in US images
Araki *et al*	CVD risk assessment using SVM and PCA based features selection algorithm in IVUS images
Araki *et al*	Stroke risk assessment using SVM based ML method in BUS images of carotid artery
Saba *et al*	Intra and interoperator reproducibility in LD measurement of CCA
Bots *et al*	Relation between cIMT and CVD
Ikeda *et al*	Carotid bulb detection in low quality videos
Saba *et al*	Cloud based cIMT measurement method and stroke risk assessment
Saba *et al*	Stroke risk assessment using plaque tissue characterization in ML based framework and PCA based feature selection algorithm
Araki *et al*	CAD risk assessment using coronary IVUS images after features extraction and classification using SVM
Acharya *et al*	Plaque tissue characterization in 160 patients using 32 features
Acharya *et al*	Atherosclerotic olaque measurement using only 3 texture features and classifying them using SVM classifier
Cheng *et al*	Plaque segment in 3D US images
Banchhor *et al*	A review on calcium detection and risk stratification system in US images
Qian and Yang	Random forest and Auto context model integration for carotid plaque segmentation
Vapnik	Statistical learning theory
Shrivastava *et al*	Psoriasis disease stratification using PCA and Fisher discriminant ratio based features selection and SVM and decision tree based classifier
Shrivastava *et al*	PCAD system for psoriasis risk stratification using 86 color features and compared with 60 grayscale and 146 color features
Shrivastava *et al*	PCA based features selection used in pCAD system with 11 HOS, 60 grayscale and 86 color features
Shrivastava *et al*	PCAD system for skin lesion classification using color feature specs with K = 5,10 and JK partition methods.
LeCunn *et al*	Recognition of objects using gradient based learning
Ilea *et al*	cIMT measurement in CCA US video
Loizou *et al*	Identification of CCA bifurcation in BUS images
Wainer *et al*	Comparison of 14 classification algorithm

Table A.4. Summary of the articles of sector 4.

Authors	Summary
Biswas *et al*	Atheroedge™: CNN based system based on encoder and decoder for cIMT measurement in CCA images
Sudha *et al*	CCA based method for cIMT measurement
Biswas *et al*	CNN based method for LD measurement in diabetic patients
Tandel *et al*	Review of the studies on brain cancer classification
Ker *et al*	Study of the deep learning applications in medical image analysis
Klang *et al*	Study of the deep learning in medical image analysis
Vieira *et al*	Psychiatric and neurological disorder assessment using deep learning method
Gibson *et al*	NiftyNet platform for medical image analysis
Hoo-Chang *et al*	Deep convolutional neural networks for computer-aided detection
Liu *et al*	A review of the deep learning applications in ultrasound images
Lundervold *et al*	A review of the deep learning applications in MRI images
Meyer *et al*	A review of the deep learning applications in radiotherapy
Razzak *et al*	Challenges of deep learning present and future aspects
Menchon-Lara *et al*	Atherosclerotic plaque characterization in US images using DL system
Zreik *et al*	RCNN application in coronary plaque detection in CTA images
Saba *et al*	Present and future aspects of deep learning in medical imaging
Dong *et al*	Carotid plaque tissue characterization in MRI images using CNN based method
Jones *et al*	Deep learning application in computational biology
Ren *et al*	FRCNN based method with region proposals for object detection
Girshik R.	FRCNN based method
He *et al*	ImageNet or AlexNet: A CNN based method for object detection
Lekadir *et al*	Plaque tissue characterization in CCA BUS images using CNN
Shrivastav *et al*	Deep CNN training
Bianchini *et al*	Comparison between shallow and deep CNN architectures
Yu *et al*	CNN application speech recognition
Ciresan *et al*	Combining multiple CNN in parallel to form MCDNN for image classification
Avendi *et al*	CNN deformable model for segmentation of left ventricle in MRI images
Havaei *et al*	Brain tumor segmentation from MRI images using CNN
Tajbaksh *et al*	A review on selection of the best CNN among developed from scratch and pre-trained fine-tuned CNN
Alom *et al*	A review of deep learning systems
Bianco *et al*	Benchmark analysis of the CNN
Szegedy *et al*	A review of the deep neural network
Simonyan *et al*	Deep CNN for large scale image recognition
Smistad *et al*	Review of deep learning applications in medical images

Appendix B Some activation functions

(1) Rectified linear unit (RELU): is a simplest and most widely used activation function in DL systems. The reason of wide acceptance is its' computational efficiency over other activation functions. It selects only the positive values in from the input signal and rectifies vanishing gradient problem. Therefore, it is most preferred over the sigmoid and tanh function [93].

$$\mathrm{ReLU}(x) = max(0, x) \tag{B.1}$$

(2) Sigmoid function: Also known as logistic function and ranges between 0 and 1. This 'S' shape function has output is centered to zero. However it has slow convergence and vanishing gradient problem [111].

$$\mathrm{Sigmoid}(x) = \frac{1}{1 + \mathrm{e}^{-x}} \tag{B.2}$$

(3) Hyperbolic tangent (tanh): This hyperbolic function has output mapping between the range −1 to 1. The optimization. Mathematically it is related to the sigmoid function and can be derived from it [142].

$$\tanh(x) = \frac{2}{1 + \mathrm{e}^{-2x}} - 1 \tag{B.3}$$

$$\tanh(x) = 2 \times \mathrm{sigmoid}(-2x) - 1 \tag{B.4}$$

(4) Leaky ReLU: Rectified liner unit in another form is used as Leaky ReLU, with enhanced range. As shown in figure B.1, leaky ReLU has the values of negative slope $m = 0.01$, which computes the values as [99]

$$\mathrm{LeakyReLU}(x) = 1(x < 0)(mx) + 1(x - 0)(x) \tag{B.5}$$

when $m = 0.01$.

Mean Squared Error (MSE): takes the average of the square of the difference between the original values and the predicted values. This error focuses on larger errors as the square emphasises the larger errors more compared to small errors [142].

$$\mathrm{MSE} = \frac{1}{n}\sum_{i=1}^{n}(Y_i - \hat{Y}_i)^2 \tag{B.6}$$

Mean Absolute Error: is the average of the difference between the original values and the predicted values. It gives us the measure of how far the predictions were from the actual output [116].

$$\mathrm{MAE} = \frac{1}{n}\sum_{i=1}^{n}(Y_i - \hat{Y}_i) \tag{B.7}$$

Log loss (Binary cross entropy): The cross-entropy loss for output label y (can take values 0 and 1) and predicted probability p is defined as [144]:

$$L = -y \times \log(p) - (1 - p) \times \log(1 - p) = -\log(1 - p) \text{ if } y = 0 \tag{B.8}$$

$$= -\log(p) \text{ if } y = 1$$

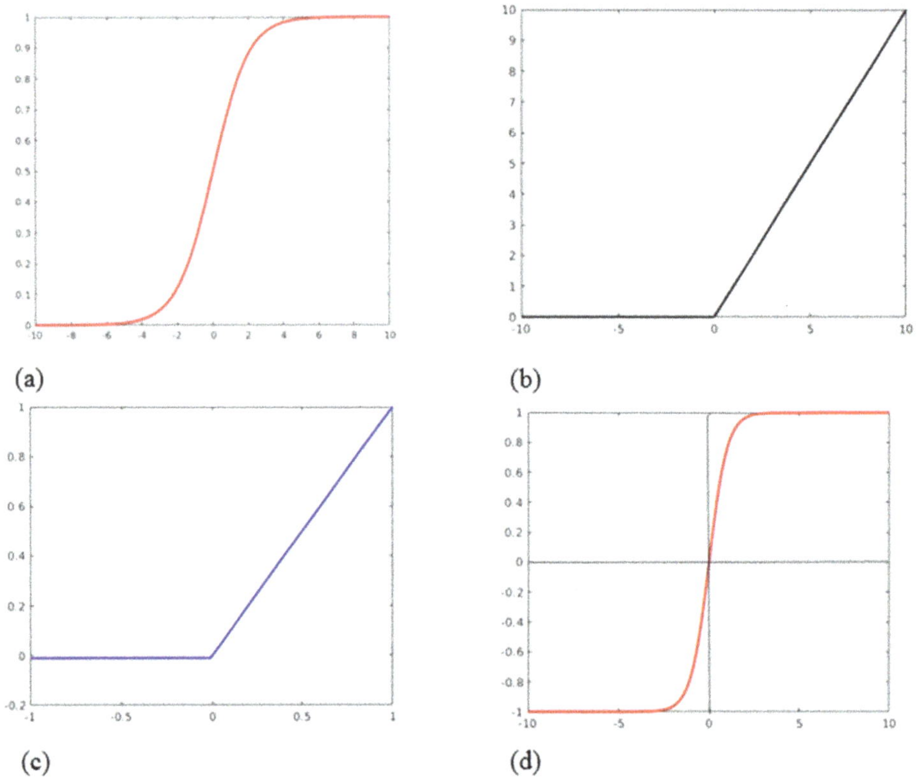

Figure B.1. Activation functions used in CNN (a) Sigmoid (b) ReLU (c) Leaky ReLU (d) tanh error functions.

to calculate the value of probability p above differential equation is converted into it's z-transform and shape of function is sigmoidal $S(z) = 1/(1+\exp(-z))$.

References

[1] Benjamin E J, Muntner P, Alonso A, Bittencourt M S, Callaway C W and Carson A P *et al* 2019 Heart disease and stroke statistics—2019 update: a report from the american heart association *Circulation* **139** 897–9

[2] WHO Reference https://who.int/topics/cerebrovascular_accident/en/

[3] Kamalakannan S, Gudlavalleti A V, Gudlavalleti V M, Goenka S and Kuper H 2017 Incidence and prevalence of stroke in india: a systematic review *Indian J. Med. Res.* **146** 175

[4] Chauhan S A B T 2015 The rising incidence of cardiovascular disease in India *J. Prev. Cardiol* **4** 735–40 http://journalofpreventivecardiology.com/pdf/Last-pdp/rising_incidence_4_2015.pdf

[5] Prabhakaran D, Jeemon P and Roy A 2016 Cardiovascular diseases in India: current epidemiology and future directions *Circulation* **133** 1605–20

[6] Benjamin E J, Virani S S, Callaway C W, Chamberlain A M, Chang A R and Cheng S *et al* 2018 Heart disease and stroke statistics – 2018 update: a report from the American Heart Association. *Circulation* **137** 67 492

[7] Suri J S 2011 *Atherosclerosis Disease Management* ed J S Suri, C Kathuria and F Molinari (New York: Springer)

[8] Molinari F, Zeng G and Suri J S 2010 A state of the art review on intima-media thickness (IMT) measurement and wall segmentation techniques for carotid ultrasound *Comput. Methods Programs Biomed.* **100** 201–21

[9] Park T H 2016 Evaluation of carotid plaque using ultrasound imaging *J. Cardiovasc. Ultrasound* **24** 91–5

[10] Togay-Işikay C, Kim J, Betterman K, Andrews C, Meads D and Tesh P *et al* 2005 Carotid artery tortuosity, kinking, coiling: stroke risk factor, marker, or curiosity? *Acta Neurol. Belg.* **105** 68–72 http://ncbi.nlm.nih.gov/pubmed/16076059

[11] Tracqui P, Broisat A, Toczek J, Mesnier N, Ohayon J and Riou L 2011 Mapping elasticity moduli of atherosclerotic plaque *in situ* via atomic force microscopy *J. Struct. Biol.* **174** 115–23

[12] Teng Z, Zhang Y, Huang Y, Feng J, Yuan J and Lu Q *et al* 2014 Material properties of components in human carotid atherosclerotic plaques: a uniaxial extension study *Acta Biomater.* **10** 5055–63

[13] Patel A K, Suri H S, Singh J, Kumar D, Shafique S and Nicolaides A *et al* 2016 A review on atherosclerotic biology, wall stiffness, physics of elasticity, and its ultrasound-based measurement *Curr. Atheroscler Rep.* (Berlin: Springer)18 83

[14] Polak J F and O'Leary D H 2016 Carotid intima-media thickness as surrogate for and predictor of CVD *Glob. Heart* **11** 295–312

[15] Rothwell P M and Warlow C P 2011 Low risk of ischemic stroke in patients with reduced internal carotid artery lumen diameter distal to severe symptomatic carotid stenosis *Stroke* **31** 622–30

[16] Kumar P K, Araki T, Rajan J, Laird J R, Nicolaides A and Suri J S 2018 State-of-the-art review on automated lumen and adventitial border delineation and its measurements in carotid ultrasound *Comput. Methods Programs Biomed.* **163** 155–68

[17] Araki T, Ikeda N, Shukla D, Jain P K, Londhe N D and Shrivastava V K *et al* 2016 PCA-based polling strategy in machine learning framework for coronary artery disease risk assessment in intravascular ultrasound: a link between carotid and coronary grayscale plaque morphology *Comput. Methods Programs Biomed.* **128** 137–58

[18] Araki T, Jain P K, Suri H S, Londhe N D, Ikeda N and El-Baz A *et al* 2017 Stroke risk stratification and its validation using ultrasonic echolucent carotid wall plaque morphology: a machine learning paradigm *Comput. Biol. Med.* **80** 77–96

[19] Saba L, Banchhor S K, Araki T, Viskovic K, Londhe N D and Laird J R *et al* 2018 Intra- and inter-operator reproducibility of automated cloud-based carotid lumen diameter ultrasound measurement *Indian Heart J.* **70** 649–64

[20] Bots M L, Baldassarre D, Simon A, De Groot E, O'Leary D H and Riley W *et al* 2007 Carotid intima-media thickness and coronary atherosclerosis: weak or strong relations? *Eur. Heart J.* **28** 398–406

[21] Ikeda N, Araki T, Dey N, Bose S, Shafique S and El-Baz A *et al* 2014 Automated and accurate carotid bulb detection, its verification and validation in low quality frozen frames and motion video *Int. Angiol.* **33** 573–89 http://ncbi.nlm.nih.gov/pubmed/24658129

[22] Biswas M, Kuppili V, Araki T, Edla D R, Godia E C and Saba L *et al* 2018 Deep learning strategy for accurate carotid intima-media thickness measurement: an ultrasound study on Japanese diabetic cohort *Comput. Biol. Med.* **98** 100–17

[23] Sudha S, Jayanthi K B, Rajasekaran C, Madian N and Sunder T 2018 Convolutional neural network for segmentation and measurement of intima media thickness *J. Med. Syst.* **42**
[24] Molinari F, Meiburger K M, Saba L, Zeng G, Acharya U R and Ledda M *et al* 2012 Fully automated dual-snake formulation for carotid intima-media thickness measurement: a new approach *J. Ultrasound Med* **31** 1123–36
[25] Molinari F, Meiburger K M, Nicolaides A and Suri J S 2012 CAUDLES-EF: carotid automated ultrasound double line extraction system using edge flow *Ultrasound Imaging Adv. Appl.* **24** 129–62
[26] Molinari F, Pattichis C S, Guang Z, Saba L, Acharya U R and Sanfilippo R *et al* 2012 Completely automated multiresolution edge snapper—a new technique for an accurate carotid ultrasound imt measurement: clinical validation and benchmarking on a multi-institutional database *IEEE Tran.s Image Process.* **21** 1211–22
[27] Kim G H and Youn H J 2017 Is carotid artery ultrasound still useful method for evaluation of atherosclerosis? *Korean Circ. J.* **47** 1–8
[28] Naim C, Douziech M, Therasse É, Robillard P, Giroux M F and Arsenault F *et al* 2014 Vulnerable atherosclerotic carotid plaque evaluation by ultrasound, computed tomography angiography, and magnetic resonance imaging: an overview *Can. Assoc. Radiol. J.* **65** 275–86
[29] Naim C, Cloutier G, Mercure E, Destrempes F, Qin Z and El-Abyad W *et al* 2013 Characterisation of carotid plaques with ultrasound elastography: feasibility and correlation with high-resolution magnetic resonance imaging *Eur. Radiol* **23** 2030–41
[30] Lee W 2014 General principles of carotid doppler ultrasonography *Ultrason (Seoul, Korea)* **33** 11–7
[31] Christopher B 2006 *Pattern Recognition and Machine Learning* (New York: Springer) 1st
[32] Saba L, Araki T, Krishna Kumar P, Rajan J, Lavra F and Ikeda N *et al* 2016 Carotid inter-adventitial diameter is more strongly related to plaque score than lumen diameter: an automated tool for stroke analysis. *J. Clin. Ultrasound* **44** 210–20
[33] Saba L, Banchhor S K, Suri H S, Londhe N D, Araki T and Ikeda N *et al* 2016 Accurate cloud-based smart IMT measurement, its validation and stroke risk stratification in carotid ultrasound : a web-based point-of-care tool for multicenter clinical trial. *Comput. Biol. Med.* **75** 217–34
[34] Molinari F, Acharya R U, Zeng G, Meiburger K M, Rodrigues P S and Saba L *et al* 2011 CARES 2.0: completely automated robust edge snapper for CIMT measurement in 300 ultrasound images—a two stage paradigm *J. Med. Imaging Heal Inform.* **1** 150–63
[35] Biswas M, Kuppili V, Saba L, Edla D R, Suri H S and Sharma A *et al* 2019 Deep learning fully convolution network for lumen characterization in diabetic patients using carotid ultrasound: a tool for stroke risk *Med. Biol. Eng. Comput.* **57** 543–64
[36] Cootes T F, Taylor C J, Cooper D H and Graham J 1995 Active shape models-their training and application *Comput. Vis. Image Underst.* **61** 38–59
[37] Yang X, Jin J, Xu M, Wu H, He W and Yuchi M *et al* 2013 Ultrasound common carotid artery segmentation based on active shape model. *Comput. Math. Methods Med* **2013** 1–11
[38] Araki T, Ikeda N, Dey N, Acharjee S, Molinari F and Saba L *et al* 2015 Shape-based approach for coronary calcium lesion volume measurement on intravascular ultrasound imaging and its association with carotid intima-media thickness *J. Ultrasound Med* **34** 469–82

[39] Loizou C P, Pattichis C S, Pantziaris M, Tyllis T and Nicolaides A 2007 Snakes based segmentation of the common carotid artery intima media *Med. Biol. Eng. Comput.* **45** 35–49
[40] Saba L, Jain P K, Suri H S, Ikeda N, Araki T and Singh B K *et al* 2017 Plaque tissue morphology-based stroke risk stratification using carotid ultrasound: a polling-based PCA learning paradigm *J. Med. Syst.* **41**
[41] Araki T, Ikeda N, Shukla D, Saba L, Nicolaides A and Shafique S *et al* 2015 A new method for IVUS-based coronary artery disease risk stratification: a link between coronary and carotid ultrasound plaque burdens *Comput. Methods Programs Biomed* **124** 161–79
[42] Tandel G S, Biswas M, Kakde O G, Tiwari A, Suri H S and Turk M *et al* 2019 A review on a deep learning perspective in brain cancer classification *Cancers (Basel)* **11** 111
[43] Klang E 2018 Deep learning and medical imaging *J. Thorac. Dis* **10** 1325–8
[44] Vieira S, Pinaya W H L and Mechelli A 2017 Using deep learning to investigate the neuroimaging correlates of psychiatric and neurological disorders: methods and applications. *Neurosci. Biobehav. Rev.* **74** 58–75
[45] Gibson E, Li W, Sudre C, Fidon L, Shakir D I and Wang G *et al* 2018 NiftyNet: a deep-learning platform for medical imaging *Comput. Methods Programs Biomed* **158** 113–22
[46] Hoo-Chang Member S, Roth hoochangshin H R, Gao M, Lu Senior Member L, Xu Z and Nogues I *et al* 2016 Deep convolutional neural networks for computer-aided detection: CNN architectures, dataset characteristics and transfer learning and daniel mollura are with center for infectious disease imaging HHS public access *IEEE Trans. Med. Imag* **35** 1285–98
[47] Lundervold A S and Lundervold A 2019 An overview of deep learning in medical imaging focusing on MRI *Z Med. Phys.* **29** 102–27
[48] Meyer P, Noblet V, Mazzara C and Lallement A 2018 Survey on deep learning for radiotherapy *Comput. Biol. Med.* **98** 126–46
[49] Razzak M I, Naz S and Zaib A 2018 Deep learning for medical image processing: overview, challenges and the future *Classification in BioApps* (Berlin: Springer) 323–50
[50] Ker J, Wang L, Rao J and Lim T 2017 Deep learning applications in medical image analysis *IEEE Access* **6** 9375–9
[51] Menchón-Lara R M, Sancho-Gómez J L and Bueno-Crespo A 2016 Early-stage atherosclerosis detection using deep learning over carotid ultrasound images *Appl. Soft. Comput. J.* **49** 616–28
[52] Zreik M, van Hamersvelt R W, Wolterink J M, Leiner T, Viergever M A and Isgum I 2018 A Recurrent CNN for automatic detection and classification of coronary artery plaque and stenosis in coronary CT angiography. arXiv:1804.04360
[53] Liu S, Wang Y, Yang X, Lei B, Liu L and Li S X *et al* 2019 Deep learning in medical ultrasound analysis: a review *Engineering* **5** 261–75
[54] Saba L, Biswas M, Kuppili V, Cuadrado Godia E, Suri H S and Edla D R *et al* 2019 The present and future of deep learning in radiology *Eur. J. Radiol.* **114** 14–24
[55] Libby P 2002 Inflammation in atherosclerosis *Nature* **420** 868–74
[56] Naghavi M, Libby P, Falk E, Casscells S W, Litovsky S and Rumberger J *et al* 2003 From vulnerable plaque to vulnerable patient. *Circulation* **108** 1664–72
[57] Ross R 1995 Cell biology of atherosclerosis *Annu. Rev. Physiol.* **57** 791–804
[58] Hopkins P N 2013 Molecular biology of atherosclerosis *Physiol. Rev.* **93** 1317–542

[59] Mannarino E and Pirro M 2008 Molecular biology of atherosclerosis *Clin. Cases Miner Bone Metab.* **5** 57–62 http://ncbi.nlm.nih.gov/pubmed/22460847

[60] Lo J and Plutzky J 2012 The biology of atherosclerosis: general paradigms and distinct pathogenic mechanisms among HIV-infected patients *J. Infect. Dis.* **205** S368–74

[61] Mohebali J, Patel V I, Romero J M, Hannon K M, Jaff M R and Cambria R P *et al* 2015 Acoustic shadowing impairs accurate characterization of stenosis in carotid ultrasound examinations presented at the plenary session of the 2014 joint annual meeting of the New England society for vascular surgery and Eastern Vascular Society, Boston, Mass. *J. Vasc. Surg.* **62** 1236–44

[62] Chiu B, Shamdasani V, Entrekin R, Yuan C and Kerwin W S 2012 Characterization of carotid plaques on 3-dimensional ultrasound imaging by registration with multicontrast magnetic resonance imaging *J. Ultrasound Med* **31** 1567–80

[63] Pedro L M, Sanches J M, Seabra J, Suri J S, Fernandes E and Fernandes J 2014 Asymptomatic carotid disease: a new tool for assessing neurological risk *Echocardiography* **31** 353–61

[64] Ho S S Y 2016 Current status of carotid ultrasound in atherosclerosis *Quant. Imaging Med. Surg* **6** 285–96

[65] Mirek A M and Wolińska-Welcz A 2013 Is the lumen diameter of peripheral arteries a good marker of the extent of coronary atherosclerosis? *Kardiol. Pol* **71** 810–7

[66] Nambi V, Brunner G and Ballantyne C M 2013 Ultrasound in cardiovascular risk prediction: don't forget the plaque! *J. Am. Heart Assoc.* **2** 1–3

[67] Picano E and Paterni M 2015 Ultrasound tissue characterization of vulnerable atherosclerotic plaque *Int. J. Mol. Sci.* **16** 10121–33

[68] Cuadrado-Godia E, Dwivedi P, Sharma S, Ois Santiago A, Roquer Gonzalez J and Balcells M *et al* 2018 Cerebral small vessel disease: a review focusing on pathophysiology, biomarkers, and machine learning strategies *J. Stroke* **20** 302–20

[69] De Korte C L, Fekkes S, Nederveen A J, Manniesing R and Hansen H R H G 2016 Review: mechanical characterization of carotid arteries and atherosclerotic plaques *IEEE Trans. Ultrason. Ferroelectr. Freq. Control* **63** 1613–23

[70] Acharya U R 2012 Plaque tissue characterization and classification in ultrasound carotid scans: a paradigm for vascular *IEEE Trans. Instrum. Meas.* **62** 392–400

[71] Acharya U R 2012 Atherosclerotic risk stratification strategy for carotid arteries using texture-based features *Ultrasound Med. Biol.* **38** 899–915

[72] Cheng J, Yu Y and Chiu B 2016 Direct 3D segmentation of carotid plaques from 3D ultrasound images *Proc. 2016 IEEE Biomed. Circuits Syst. Conf. BioCAS 2016* pp 123–6

[73] Dong Y, Pan Y, Zhao X, Li R, Yuan C and Xu W 2017 Identifying carotid plaque composition in MRI with convolutional neural networks *2017 IEEE Int. Conf. Smart Comput. SMARTCOMP* 2017

[74] Jones W, Alasoo K, Fishman D and Parts L 2017 Computational biology: deep learning *Emerg. Top. Life Sci.* **1** 257–74

[75] Turan T N, Lematty T, Martin R, Chimowitz M I, Rumboldt Z and Spampinato M V *et al* 2015 Characterization of intracranial atherosclerotic stenosis using high-resolution MRI study – rationale and design *Brain Behav* **5** 1–9

[76] Gupta A, Kesavabhotla K, Baradaran H, Kamel H, Pandya A and Giambrone A E *et al* 2015 Plaque echolucency and stroke risk in asymptomatic carotid stenosis *Stroke* **46** 91–7

[77] Hunt K J, Evans G W, Folsom A R, Sharrett A R, Chambless L E and Tegeler C H *et al* 2001 Acoustic shadowing on B-mode ultrasound of the carotid artery predicts ischemic stroke. The Atherosclerosis Risk in Communities (ARIC) study *Stroke* **32** 1120–6

[78] Araki T, Banchhor S K, Londhe N D, Ikeda N, Radeva P and Shukla D *et al* 2016 Reliable and accurate calcium volume measurement in coronary artery using intravascular ultrasound videos *J. Med. Syst.* **40** 1–20

[79] Chu B, Kampschulte A, Ferguson M S, Kerwin W S, Yarnykh V L and O'Brien K D *et al* 2004 Hemorrhage in the atherosclerotic carotid plaque: a high-resolution MRI study *Stroke* **35** 1079–84

[80] Kamenskiy A V, Pipinos I I, Carson J S, Mactaggart J N and Baxter B T 2015 Age and disease-related geometric and structural remodeling of the carotid artery *J. Vasc. Surg.* **62** 1521–8

[81] Londhe N D and Suri J S 2016 Superharmonic imaging for medical ultrasound: a review *J. Med. Syst.* **40** 279

[82] Grønholdt M L M, Nordestgaard B G, Schroeder T V, Vorstrup S and Sillesen H 2001 Ultrasonic echolucent carotid plaques predict future strokes *Circulation* **104** 68–73

[83] Jashari F, Ibrahimi P, Bajraktari G, Grönlund C, Wester P and Henein M Y 2016 Carotid plaque echogenicity predicts cerebrovascular symptoms: a systematic review and meta-analysis *Eur. J. Neurol.* **23** 1241–7

[84] Madore B and Meral F C 2012 Reconstruction algorithm for improved ultrasound image quality *IEEE Trans. Ultrason. Ferroelectr. Freq. Control* **59** 217–30

[85] Remington L A 2012 Orbital blood supply *Clinical Anatomy and Physiology of the Visual System* (Amsterdam: Elsevier) 202–17

[86] Binning M J 2018 Internal carotid artery aneurysms introduction *Intracranial Aneurysms* (Amsterdam: Elsevier) 479–81

[87] Webb W G 2017 Organization of the nervous system II *Neurology for the Speech-Language Pathologist* (Amsterdam: Elsevier) 44–73

[88] Love B B and Biller J 2007 Neurovascular system *Textbook of Clinical Neurology* (Amsterdam: Elsevier) 405–34

[89] Hong J T, Kim T H, Kim I S, Yang S H, Sung J H and Son B C *et al* 2010 The effect of patient age on the internal carotid artery location around the atlas *J. Neurosurg. Spine* **12** 613–8

[90] Banchhor S K, Londhe N D, Araki T, Saba L, Radeva P and Khanna N N *et al* 2018 Calcium detection, its quantification, and grayscale morphology-based risk stratification using machine learning in multimodality big data coronary and carotid scans: a review *Comput. Biol. Med.* **101** 184–98

[91] Biswas M, Kuppili V, Saba L, Edla D R, Suri H S and Cuadrado-Godia E *et al* 2019 State-of-the-art review on deep learning in medical imaging *Front. Biosci. (Landmark Ed)* **24** 392–426

[92] Ren S, He K, Girshick R and Sun J 2015 Faster R-CNN: Towards real-time object detection with region proposal networks arXiv:1506.01497

[93] Girshick R 2015 *Fast R-CNN IEEE Int. Conf. on Computer Vision (ICCV)* 2015 (IEEE) 1440–8

[94] He K, Zhang X, Ren S and Sun J 2015 Delving deep into rectifiers: surpassing human-level performance on imagenet classification arXiv:1502.01852

[95] Lal B K, Hobson R W, Pappas P J, Kubicka R, Hameed M and Chakhtura E Y *et al* 2002 Pixel distribution analysis of B-mode ultrasound scan images predicts histologic features of atherosclerotic carotid plaques *J. Vasc. Surg.* **35** 1210–7

[96] Broomhead D and David L 1988 Multivariable Functional Interpolation and Adaptive Networks *Complex Systems* **2** 321–55

[97] Gonzalez R C and Woods R E 2002 *Digital Image Processing* (Hoboken, NJ: Prentice-Hall)

[98] Qian C and Yang X 2018 Computer methods and programs in biomedicine An integrated method for atherosclerotic carotid plaque segmentation in ultrasound image *Comput. Methods Programs Biomed.* **153** 19–32

[99] Lekadir K, Galimzianova A, Betriu A, Del Mar Vila M, Igual L and Rubin D L *et al* 2017 A convolutional neural network for automatic characterization of plaque composition in carotid ultrasound *IEEE J. Biomed. Heal Inform.* **21** 48–55

[100] Barnett H J M, Taylor D W, Eliasziw M, Fox A J, Ferguson G G and Haynes R B *et al* 1998 Benefit of carotid endarterectomy in patients with symptomatic moderate or severe stenosis *N. Engl. J. Med.* **339** 1415–25

[101] Vapnik V N 1999 An overview of statistical learning theory *IEEE Trans. Neural Networks* **10** 988–99

[102] Shrivastava V K, Londhe N D, Sonawane R S and Suri J S 2016 A novel approach to multiclass psoriasis disease risk stratification: machine learning paradigm. *Biomed. Signal. Process. Control* **28** 27–40

[103] Shrivastava V K, Londhe N D, Sonawane R S and Suri J S 2015 Exploring the color feature power for psoriasis risk stratification and classification: a data mining paradigm *Comput. Biol. Med.* **65** 54–68

[104] Shrivastava V K, Londhe N D, Sonawane R S and Suri J S 2016 Computer-aided diagnosis of psoriasis skin images with HOS, texture and color features: a first comparative study of its kind *Comput. Methods Prog. Biomed.* **126** 98–109

[105] Shrivastava V K, Londhe N D, Sonawane R S and Suri J S 2015 Reliable and accurate psoriasis disease classification in dermatology images using comprehensive feature space in machine learning paradigm *Expert. Syst. Appl.* **42** 6184–95

[106] Srivastava R K, Greff K and Schmidhuber J 2015 Training very deep networks arXiv:1507.06228

[107] Bianchini M and Scarselli F 2014 On the complexity of neural network classifiers: a comparison between shallow and deep architectures *IEEE Trans. Neural. Networks Learn. Syst.* **25** 1553–65

[108] Yu D, Seltzer M L, Li J, Huang J T and Seide F 2013 Feature learning in deep neural networks – studies on speech recognition tasks arXiv:1301.3605

[109] Ciresan D, Meier U and Schmidhuber J 2012 Multi-column deep neural networks for image classification *IEEE Conf. on Computer Vision and Pattern Recognition* 2012 (IEEE) 3642–9

[110] Simonyan K and Zisserman A 2014 Very deep convolutional networks for large-scale image recognition arXiv:1409.1556

[111] He K, Zhang X, Ren S and Sun J 2015 Deep residual learning for image recognition arXiv:1512.03385

[112] Alex K and Ilya S G E H 2012 ImageNet classification with deep convolutional neural networks. *Advances in Neural Information Processing Systems* **25** pp 1097–105 https://cs.toronto.edu/~fritz/absps/imagenet.pdf

[113] Szegedy C, Liu W, Jia Y, Sermanet P, Reed S and Anguelov D *et al* 2014 Going deeper with convolutions arXiv:1409.4842
[114] Lecun Y, Bottou L, Bengio Y and Haffner P 1998 Gradient-based learning applied to document recognition *Proc IEEE* **86** 2278–324
[115] Avendi M R, Kheradvar A and Jafarkhani H 2016 A combined deep-learning and deformable-model approach to fully automatic segmentation of the left ventricle in cardiac MRI *Med. Image Anal.* **30** 108–19
[116] Havaei M, Davy A, Warde-Farley D, Biard A, Courville A and Bengio Y *et al* 2017 Brain tumor segmentation with deep neural networks *Med. Image Anal.* **35** 18–31
[117] Tajbakhsh N, Shin J Y, Gurudu S R, Hurst R T, Kendall C B and Gotway M B *et al* 2016 Convolutional neural networks for medical image analysis: full training or fine tuning? *IEEE Trans. Med. Imaging* **35** 1299–312
[118] Alom M Z, Taha T M, Yakopcic C, Westberg S, Sidike P and Nasrin M S *et al* 2018 The history began from AlexNet: a comprehensive survey on deep learning approaches arXiv:1803.01164
[119] Bianco S, Cadene R, Celona L and Napoletano P 2018 Benchmark analysis of representative deep neural network architectures *IEEE Access* **6** 64270–7
[120] LeCun Y, Haffner P, Bottou L and Bengio Y 1999 Object recognition with gradient-based learning *Shape, Contour and Grouping in Computer Vision* (Berlin: Springer) 319–45
[121] He K, Zhang X, Ren S and Sun J 2014 Spatial pyramid pooling in deep convolutional networks for visual recognition *Computer Vision – ECCV 2014* (Springer) 346–61
[122] Özdemir H, Artaş H and Serhatlioğlu S 2006 Oğur E. Effects of overweight on luminal diameter, flow velocity and intima-media thickness of carotid arteries *Diagnost. Interv. Radiol* **12** 142–6
[123] Lloyd K D, Barinas-Mitchell E, Kuller L H, Mackey R H, Wong E A and Sutton-Tyrrell K 2012 Common carotid artery diameter and cardiovascular risk factors in overweight or obese postmenopausal women *Int. J. Vasc. Med* **2012** 1–7
[124] Krejza J, Arkuszewski M, Kasner S E, Weigele J, Ustymowicz A and Hurst R W *et al* 2006 Carotid artery diameter in men and women and the relation to body and neck size *Stroke* **37** 1103–5
[125] Ruan L, Chen W, Srinivasan S R, Sun M, Wang H and Toprak A *et al* 2009 Correlates of common carotid artery lumen diameter in black and white younger adults: the Bogalusa heart study *Stroke* **40** 702–7
[126] Mancini G B J 2004 Surrogate markers for cardiovascular disease: structural markers *Circulation* **109** IV-22–30
[127] Cohn J N 2004 Surrogate markers for cardiovascular disease: functional markers *Circulation* **109** IV-31–46
[128] Amato M, Montorsi P, Ravani A, Oldani E, Galli S and Ravagnani P M *et al* 2007 Carotid intima-media thickness by B-mode ultrasound as surrogate of coronary atherosclerosis: correlation with quantitative coronary angiography and coronary intravascular ultrasound findings *Eur. Heart J.* **28** 2094–101
[129] O'Leary D H, Polak J F, Wolfson S K, Bond M G, Bommer W and Sheth S *et al* 2011 Use of sonography to evaluate carotid atherosclerosis in the elderly. the cardiovascular health study. CHS collaborative research group *Stroke* **22** 1155–63
[130] Smistad E 2016 Deep learning and data labeling for medical applications. *First International Workshop, LABELS 2016, and Second International Workshop, DLMIA*

2016, Held in Conjunction with MICCAI 2016, Athens, Greece, October 21, 2016, Proceedings 10008 (Springer) 30–8

[131] Bastida-Jumilla M C, Menchón-Lara R M, Morales-Sánchez J, Verdú-Monedero R, Larrey-Ruiz J and Sancho-Gómez J L 2015 Frequency-domain active contours solution to evaluate intima–media thickness of the common carotid artery *Biomed. Signal Process. Control* **16** 68–79

[132] Ilea D E, Duffy C, Kavanagh L, Stanton A and Whelan P F 2013 Fully automated segmentation and tracking of the intima media thickness in ultrasound video sequences of the common carotid artery. *IEEE Trans. Ultrason. Ferroelectr. Freq. Control* **60** 158–77

[133] Loizou C P, Kasparis T, Spyrou C and Pantziaris M 2013 Integrated system for the complete segmentation of the common carotid artery bifurcation in ultrasound images. *AIAI 2013: Artificial Intelligence Applications and Innovations* (Springer) 292–301

[134] Razzak M I, Naz S and Zaib A 2018 Deep learning for medical image processing: overview, challenges and the future *Lect. Notes Comput. Vis. Biomech* **26** 323–50

[135] Agarap A F 2017 An architecture combining convolutional neural network (CNN) and support vector machine (SVM) for image classification arXiv:1712.03541

[136] Lipton Z C 2016 The mythos of model interpretability arXiv:1606.03490

[137] Kane A G, Dillon W P, Barkovich A J, Norman D, Dowd C F and Kane T T 1996 Reduced caliber of the internal carotid artery: a normal finding with ipsilateral absence or hypoplasia of the A1 segment *Am. J. Neuroradiol* **17** 1295–301

[138] Ascher E, Markevich N, Hingorani A and Kallakuri S 2002 Pseudo-occlusions of the internal carotid artery: a rationale for treatment on the basis of a modified carotid duplex scan protocol *J. Vasc. Surg.* **35** 340–5

[139] Bartlett E S, Symons S P and Fox A J 2006 Correlation of carotid stenosis diameter and cross-sectional areas with CT angiography *AJNR Am. J. Neuroradiol.* **27** 638–42 http://ncbi.nlm.nih.gov/pubmed/16552008

[140] Hyde D E, Fox A J, Gulka I, Kalapos P, Lee D H and Pelz D M *et al* 2004 Internal carotid artery stenosis measurement: comparison of 3D computed rotational angiography and conventional digital subtraction angiography *Stroke* **35** 2776–81

[141] Baradaran X H, Patel P, Gialdini G, Al-Dasuqi K, Giambrone A and Kamel H *et al* 2017 Quantifying intracranial internal carotid artery stenosis on MR angiography *Am. J. Neuroradiol* **38** 986–90

[142] Lu N, Wu Y, Feng L and Song J 2019 Deep learning for fall detection: three-dimensional CNN combined with LSTM on video kinematic data *IEEE J. Biomed. Heal Informatics* **23** 314–23

[143] Savioli N, Visentin S, Cosmi E, Grisan E, Lamata P and Montana G 2018 Temporal convolution networks for real-time abdominal fetal aorta analysis with ultrasound 1–10 arXiv:1807.04056

[144] Babalyan K, Sultanov R, Generozov E, Sharova E, Kostryukova E and Larin A *et al* 2018 LogLoss-BERAF: an ensemble-based machine learning model for constructing highly accurate diagnostic sets of methylation sites accounting for heterogeneity in prostate cancer; A Elofsson *PLoS One* **13** e0204371

IOP Publishing

Multimodality Imaging, Volume 2

Heart, lungs and peripheral organs

Mainak Biswas and Jasjit S Suri

Chapter 8

Deep learning framework for ultrasound-based carotid artery disease management: a review

Mainak Biswas and Jasjit S Suri

Cardiovascular diseases (CVDs) are responsible for 30.8% deaths worldwide, and are termed global killers. One of the major causes of CVDs is build-up of plaque within the common carotid arteries (CCAs). Routine check-ups can help in early detection of CVDs and can play an instrumental role in recovery and well-being of the patient. The measurements related to ultrasound (US) CCA images, such as lumen diameter (LD) and carotid intima-media thickness (cIMT) can help in quantifying the stroke risk of the patient and pave the way for early treatment and recovery. The study on characterization of the nature of the plaque i.e., asymptomatic or symptomatic, calcified or fibrous, vulnerable or non-vulnerable, from ultrasound CCA images can also help in early diagnosis of CVDs. In this paper, we focus on various automated techniques for CCA related measurements, plaque characterization and quantification of plaque vulnerability. The automated techniques are divided into two classes: machine learning (ML) and deep learning (DL). ML techniques represent traditional techniques of feature extraction and application of ML algorithms on these features for plaque characterization or CCA measurements. Unlike ML, DL is capable of automated high-level feature extraction and is capable of both characterization and segmentation. The application of many layers of abstraction within DL-based models allow extraction of highly accurate features from ultrasound CCA images, which are independent of hand-made features. These DL features are used for both CCA measurements and plaque characterization. We do a comprehensive analysis covering strengths and weaknesses of both DL and ML techniques.

8.1 Introduction

The National Health and Nutrition Examination Survey (NHANES) [1] from 2011 to 2014 estimated that 92.1 million Americans suffer from some form of

doi:10.1088/978-0-7503-2352-9ch8

cardiovascular diseases (CVDs). Among them 46.7 million are over 60 years of age. A total of 11.5% (27.6 million) of the adult population of the USA have heart disease [2]. By 2030, it is projected that 43.9% of US population will have some form of CVD. The mortality rate due to CVDs is the highest, contributing to 30.8% of all deaths, accounting for approximately 2200 deaths each day or one death every 40 seconds [3]. Further, the direct and indirect medical costs in the USA are projected to rise from the approximate 316.1 billion dollars spent in the year 2012–13 to 1.094 trillion dollars in the year 2030 [4], putting the heavy financial burden on family and society.

The main cause of stroke is the disease called atherosclerosis [5, 6]. This occurs when the plaque deposited with time within the carotid artery starts to harden, leading to restricted blood flow to the brain and ultimately stroke. A small discussion on anatomy of the common carotid artery (CCA) is required before we proceed further. The CCA rises from the arch of aorta and supplies oxygenated blood to the face and brain [7, 8]. There are two CCAs *i.e.*, the left CCAand right CCA. The left CCA rises directly from the aortic arch and proceed towards the head. However, the right CCA rises from the brachiocephalic artery. The corresponding anatomy is shown diagrammatically in figure 8.1(a). The left and the right CCA each bifurcate into internal and external CCAs near the larynx. The internal CCA supplies blood to the brain while the external CCA supplies blood to the face and scalp. It is diagrammatically shown in figure 8.1(b). The most common area of plaque development is near this bifurcation. There are many different kinds of forces which act on the vessel walls due to blood flow. One force that acts outward in a direction perpendicular to the blood flow and is called tensile stress. The other is the frictional force that acts in parallel opposite to the direction of blood flow and is called wall shear stress (WSS). Laminar flow is characterized by smooth and streamlined passage of blood through the artery while the disturbed flow is characterized by departure from this streamline, separation, recirculation and reattachment to the forward flow of blood (shown in figure 8.2). The laminar velocity increases beyond a critical point where it becomes turbulent where inertial forces precede over viscous forces. Near the bifurcation of the CCA, the disturbed and turbulent flow happens which results in prolonged resident time of atheromatous material where it transports these materials to the vessel wall, resulting in formation of plaque [9]. It is also important to understand the nature of plaque deposited along the arteries. In this regard, the echogenicity of plaque is categorized as hypoechoic or hyperechoic. The hypoechoic plaque constitutes lipid and thrombus while the hyperechoic plaque is made of fibrous tissue or calcium. Again, we make a distinction between stable and vulnerable plaque. The stable plaque is composed of fibrous and calcified tissue while the vulnerable plaque constitutes a lipid-rich necrotic core, a thin fibrous cap, spotty calcium and neovascularization. Neovascularization is a phenomenon of formation of new blood vessels which supply lesion components, metabolic substrates, to the plaque aiding in its enhancement. A stable or small plaque does not affect the blood flow, however, a vulnerable plaque with a large lipid-rich necrotic core can rupture the thin fibrous cap resulting in a thrombogenic reaction restricting or stopping blood flow, resulting

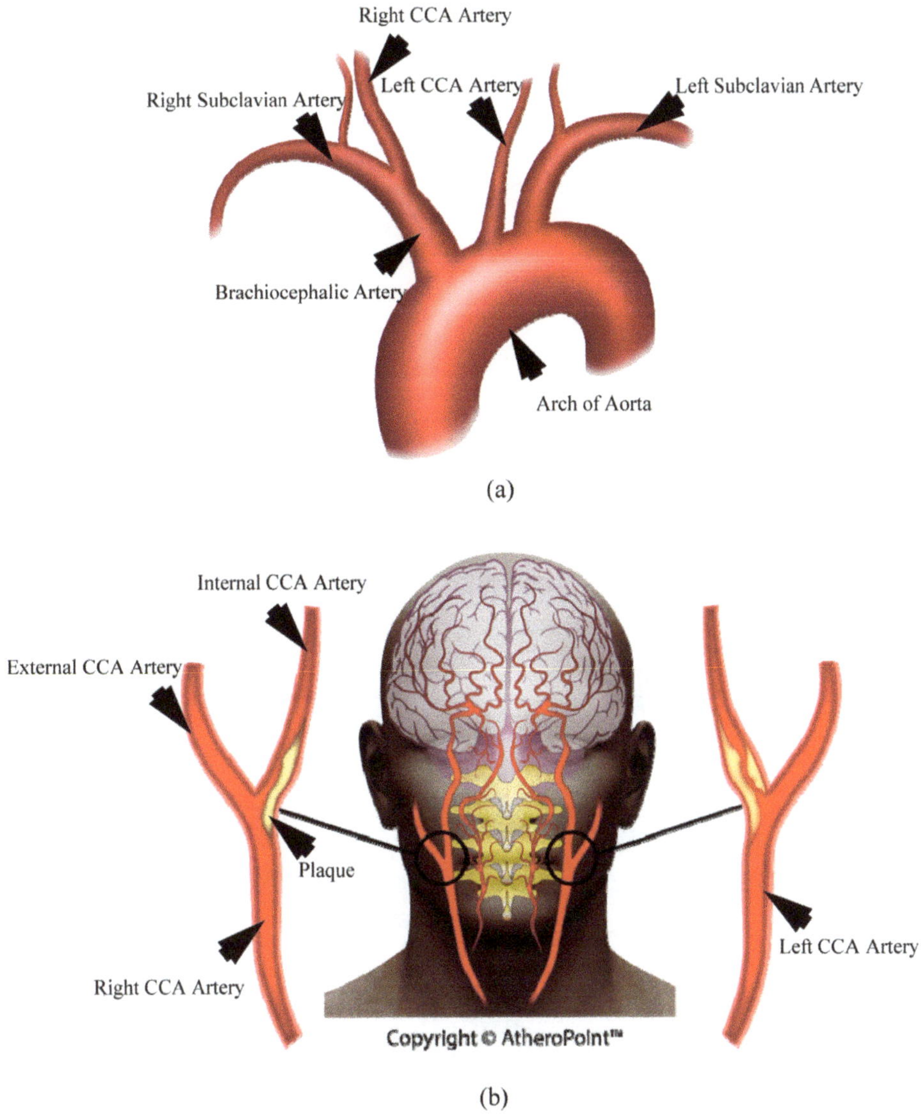

Figure 8.1. Plaque formation geometrical locations (courtesy of AtheroPoint™).

in ischemic stroke, myocardial infarction, as well as other CVD events. This process is shown in figure 8.3.

Early detection and diagnosis of CVDs is a key factor in treatment, lessening the financial impact on the family and society and enabling speedy recovery of the patient. Medical imaging is a non-invasive and painless technique widely used for plaque detection in carotid arteries and risk assessment. The different types of medical imaging are ultrasound (US) [10], magnetic resonance imaging (MRI) [11] or computed tomography (CT) [12]. Medical ultrasound imaging uses US waves for capturing images of internal organs of the body. Similarly, MRI uses magnetic

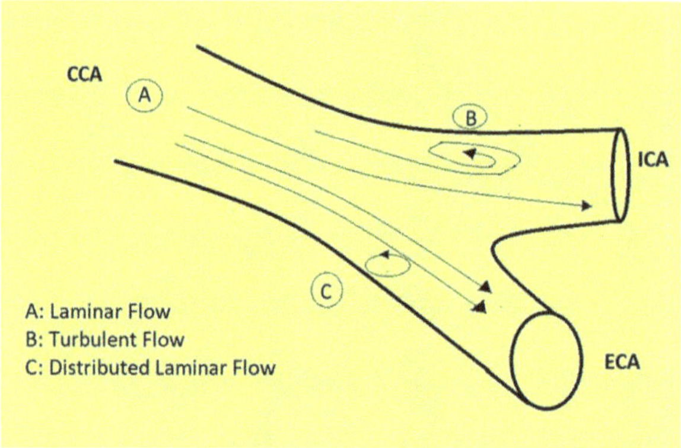

Figure 8.2. Laminar, turbulent and disturbed blood flow near the bifurcation of CCA.

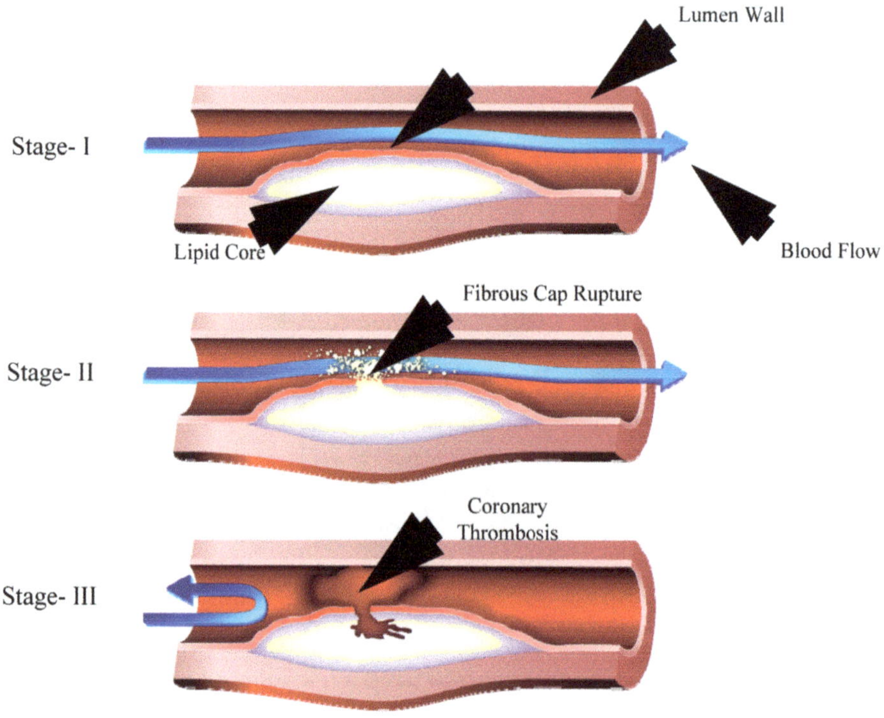

Figure 8.3. Rupture of fibrous cap (courtesy of AtheroPoint™).

fields, and CT uses x-rays for capturing internal images. Among these three, US is the most widely used medical imaging technique. US is the safest among them as it does not use ionizing radiation thereby, minimizing cellular damage. US can also quickly provide images in real time. Further, equipment cost of US is far lower compared to MRI and CT imaging techniques [5, 13–15]. However, US waves cannot penetrate bones and therefore in cases where information beyond bone structure is required, MRI and CT scans are recommended for better results. The two most prevalent US techniques are Doppler and B-mode US. Doppler-US mainly used in measuring and visualizing blood flow while B-mode US scans a plane of the body resulting in a 2D image. In this study, we focus on static B-mode US carotid scan study.

Over the decades, radiologists had had to study these medical images and prepare the diagnostic reports. However, recently, with the advent of high power computers and image processing techniques, computer aided diagnosis (CAD) is rapidly making improvements. In addition to image processing techniques, computers are specifically trained on diagnosed images to make learned diagnosis of unseen images, which is also known as machine learning (ML) [16]. The ML techniques extract hand-made features based on various image modalities and apply ML techniques such as support vector machines (SVM) [17], extreme learning machines (ELM) [18], artificial neural networks (ANNs) [19, 20] etc, for enabling tissue characterization. Recently, DL [21, 22] is making rapid strides in medical imaging which is evident from the volume of literature in this field. DL, unlike ML, does not require hand-made features. The features are generated automatically for DL and thus bypassing the feature extraction stage. The DL systems such as convolution neural networks (CNNs) mimic the human visual cortex while extracting features and therefore these features are more distinct than the hand-made ones.

In the case of CVD risk measurement, all the techniques developed so far can be divided into two branches of study: (1) segmentation, measurement and risk assessment and (2) plaque characterization and risk assessment. In segmentation and measurement, several measurements are derived:

(a) Lumen diameter and stenosis measurement;
(b) Inter-adventitial diameter measurement;
(c) Intima-medial thickness measurement; and
(d) Total plaque area measurement and then assessing stroke risk using the above.

In the case of plaque characterization, the region of interest covering the plaque is taken out and the nature of the plaque is found out, whether the plaque is symptomatic or asymptomatic, hyperechoic or hypoechoic, stable or vulnerable, etc, so that a proper diagnosis is made. Plaque rupture is one of the frequent causes of carotid/coronary thrombosis [23–25]. The process of plaque rupture and its consequent reaction is shown in figure 8.3. It is shown in three stages: stage-I represents the normal blood flow through the lumen, with lesions representing the lipid core covered by a fibrous cap. Stage-II is where the fibrous cap ruptures. In stage-II due to rupture, carotid thrombosis occurs resulting in blocking of blood flow and ultimately myocardial infarction. A detailed explanation on plaque rupture is given in the Discussion section. In this chapter, we cover various technologies that have been implemented for segmentation and measurement and plaque characterization. This study will be unique in its approach as it covers all major technologies developed so far for these two branches of CVD risk assessment. This chapter is further divided into five sections. Section 1 gives the introduction, section 2 covers segmentation and measurement techniques using ML/DL, section 3 describes plaque characterization using ML/DL, section 4 provides the discussion and finally the conclusion is provided in section 5.

8.2 Arterial segmentation using deep learning framework

In the past few decades, there has been an exponential rise in literature on automated techniques for US carotid segmentation. This rise can be attributed to the lowering of computer hardware costs and upsurge in new automated techniques such as ML and DL. There are several benefits of using automated techniques when compared with human operators, i.e., it allows for multiple original equipment machine (OEM) comparison studyies, machine-based automated technique usage can be scaled to multicentric and multi-ethnic large database studies. Further, it allows for more accurate design systems and removal of laborious human operator dependency, thereby removing inter- and intra-observer and inter- and intra-operator variability. However, there are multiple challenges to automated study of CCAs. The foremost among them is noise. This is due to inherent noise in US scans, which is due to blood scattering, patient movement, breathing and the physics of image acquisition. The second type of challenge is the gaps in the far wall due to shadows. These shadows are due to the presence of calcium in the near wall. The third challenge in the arterial imaging is oversaturation of the brightness in the scans and one has to sometimes customize the window-level operation for all scans. The fourth challenge occurs when imaging the bulb of the carotid artery, which has a cup-like structure. During this process, there is a difficulty in displaying plaque while trying to image the distal and proximal segments of the bulb region. This is because the probe is not wide enough to capture all the segments at the same time. The worst part is the scanning of the internal carotid artery, which is most challenging as it is buried near the brain entrance, narrow in diameter compared to the common carotid artery. Not only this, but the arteries have different curvatures throughout when scanning bulb and internal carotid arteries. Even the common carotid arteries are fully curved like a necklace, either a convex shape or concave shape. There are structural challenges too. These kinds of challenge are the interference of the jugular vein in the distal end of the artery (figures 8.4(a) and (b)) and bright muscles (figures 8.4(c) and (d)).

In the following subsections, we will look at several automated techniques in the ML and DL domain for CCA segmentation and measurement and how they deal with these challenges.

8.2.1 ML-based CCA segmentation techniques

ML-based CCA segmentation techniques involve a one or two-stage process. In the one-stage model, lumen borders are extracted by using various image features such as brightness, contrast, etc. In a two-stage model, first the ROI is extracted. In the next stage, a border is delineated from the ROI using level sets or machine learning techniques based on various image features. Both the one and two-stage ML-based techniques are given in sections 8.2.1.1 and 8.2.1.2, respectively.

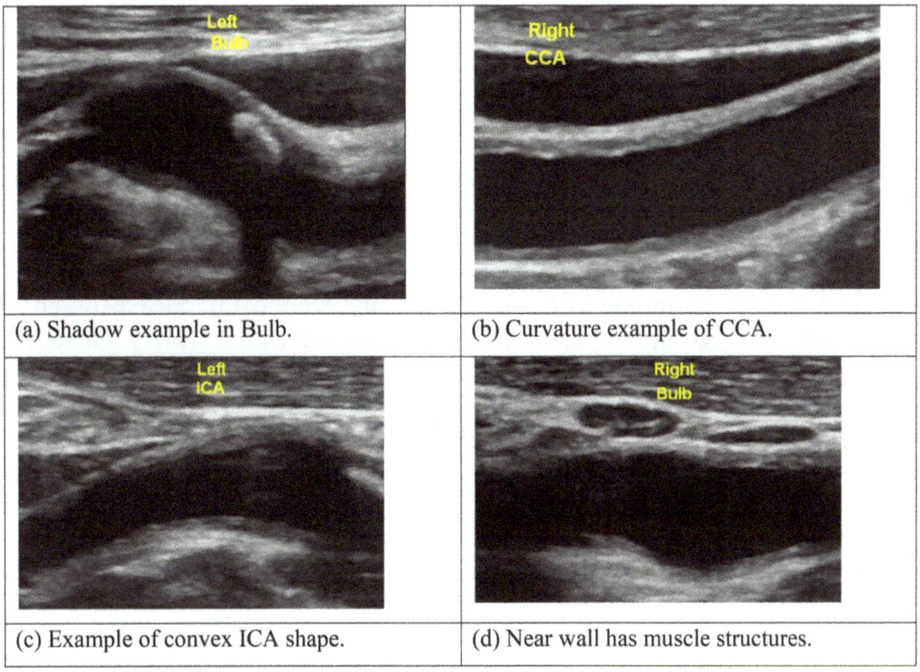

(a) Shadow example in Bulb.

(b) Curvature example of CCA.

(c) Example of convex ICA shape.

(d) Near wall has muscle structures.

Figure 8.4. (a) Noise due to shadow, (b) Noise due to curvature, (c) Noise due to convex ICA shape, (d) Noise due to near wall muscle structures.

8.2.1.1 Completely automated layers extraction (CALEX)

CALEX [26, 27] is based on integrated technique consisting of feature extraction, line fitting and classification. It is inspired from the fact that far wall pixels have the high intensity in ultrasound CCA images. The CALEX system constitutes two stages: stage-I for feature extraction and stage-II for fuzzy *K*-means [28] classification. The stage-I of the CALEX system employs feature extraction leading to determination far adventitia border. In this process each column of the CCA US scan is analysed with a linear discriminator to find those points with highest intensity. These points are called seed points, which are later linked to form the line segment. Validation probability is computed to remove the false line segments which is given as:

$$P(X_{val} \mid l_i) = \exp\{\gamma(X_{val} \mid l_i)\} \tag{8.1}$$

where X_{val} denotes the event where line segment l_i is valid, $\gamma(.)$ is the energy function, which is dependent on two properties of the line segment, i.e., width stability $f_1(l_i)$ and support $f_2(l_i)$, such that:

$$\gamma(X_{val} \mid l_i) = c_1.\ f_1(l_i) + c_2.\ f_2(l_i) \tag{8.2}$$

After validation, two line segments l_i and l_j are connected based on the connection probability based on proximity $f_3(l_i,\ l_j)$ and alignment $f_4(l_i,\ l_j)$, which are given by:

$$P(X_{con}|l_i, l_j) = \exp\{\gamma(X_{con}|l_i, l_j)\} \tag{8.3}$$

$$\gamma(X_{con}|l_i, l_j) = c_3. f_3(l_i) + c_4. f_4(l_i) \tag{8.4}$$

where X_{con} is the event that l_i and l_j are connected, and c_1, c_2, c_3 and c_4 are weights determined by training data. After the far wall is segmented, fuzzy *K*-means classification is applied in stage-II to cluster the seed points obtained from stage-I into three groups. The intermediate pixel between the first and second group is designated the lumen intima (LI) point while the second interface between the second and third group is taken as the media adventitia (MA) point. The segmentation was performed on 665 multi-institutional US CCA scan images. Absolute IMT error was found to be 0.191 $\pm$ 0.217 mm. The average distances and variability between far adventitia (AF) and LI boundaries are 0.79 $\pm$ 0.77 mm and 0.25 $\pm$ 0.26 mm, respectively. Similarly, the average distances and variability between far adventitia (AF) and MA boundaries are 0.48 $\pm$ 0.59 and 0.26 $\pm$ 0.25 mm, respectively. In addition CALEX showed a recognition ability of the CCA to be 100%. A segmented image by using CALEX is given in figure 8.5.

8.2.1.2 Lumen segmentation using region and boundary-based approaches

Suri and his team [29] used dual approaches of region- and boundary-based techniques for lumen segmentation. The region-based technique is inspired from the fact that the lumen region has the lowest intensity. The boundary-based methodology uses the concept of deformable models for CCA segmentation. In the region-based strategy, a global pixel identification strategy is followed using a row-wise intensity distribution model from histograms to segment the lumen from the ultrasound CCA images. The region-based strategy is independent of gradient information, making segmentation robust. In the boundary-based methodology the parametric or geometric lumen curve is followed. The model used two stages; stage-I for global shape or ROI extraction from the CCA image and stage-II employs region- and boundary-based strategies separately for segmenting the lumen. In stage-I, the ROI is extracted using a combination of scale space [30] and spectral analysis [30]. This ROI is fed into stage-II constituting a region-based and boundary-based algorithm. The ROI consists of three regions: (1) low-intensity lumen, (2) high-intensity adventitia region and (3) medium-intensity plaque region. A k-means classifier is adapted in the stage-II of the region-based strategy to yield the three regions. The lumen is extracted as the largest region in the binary form. The far and near walls are delineated from the binary lumen and polyline distance metric [31] is applied to obtain the LD. In the boundary-based approach of stage-II, geometric models i.e., level sets are used to detect the near and far lumen boundaries. The level set methodology used in this paper is distance regularized level set evolution (DRLSE) [32]. Hough's transform is applied before DRLSE. The mathematical formulation of the DRLSE is described as follows:

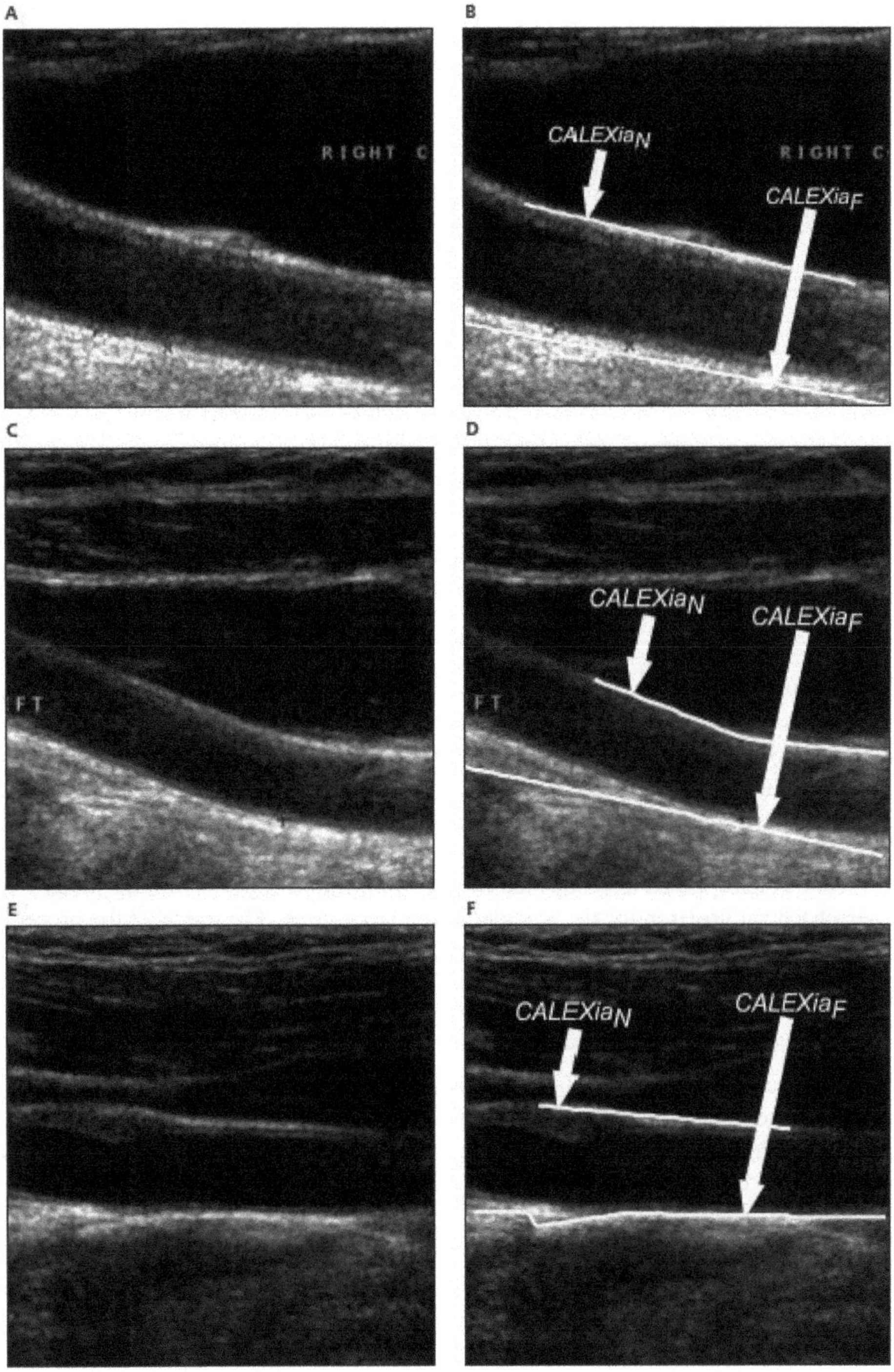

Figure 8.5. CCA segmentation using CALEX (image source: [27]).

Let P be the image in the domain δ. The edge indicator function E is defined as:

$$E = \frac{1}{1 + \left|\nabla g_\sigma * P\right|^2} \tag{8.5}$$

where ∇g_σ denotes the Gaussian kernel with standard deviation σ. The energy function $\varepsilon(\theta)$ is given by:

$$\varepsilon(\theta) = \alpha R(\theta) + \beta L(\theta) + \gamma A(\theta) \tag{8.6}$$

where α is the coefficient of the distance regularization term $R(\theta)$, and β and γ are the coefficients of the energy functional $L(\theta)$ and $A(\theta)$, which are given by:

$$L(\theta) = \int EK(\theta)|\nabla\theta| \, dx \tag{8.7}$$

$$A(\theta) = \int EH(-\theta)dx \tag{8.8}$$

where K and H are the Dirac Delta function and Heaviside function, respectively. The iteration process is given by:

$$\theta_{i,j}^{k+1} = \theta_{i,j}^{k} + \Delta t L(\theta_{i,j}^{k}) \tag{8.9}$$

Two datasets were used, one from Japan consisting of 404 images and one from Hong Kong consisting of 300 images. The LD based on the region-based method obtained from the Japanese and the Hong Kong dataset was 6.35 ± 0.95 and 6.20 ± 1.35 mm, respectively. Similarly, the LD based on the boundary-based method obtained from the Japanese and the Hong Kong dataset was 5.90 ± 0.97 mm and 5.43 ± 1.24 mm, respectively. The results showed that for 90% of images, the LD error for the region-based technique fell within 1mm, whereas, for 80% of the images, the LD error fell within 1mm for the boundary-based technique. The model of the process is shown in figure 8.6. The delineated borders obtained from the region-based and boundary-based methodology is given in figure 8.7.

8.2.2 DL-based CCA segmentation

The DL-based CCA segmentation involves two stages. The first stage involves pre-processing the images. In the second stage CCA is segmented based on a DL model trained on radiologist-delineated ground images. Two CCA segmentation and measurement techniques are given in sections 8.2.2.1 and 8.2.2.2, respectively.

8.2.2.1 Lumen diameter measurement using deep learning

Deep learning [33] is currently one of the major research areas in medical imaging. The major attraction point of DL that separates it from the ML domain is its automated feature extraction ability, which can be both modified for characterization and segmentation purposes. Convolutional neural networks (CNNs) are the most commonly used DL model for feature extraction and classification. Among the DL models, fully convolutional networks (FCNs) [34] are used for image

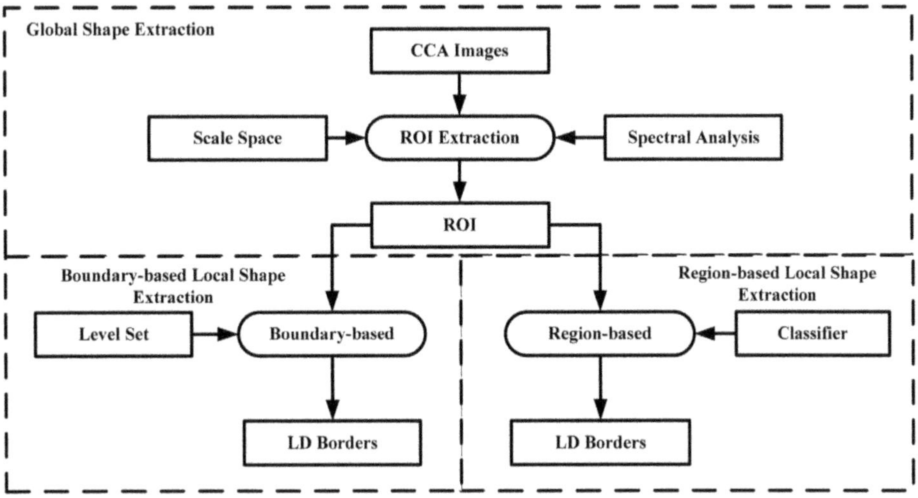

Figure 8.6. Region and boundary-based model for CCA segmentation.

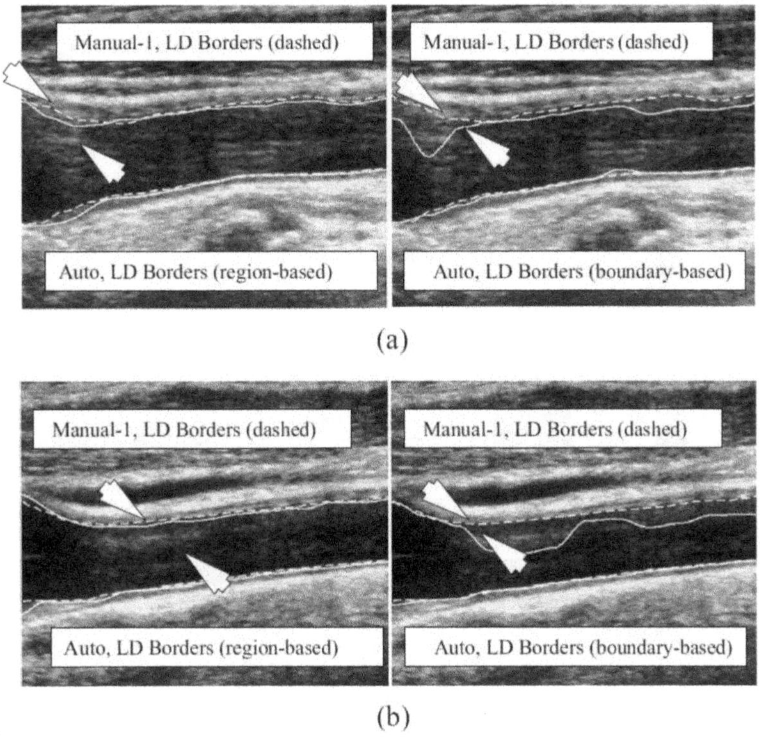

Figure 8.7. Results from the region- and boundary-based model (image source: [29]).

segmentation. CNNs use convolution filters in multiple layers to extract feature maps from the original image. It also uses pooling layers for dimensionality reduction. FCNs on the other hand use a series of upsample layers and a skipping operation for segmenting the image.

Suri and his team [35] used a combination of CNNs and FCNs for lumen segmentation from CCA images. The model used consisted of three phases: image pre-processing, DL and performance evaluation as shown in figure 8.8. The pre-processing multiresolution framework consists of four stages: cropping, binarization, reduction and downsampling. In the cropping stage all background information is removed. In the binarization stage ground truth (GT) information in the form of manually delineated lumen border coordinates from GT images are collected for training. During the US probe usage, the best information is collected from the centre of the probe due to full contact with the neck region. However, the image information collected at the sides of probe become blurry due to partial contact. Therefore, 10% reduction is done from the sides of the CCA images to retrieve only the crisp information. Further, 50% downsampling is done for speedy computation. The images are input to the DL phase. The DL phase consists of two stages: an encoder for feature extraction and a decoder for segmentation. The encoder constitutes the first 13 layers (13 convolution layers with intermediate five max pooling layers) of the VGG16 [36] CNN. The features extracted from the encoder are fed into the decoder. The decoder consists of the three upsample layers of the FCN. It uses two skipping operations to merge outputs from the last two max pool layers with outputs of the first two

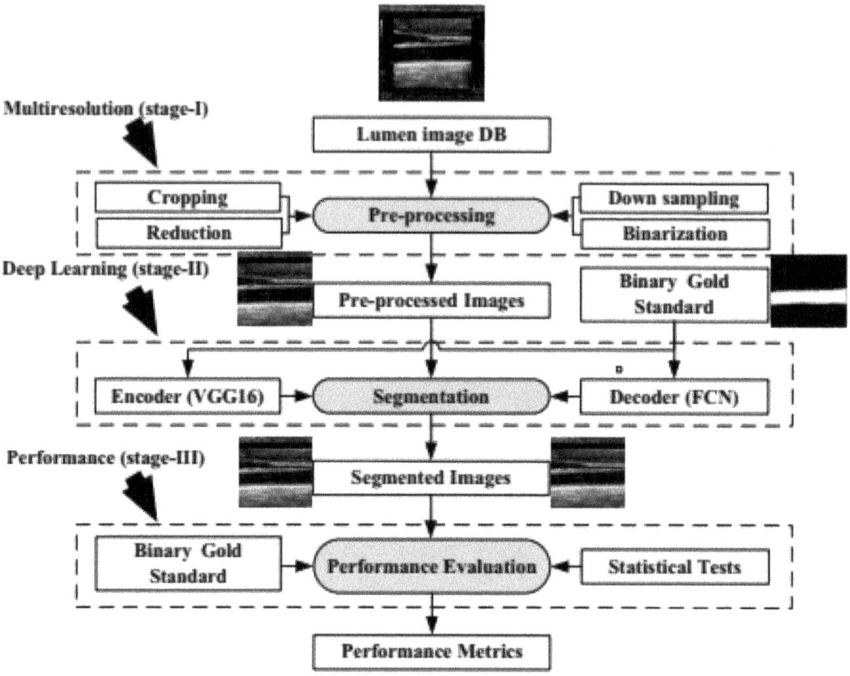

Figure 8.8. Deep learning model for LD measurement from US CCA images (image source: [35]).

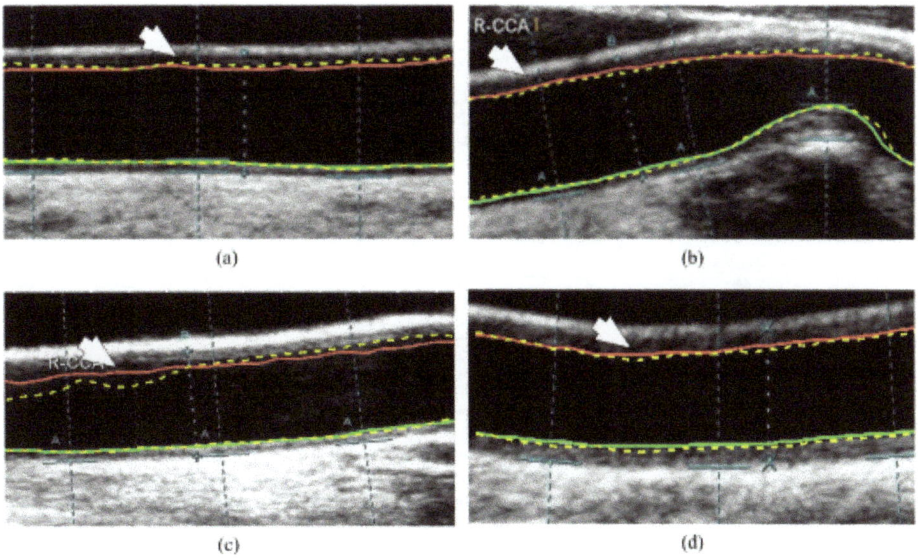

Figure 8.9. DL segmentation is shown in red and green, GT is shown in dashed yellow (image source: [35]).

upsample layers, respectively. The decoder segments the images, which are then fed into the performance evaluation stage for computing the statistics.

Altogether, three DL models were used for three GTs from three radiologists of different experience levels. A total of 407 US CCA images were used for the experiment. The mean LD for DL1 and GT1 was 6.09 ± 0.94 and 6.06 ± 0.91 mm, respectively. For DL2 and GT2, the corresponding mean LD was 6.05 ± 0.91 and 5.91 ± 0.88 mm, respectively. Finally, for DL3 and GT3, the mean LD was 6.09 ± 0.94 and 6.00 ± 0.90 mm, respectively. The mean LD error for DL1, DL2, and DL3 was 0.19 ± 0.27, 0.23 ± 0.23, and 0.21 ± 0.19, respectively. The diagrammatic representation of segmentation is shown in figure 8.9.

8.2.2.2 Carotid intima segmentation measurement using deep learning

Suri and his team [37] used a four-phase process for cIMT measurement from US CCA images. The four phases are: multiresolution, deep learning, boundary extraction and performance measurement. The multiresolution phase function is similar to that discussed earlier in section 8.2.3. It also constitutes cropping, reduction, downsampling and binarization stages. However, the GT images are divided into two groups. The first group is the MA delineated GT images and the second group is the lumen border-delineated GT images. Two DL models of the same configuration as section 8.2.3 are used in the second phase. The first DL model is used for MA border segmentation (DL-MA) and the second model is used for lumen segmentation (DL-L). The MA delineated GT images and lumen border-delineated GT images are fed into DL-MA and DL-L models for segmentation. Finally, the MA segmented and lumen segmented images are fed into the boundary extraction phase. It constitutes of two stages: border extraction and calibration.

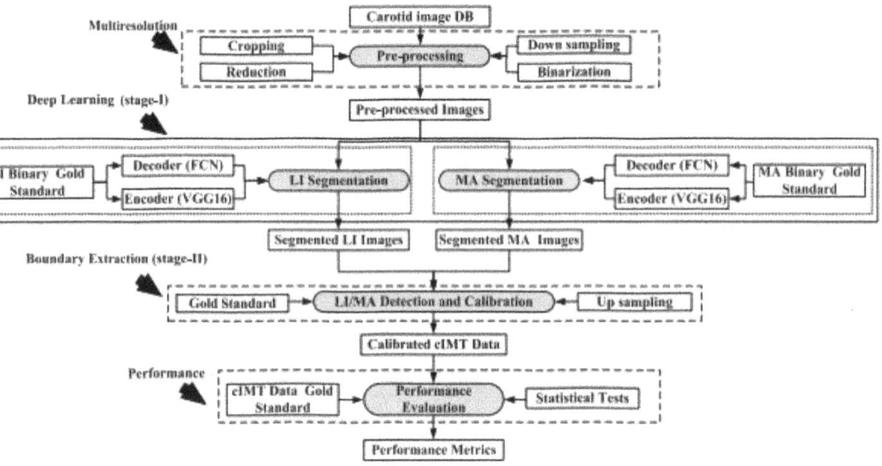

Figure 8.10. DL model for cIMT measurement (image source: [37]).

The LI-far wall and MA-far wall are delineated in the border extraction phase. These LI-far and MA-far walls are then calibrated using a least-squares model [38]. Finally, cIMT is measured between LI-far and MA-far wall using polyline distance method. Two GTs from two different radiologists were used for training. A total of 396 US CCA images were used. The cIMT error for DL1 and DL2 was 0.126 ± 0.134 and 0.126 ± 0.100 mm, respectively. The diagrammatic representation of the model is given in figure 8.10. The results for segmentation is shown in figure 8.11.

8.3 Carotid plaque characterization using machine learning and deep learning approaches

Carotid plaque is made of lipids, cholesterol, calcium and other tissues. In terms of risk, carotid plaque is characterized into two categories: symptomatic and asymptomatic. Symptomatic plaque is the case where the patient shows symptoms of disease. In the case of asymptomatic class, the patient does not show any symptoms, but regular monitoring is required. An example of symptomatic and asymptomatic plaque is given in figure 8.12. Earlier, the grayscale median (GSM) [39], which is the median value of all pixels in the plaque, was used to characterize high-risk and low-risk plaque. It is observed that a GSM value of 32 or less corresponds to high-risk plaque. However, GSM value may vary in images from different OEMs. Diagnosis of symptomatic and asymptomatic plaques using ML/DL techniques can help clinicians categorize patients into high risk and low risk. Based on this diagnosis, better decision making regarding disease and its treatment can be made. There are several computer-based techniques for plaque characterization, which can characterize plaque into symptomatic or asymptomatic, fibrous or non-fibrous etc. These techniques are again divided into two classes: ML and DL. A general discussion is made on these two techniques for plaque characterization followed by examples.

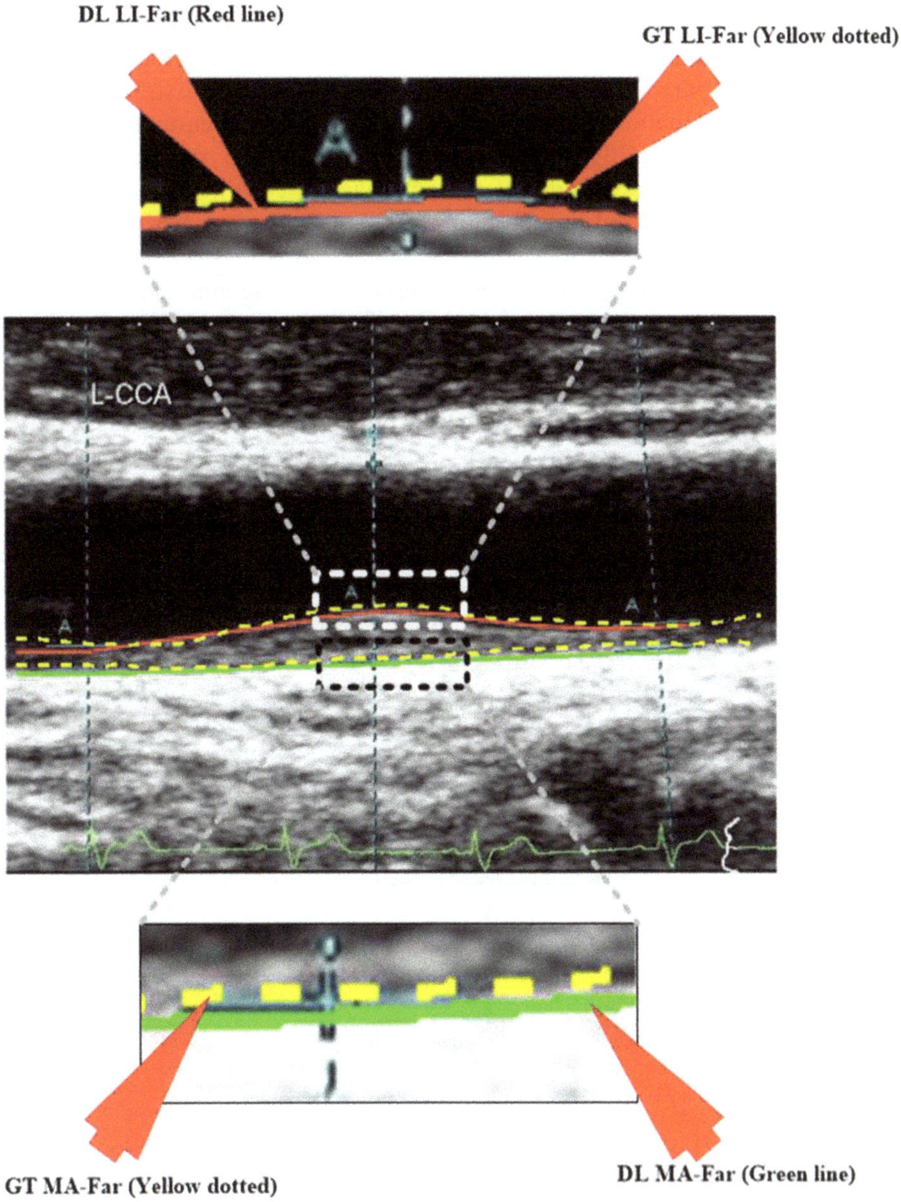

Figure 8.11. DL segmentation is shown in red and green, GT is shown in dashed yellow (image source: [37]).

Figure 8.12. Row (a) symptomatic plaque; Row (b) asymptomatic plaque (image source: [39]).

8.3.1 ML-based technique for plaque characterization

The ML-based techniques consist of two stages: feature extraction and training/testing in the ML paradigm. The ROI consisting of plaque is extracted from the carotid image either manually or via automated techniques. Features are extracted from plaque ROI. ML algorithms are applied on these plaque features for characterization. A process model of this approach is shown in figure 8.13. Two techniques using ML methods are given in the next two sections 8.3.1.1 and 8.3.1.2.

8.3.1.1 Plaque classification using support vector machine and AdaBoost

Suri and his team [40] used two ML-based classifiers, AdaBoost [41] and SVM [17], for plaque characterization. 346 carotid plaques images were collected. Plaque ROI was extracted based on echogenic measures. Texture-based features were extracted from these plaque images based on statistical properties of pixels such as standard deviation, entropy, symmetry, run percentage etc. Other measurements based on the grey level co-occurrence matrix and run-length matrix are also collected such as: symmetry, entropy etc. The features extracted are fed into the AdaBoost and SVM classifier. AdaBoost is an ensemble classifier which achieves results by combining accuracy from different classifiers. Some of the classifiers are a least-square (LS) method [42], a maximum likelihood (ML) [43] algorithm, a normal density discriminant function (NDDF) [44], Pocket [45] and Stumps [46]. The SVM classifier used linear, poly and radial basis kernels for computing the separating hyper plane between two classes. SVM recorded better accuracy, sensitivity and specificity results, which are: 82.4%, 82.9% and 82.1%, respectively. AdaBoost recorded

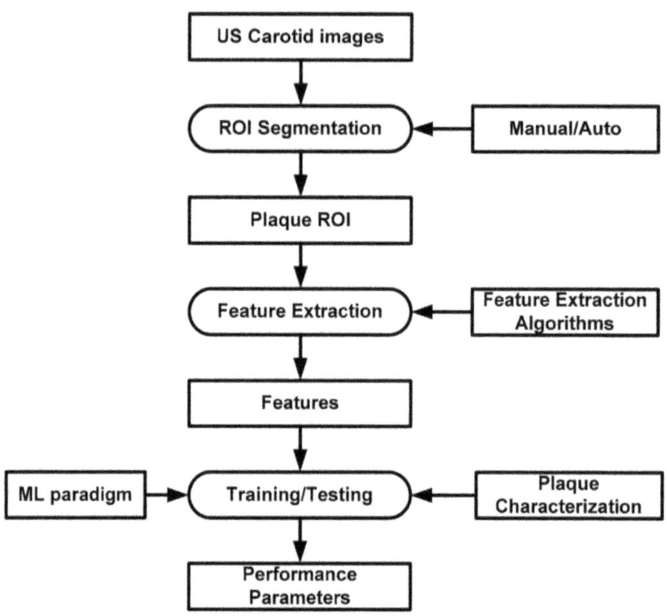

Figure 8.13. Process model showing plaque characterization using ML-based methods.

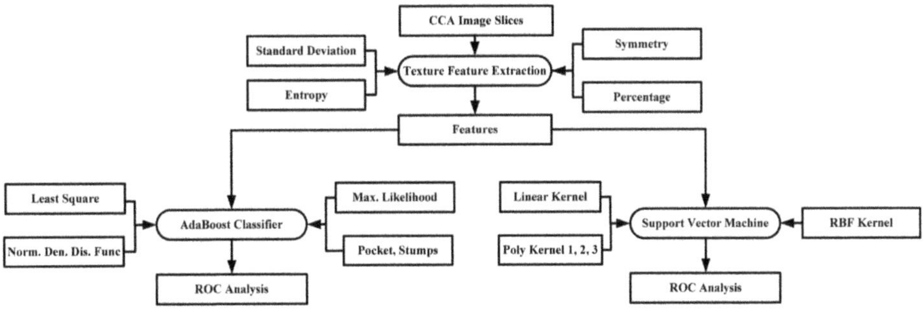

Figure 8.14. Ml model for plaque characterization.

accuracy, sensitivity and specificity of 81.7%, 82.3% and 81.9%, respectively. A model of the methodology is given in figure 8.14.

8.3.1.2 Automated plaque classification using ML-based technique

For the purpose of plaque characterization [47], US CCA images were collected from two datasets: UK and Portugal (146 + 346 = 492). Plaque was extracted using echogenic measures. Grayscale texture features were extracted using GLCM and fuzzy run-length matrix (FRLM) [48]. The classification algorithms which were applied were: SVM, Gaussian mixture model (GMM) [49], radial basis probabilistic neutral network (RBPNN) [50], decision tree [51], K-Nearest Neighbour (KNN) [52], Naïve Bayes Classifier (NBC) [53] and the Fuzzy Classifier model (FCM) [54]. Among all the classifiers, FCM gave the highest accuracy, sensitivity and specificity at 93.1%, 99% and 80%, respectively. The training and testing model is given in figure 8.15.

8.3.2 Plaque characterization using DL model

The DL is an evolving field in the ML domain. The key difference between DL and ML techniques is that DL is capable of automated feature extraction and thus bypassing the hand-made feature extraction model used by ML-based techniques. The US plaque images are fed into the DL model after pre-processing. The DL model trains and tests on labelled plaque images to extract high-level features and perform characterization over them. The process model of the DL plaque characterization system is shown in figure 8.16. An example of usage of a CNN model for plaque characterization is given in section 8.3.3.1.

8.3.2.1 Plaque characterization using CNN

In this work [55], CNNs have been used for plaque classification. For this purpose, 56 images were collected. A patch-based approach was used where these 56 images were translated to 90,000 images for better plaque patch classification. The plaque tissue was divided into three classes: lipid core, fibrous tissue, and calcified tissue. The CNN architecture consisted of four convolution layers with a kernel size of

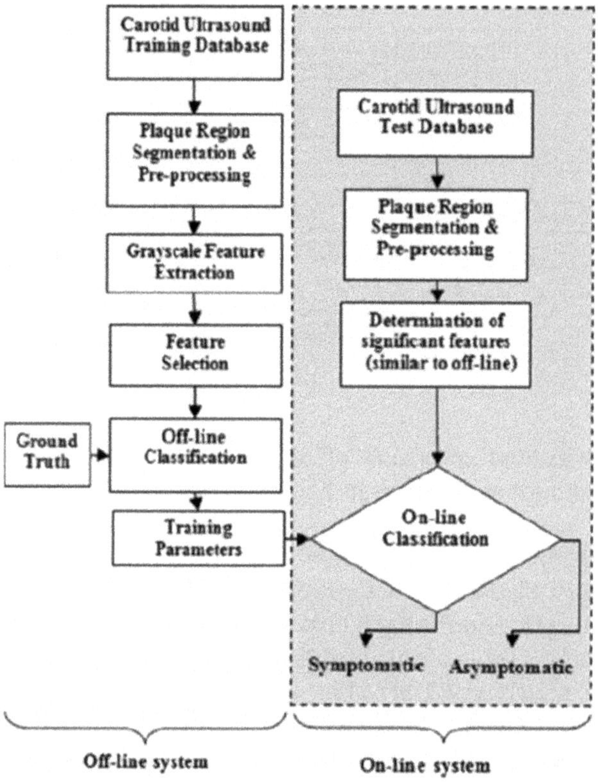

Figure 8.15. Ml training/testing model for plaque characterization (image source: [47]).

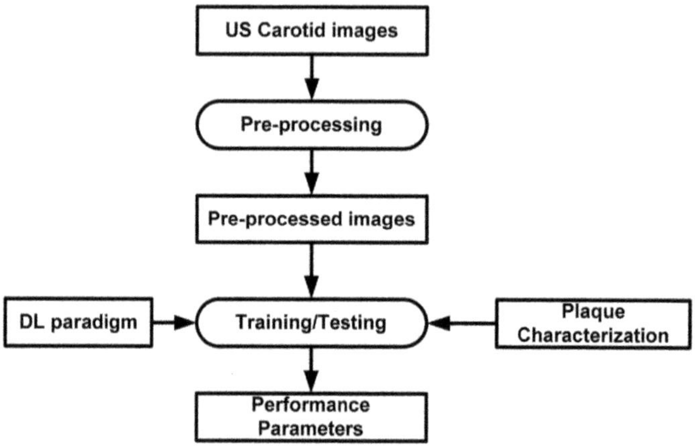

Figure 8.16. DL-based plaque characterization model.

dimension 3×3. It was followed three, fully connected layers, followed by leaky ReLu layer [56]. In addition to CNN, SVM was also used for classification. However, CNN gave a high accuracy of 78.5% when compared with SVM which gave an accuracy of 14.3%.

8.4 Discussion

The main focus of this paper is stroke risk estimation from CCA segmentation and measurement and plaque characterization. Both of methods are based on either ML or DL-based techniques. A thorough investigation is made on each of the methods by ML- and DL-based techniques. The ML-based techniques apply characterization algorithms on features extracted from US CCA images. The DL-based techniques learn directly from the images, skipping the feature extraction part. A comparison on both methods using ML and DL techniques is made in the following subsections.

8.4.1 Benchmarking on CCA segmentation and measurement

The benchmarking of several ML and DL techniques for CCA segmentation and measurement is given in table 8.1.

8.4.2 Benchmarking on CCA plaque characterization

The benchmarking table of ML and DL methods for plaque characterization is given in table 8.2.

A short note on vulnerable plaque

The most common reason for myocardial infarction or sudden death is occurrence of thrombosis in the coronary artery or CCA [57, 58]. Plaque rupture is considered 60%–65% responsible for the thrombosis [59]. There are other causes of coronary or

Table 8.1. Benchmarking table for CCA segmentation and measurement.

S. No.	Method	Features	No. of images	Performance
1	CALEX, fuzzy k-means [26, 27]	Intensity seed points	665	cIMT error: 0.191 ± 0.217 mm
2a	Region-based [29]	Scale-space	404 (JP*) 300 (HK*)	LD (JP*): 6.35 ± 0.95 mm LD (HK*): 6.20 ± 1.35 mm
2b	Boundary-based [29]	Scale-space	404 (JP*) 300 (HK*)	LD (JP*): 5.90 ± 0.97 mm LD (HK*): 5.43 ± 1.24 mm
3	DL for lumen segmentation [35]	Deep features	407	Mean LD error DL1: 0.19 ± 0.27 mm DL2: 0.23 ± 0.23 mm DL3: 0.21 ± 0.19 mm
4	DL for cIMT measurement [37]	Deep features	396	Mean cIMT error DL1: 0.126 ± 0.134 mm DL2: 0.126 ± 0.100 mm

*JP: Japan; HK: Hong Kong.

Table 8.2. Benchmarking table for CCA plaque characterization.

S. No.	Method	Features	No. of images	Performance
1a	SVM [40]	GLCM*, GLRLM*	346	Acc*: 82.4% Sen*: 82.9% Spec*: 82.1%
1b	AdaBoost [40]	GLCM, GLRLM	346	Acc: 81.7% Sen: 82.3% Spec: 81.9%
2	FCM* [47]	GLCM, FRLM*	492	Acc: 93.1% Sen: 99.0% Spec: 80.0%
3	CNN [55]	Deep features	56	Acc: 78.5%

*GLCM: grey level co-occurrence matrix; GLRLM: gray level run-length matrix; Acc: accuracy; Sen: sensitivity, Spec: specificity; FRLM: fuzzy run-length matrix; FCM: fuzzy classifier.

CCA thrombosis, i.e., mental or physical stress, increase in coagulability, etc [60]. The descriptive term to define a vulnerable plaque which is about to rupture (or ruptured plaque) is thin-cap fibroatheroma (TCF). TCF is characterized by a large lipid-rich necrotic core which is separated from the blood stream by a thin layer of fibrous tissue i.e., fibrous cap (FC).

The lipid core is composed of free cholesterol, cholesterol crystals, cholesterol esters derived from lipids, and also lipids from death of foam cells, interplaque haemorrhage, etc [61]. Further, this lipid core also contains prothrombotic oxidized lipids making it highly thrombotic to the circulating blood, making the plaque highly vulnerable. It has been seen that plaque area-wise, the necrotic core is significantly higher (34 ± 17 %) for ruptured plaque than TCF (23 ± 17 %). Also, the mean calcification score of ruptured plaque is significantly higher than TCF. The FC is characterized into three classes. Intact FC is associated with low-risk plaque rupture, thin FC is associated with mild-risk rupture and fissured FC is associated with high-risk rupture. Vulnerable plaques are characterized by thin FC whose size is less than 65 μm.

The presence, number, identification and risk characterization of TCF in coronary and carotid artery can be done by both invasive and non-invasive imaging modalities. For coronary arteries, TCF with large plaque burden can be identified by radio frequency intra vascular ultrasound (RF-IVUS). RF-IVUS technique however, cannot accurately distinguish calcific plaque and non-calcific tissues due to acoustic shadowing. Further, the spectra that differentiates the calcium and necrotic core overlap, adding to inaccuracy. That said, however, RF-IVUS can strongly predict non-culprit lesion-related major adverse cardiovascular events at a median follow-up of 3.4 years, making RF-IVUS validated imaging modality for TCF characterization. The imaging modality used for common carotid artery TCF detection is MRI, which characterizes TF as intact, thin and fissured. The thick FC appears as a juxtaluminal band on low signal, which is identifiable. However, this dark band is absent in thin TCF. The ruptured plaque is identified by absence of the dark band and presence of a bright grey region adjacent to the lumen, which represents the plaque haemorrhage.

In this regard, ML and DL algorithms can be applied for patient risk stratification based on identification and characterization of two features of TCF: necrotic core and FC. The amount of calcification of the necrotic core can be used to characterize the plaque as vulnerable or non-vulnerable. This can be done by using image features such as brightness, contrast, and texture, which can differentiate between a calcified and non-calcified necrotic core. Also, the characterization of the FC is based on the presence or absence of the dark band as well as presence of a grey bright region signifying haemorrhage. ML and DL models can be trained using two classes of images, calcified and non-calcified necrotic cores, to identify vulnerable plaque. Similarly, ML/DL models can be trained using thick, thin and ruptured plaque images to characterize low-risk, mild-risk and high-risk patients.

References

[1] National Center for Health Statistics 2018 https://www.cdc.gov/nchs/nhanes/index.htm accessed 2 October 2018

[2] MEMBERS, WRITING GROUPEmelia J B, Michael J B, Stephanie E C, Mary C, Sandeep R D and Rajat D *et al* 2017 Heart disease and stroke statistics—2017 update: a report from the American Heart Association *Circulation* **135** e146

[3] National Center for Health Statistics 2014 Mortality multiple cause micro-data files, 2014: public-use data file and documentation: NHLBI tabulations http://cdc.gov/nchs/data_access/Vitalstatsonline.htm#Mortality_Multiple accessed 2 October 2018

[4] Heidenreich P A, Justin G T, Olga A K, Javed B, Kathleen D, Michael D E and Finkelstein E A *et al* 2011 Forecasting the future of cardiovascular disease in the United States: a policy statement from the American Heart Association *Circulation* **123** 933–44

[5] Suri J S, Kathuria C and Molinari F 2010 *Atherosclerosis Disease Management* (Berlin: Springer Science & Business Media)

[6] Libby P, Karin E B and Alan R T 2016 Atherosclerosis: successes, surprises, and future challenges *Circ. Res.* **118** 531–4

[7] https://mayfieldclinic.com/pe-carotidstenosis.htm

[8] Sato K, Ogoh S, Hirasawa A, Oue A and Sadamoto T 2011 The distribution of blood flow in the carotid and vertebral arteries during dynamic exercise in humans *J. Physiol.* **589** 2847–56

[9] Dhawan S S *et al* 2010 Shear stress and plaque development *Expert Rev. Cardiovasc. Ther.* **8** 545–56

[10] He X, Peter P C, Kischell E R and Chiang A M 2003 Ultrasound imaging system *U.S. Patent 6,638,226 issued*

[11] Punchard W F B and Robert D P 1988 Magnetic resonance imaging systems *U.S. Patent 4,733,189 issued*

[12] Hsieh J 2009 *Computed Tomography: Principles, Design, Artifacts, and Recent Advances* (Bellingham, WA: SPIE)

[13] Narayanan R, Kurhanewicz J, Shinohara K, David Crawford E, Simoneau A and Suri J S 2009 MRI-ultrasound registration for targeted prostate biopsy *Biomedical Imaging: From Nano to Macro, 2009. ISBI'09. IEEE Int. Symp. on* (Piscataway, NJ: IEEE) pp 991–4
[14] Saba L, Montisci R, Molinari F, Tallapally N, Zeng G, Mallarini G and Suri J S 2012 Comparison between manual and automated analysis for the quantification of carotid wall by using sonography. A validation study with CT *Eur. J. Radiol.* **81** 911–8
[15] El-Baz A and Jasjit S S (ed) 2011 *Lung Imaging and Computer Aided Diagnosis* (Boca Raton, FL: CRC Press)
[16] Haykin S S 2009 *Neural Networks and Learning Machines* 3 (Upper Saddle River, NJ: Pearson)
[17] Cortes C and Vladimir V 1995 Support-vector networks *Mach. Learn.* **20** 273–97
[18] Huang G B, Zhu Q Y and Siew C K 2006 Extreme learning machine: theory and applications *Neurocomputing* **70** 489–501
[19] Yegnanarayana B 2009 *Artificial Neural Networks* (Delhi: PHI Learning Pvt. Ltd)
[20] Hopfield J J 1988 Artificial neural networks *IEEE Circuits Dev. Mag.* **4** 3–10
[21] LeCun Y, Yoshua B and Geoffrey H 2015 Deep learning *Nature* **521** 436
[22] Goodfellow I, Bengio Y, Courville A and Bengio Y 2016 *Deep Learning* 1 (Cambridge, MA: MIT Press)
[23] Virmani R, Allen P B, Kolodgie F D and Farb A 2003 Pathology of the thin-cap fibroatheroma: a type of vulnerable plaque *J. Interv. Cardiol.* **16** 267–72
[24] Torvik A, Svindland A and Lindboe C F 1989 Pathogenesis of carotid thrombosis *Stroke* **20** 1477–83
[25] Roux S, Carteaux J P, Hess P, Falivene L and Clozel J P 1994 Experimental carotid thrombosis in the guinea pig *Thromb. Haemost.* **71** 252–6
[26] Molinari F, Zeng G and Suri J S 2010 A state of the art review on intima–media thickness (IMT) measurement and wall segmentation techniques for carotid ultrasound *Comput. Methods Prog. Biomed.* **100** 201–21
[27] Molinari F, Zeng G and Suri J S 2010 An integrated approach to computer-based automated tracing and its validation for 200 common carotid arterial wall ultrasound images: a new technique *J. Ultrasound Med.* **29** 399–418
[28] Huang Z and Ng M K 1999 A fuzzy k-modes algorithm for clustering categorical data *IEEE Trans. Fuzzy Syst.* **7** 446–52
[29] Araki T, Kumar P K, Harman S S, Nobutaka I, Ajay G, Luca S and Jeny R *et al* 2016 Two automated techniques for carotid lumen diameter measurement: regional versus boundary approaches *J. Med. Syst.* **40** 182
[30] Suri J S, Liu K, Reden L and Laxminarayan S 2002 A review on MR vascular image processing algorithms: acquisition and prefiltering: part I *IEEE Trans. Inform. Technol. Biomed.* **6** 324–37
[31] Suri J S, Haralick R M and Florence H S 2000 Greedy algorithm for error correction in automatically produced boundaries from low contrast ventriculograms *Pattern Anal. Appl.* **3** 39–60
[32] Li C, Xu C, Gui C and Fox M D 2010 Distance regularized level set evolution and its application to image segmentation *IEEE Trans. Image Process.* **19** 3243
[33] LeCun Y, Yoshua B and Geoffrey H 2015 Deep learning *Nature* **521** 436

[34] Long J, Shelhamer E and Darrell T 2015 Fully convolutional networks for semantic segmentation *Proc. IEEE Conf. on Computer Vision and Pattern Recognition* (IEEE) 3431–40

[35] Biswas M, Kuppili V, Saba L, Reddy Edla D, Suri H S, Sharma A, Cuadrado-Godia E, Laird J R, Nicolaides A and Suri J S 2018 Deep learning fully convolution network for lumen characterization in diabetic patients using carotid ultrasound: a tool for stroke risk *Med. Biol. Eng. Comput.* **57** 1–22

[36] Simonyan K and Zisserman A 2014 Very deep convolutional networks for large-scale image recognition arXiv preprint arXiv:1409.1556

[37] Biswas M *et al* 2018 Deep learning strategy for accurate carotid intima-media thickness measurement: an ultrasound study on Japanese diabetic cohort *Comput. Biol. Med.* **98** 100–17

[38] Molinari F, Kristen M M, Saba L, Zeng G, Rajendra Acharya U, Ledda M, Nicolaides A and Suri J S 2012 Fully automated dual-snake formulation for carotid intima-media thickness measurement: a new approach *J. Ultrasound Med.* **31** 1123–36

[39] Sharma A M, Gupta A, Krishna Kumar P, Rajan J, Saba L, Nobutaka I, Laird J R, Nicolades A and Suri J S 2015 A review on carotid ultrasound atherosclerotic tissue characterization and stroke risk stratification in machine learning framework *Curr. Atheroscler. Rep.* **17** 55

[40] Acharya R U, Faust O, Peng Chuan Alvin A, Vinitha Sree S, Molinari F, Saba L, Nicolaides A and Suri J S 2012 Symptomatic vs. asymptomatic plaque classification in carotid ultrasound *J. Med. Syst.* **36** 1861–71

[41] Hastie T, Rosset S, Zhu J and Zou H 2009 Multi-class adaboost *Stat. Interface* **2** 349–60

[42] 1963 Whittle, Peter, Peter Whittle, Peter Whittle, Nouvelle-Zélande Mathématicien, Peter Whittle, New Zealand Mathematician, and Great Britain *Prediction and Regulation by Linear Least-Square Methods* (London: English Universities Press)

[43] Chakravarti A, Laura K L and Jillian E 1991 Reefer. 'a maximum likelihood method for estimating genome length using genetic linkage data *Genetics* **128** 175–82

[44] Fraley C and Adrian E R 2002 Model-based clustering, discriminant analysis, and density estimation *J. Am. Stat. Assoc.* **97** 611–31

[45] Gallant S I and Stephen I G 1993 *Neural Network Learning and Expert Systems* (Cambridge, MA: MIT Press)

[46] Iba W and Langley P 1992 Induction of one-level decision trees *Machine Learning Proceedings* (Morgan Kaufmann) 233–40

[47] Acharya U *et al* 2013 Atherosclerotic plaque tissue characterization in 2D ultrasound longitudinal carotid scans for automated classification: a paradigm for stroke risk assessment *Med. Biol. Eng. Comput.* **51** 513–23

[48] Shi Z and Govindaraju V 2004 Line separation for complex document images using fuzzy runlength *First International Workshop on Document Image Analysis for Libraries* (Piscataway, NJ: IEEE) pp 306–12

[49] Reynolds D 2015 Gaussian mixture models *Encyclopedia of Biometrics* (Berlin: Springer) 827–32

[50] Huang D 1999 Radial basis probabilistic neural networks: model and application *Int. J. Pattern Recognit. Artif. Intell.* **13** 1083–101

[51] Safavian S, Rasoul and Landgrebe D 1991 A survey of decision tree classifier methodology *IEEE Trans. Syst. Man Cybern.* **21** 660–74

[52] Peterson L E 2009 K-nearest neighbor *Scholarpedia* **4** 1883
[53] Murphy K P 2006 *Naive Bayes Classifiers* (Vancouver: University of British Columbia) 18
[54] Kuncheva L 2000 *Fuzzy Classifier Design* 49 (Berlin: Springer Science & Business Media)
[55] Lekadir K, Alfiia G, Àngels B, Vila M, Igual L, Daniel L R, Elvira F, Petia R and Sandy N 2017 A convolutional neural network for automatic characterization of plaque composition in carotid ultrasound *IEEE J. Biomed. Health Inform.* **21** 48–55
[56] Xu B, Wang N, Chen T and Li M 2015 Empirical evaluation of rectified activations in convolutional network arXiv preprint arXiv:1505.00853
[57] Tofler G *et al* 1987 Concurrent morning increase in platelet aggregability and the risk of myocardial infarction and sudden cardiac death *New Engl. J. Med.* **316** 1514–8
[58] Falk E 1992 Why do plaques rupture? *Circulation* **86** III30–42
[59] Stone G W, Gary S M and Virmani R 2018 Vulnerable plaques, vulnerable patients, and intravascular imaging *J. Am. Coll. Cardiol.* **72** 2022–6
[60] Muller , James E, Geoffrey and Tofler H 1992 Triggering and hourly variation of onset of arterial thrombosis *Ann. Epidemiol.* **2** 393–405
[61] Saba L, Michele A, Marincola B C, Mario P, Eytan R, Pier Paolo B, Alessandro N, Lorenzo M, Carlo C and Max W 2014 Imaging of the carotid artery vulnerable plaque *Cardiovasc. Interv. Radiol.* **37** 572–85